Lecture Notes in Computer Science   15504

Founding Editors

Gerhard Goos
Juris Hartmanis

The series Lecture Notes in Computer Science (LNCS), including its subseries Lecture Notes in Artificial Intelligence (LNAI) and Lecture Notes in Bioinformatics (LNBI), has established itself as a medium for the publication of new developments in computer science and information technology research, teaching, and education.

LNCS enjoys close cooperation with the computer science R & D community, the series counts many renowned academics among its volume editors and paper authors, and collaborates with prestigious societies. Its mission is to serve this international community by providing an invaluable service, mainly focused on the publication of conference and workshop proceedings and postproceedings. LNCS commenced publication in 1973.

Jean Baratgin · Baptiste Jacquet ·
Emmanuel Brochier · Hiroshi Yama
Editors

Human and Artificial Rationalities

Advances in Cognition, Computation, and Consciousness

Third International Conference, HAR 2024
Paris, France, September 17–20, 2024
Proceedings

 Springer

Editors
Jean Baratgin 🆔
Université Paris 8
Saint-Denis Cedex, France

Baptiste Jacquet 🆔
Université Paris 8
Saint-Denis Cedex, France

Emmanuel Brochier 🆔
Facultés Libres de Philosophie et de
Psychologie (IPC)
Paris, France

Hiroshi Yama 🆔
Osaka Metropolitan University
Osaka, Japan

ISSN 0302-9743   ISSN 1611-3349 (electronic)
Lecture Notes in Computer Science
ISBN 978-3-031-84594-9   ISBN 978-3-031-84595-6 (eBook)
https://doi.org/10.1007/978-3-031-84595-6

Preface

The Human and Artificial Rationalities (HAR) conference series is focused on comparing human and artificial rationalities, investigating how they interact together in a practical sense, but also on the theoretical and ethical aspects behind rationality from three main perspectives: Philosophy, Psychology, and Computer Sciences. HAR aims to build bridges between these three fields of research.

This volume, entitled Advances in Cognition, Computation, and Consciousness, contains the papers presented at the 3rd International Conference on Human and Artificial Rationalities (HAR 2024), which was held in Paris (France), during September 17–20. This edition continued the theme initiated in the first and second conferences of bridging three different fields of study: Philosophy, psychology, and computer sciences, to share knowledge about intelligent systems, whether natural or artificial, to generate a multidisciplinary understanding of reasoning.

In this third year of the HAR conference, we received 52 submissions. For publication in the proceedings, we accepted 19 of them as full papers and 4 as short papers. All submissions were carefully reviewed by the program committee and additional reviewers in a multi-stage double-blind peer-review process, which began with a review of the abstracts. Three reviewers were assigned to each submission by the program chairs according to their area of expertise, in addition to an AI reviewer (for the abstract review stage only, if the authors accepted it). More could be assigned if reviewers did not agree on the decision. On average, there were 5 submissions reviewed per reviewer. Reviewers were assigned based on a balance between expertise in the topic of the paper and minimization of potential conflicts of interest. In the situation where a reviewer would have personal knowledge of the authors of a paper but their expertise would also have made them too invaluable to replace, we ensured that the two other reviewers did not have personal contacts with the authors. The final decision was made by the program committee chairs and general chair based on the score on the 6 criteria (Content quality, Significance, Originality, Relevance, Rigor, and Coherence with conference) and on the written reviews. Decisions could include acceptance for publication, major revisions, minor revisions, or rejection. The accepted papers cover a wide array of topics pertaining to rationality, such as reasoning, conceptual thinking, belief revision, judgments, theory of mind, anthropomorphism, language and culture on reasoning, socio-cultural aspects of artificial agents, evaluation of artificial agents, use of artificial devices in life-span approaches, neuropsychology of reasoning, and thinking with disabilities.

In addition to the list of accepted papers, the conference greatly benefited from three invited lectures by important researchers on reasoning from the field of computer sciences: Antonio Lieto, from the University of Salerno, Italy (Knowledge Representation and Reasoning, Semantic/Language Technologies, Cognitive Systems and Architectures, Persuasive Technologies, Interactive Intelligent Technologies); from the field of philosophy: Roger Pouivet, from Université de Lorraine and Honorary member of the Institut Universitaire de France Archives Henri-Poincaré - Philosophie et Recherches sur les Sciences et les Techniques (CNRS); and from the field of probability theory: Giulianella Coletti, from the University of Perugia (cognitive sciences, uncertainty models, and decision theory under partial knowledge). One additional paper has also been included in the conference proceedings of this year's conference following a mistake with the proceedings of HAR 2023, which did not include it.

We would like to thank all the members of the Program Committee, as well as the additional reviewers, who devoted their time for the reviewing process. We thank all the authors of submitted papers, the invited speakers, and the participants for their scientific contributions to the conference. Finally, we would like to thank all the Organizing Committee members (with a special thanks to Emmanuel Brochier from IPC, Facultés Libres de Philosophie et de Psychologie) for their excellent local organization, which made the HAR conference a success.

The HAR conference could not have happened without the organizational support of partner institutions: The CHArt laboratory of Université Paris 8, Facultés Libres de Philosophie et de Psychologie (IPC), and the P-A-R-I-S Reasoning Association. We would also like to particularly thank the MAIF, the BONHEUR laboratory of CY Cergy University, the CHArt Laboratory, and the EDUS4EL project from the ERASMUS+ program for their financial support.

August 2024   Jean Baratgin
                                                                    Baptiste Jacquet
                                                                  Emmanuel Brochier
                                                                      Hiroshi Yama

Organization

General Chair

Jean Baratgin   Université Paris 8, France

Program Committee Chairs

Baptiste Jacquet   Université Paris 8, France
Emmanuel Brochier   IPC, Facultés Libres de Philosophie, France
Hiroshi Yama   Osaka Metropolitan University, Japan

Honorary President

Daniel Andler   Sorbonne Université and Académie des sciences
                                 morales et politiques, France

Program Committee

Julien Bugmann   Haute école Pédagogique du Canton de Vaud,
                                 Switzerland
Sophie Charles   CY Cergy Paris Université, France
Charles El-Nouty   Université Sorbonne Paris Nord, France
Hirofumi Hashimoto   Osaka Metropolitan University, Japan
Alain Jaillet   CY Cergy Paris Université, France
Frank Jamet   CY Cergy Paris Université, France
Vassilis Komis   University of Patras, Greece
Daniel Lassiter   Edinburgh University, UK
Laura Macchi   Università degli Studi di Milano-Bicocca, Italy
Laura Martignon   Pädagogische Hochschule Ludwigsburg,
                                 Germany
Bernard N'Kaoua   Université de Bordeaux, France

Additional Reviewers

Kevin Bague
Kevin Beaufils
Maxime Bourlier
Laura Caravona
Bruno Corcos
Marion Dubois-Sage
Darya Filatova
Antoine Gazeaud
Léa Lachaud
Kaede Maeda
Anastasia Misrili
Alberto Mura
François Nollé
David Over
Davide Petturiti
Véronique Salvano-Pardieu
Kerem Tahan
Taïna Victor

Keynote Talks

Cognitively-Inspired Machine Reasoning

Antonio Lieto[iD]

Università degli Studi di Palermo, 61 Piazza Marina, 90133 Palermo, Italy

Commonsense reasoning is one of the main open problems in the field of Artificial Intelligence (AI) while, on the other hand, it seems to be a very intuitive and default reasoning mode in humans and other animals. In this lecture, I will show - via two different case studies concerning commonsense categorization and knowledge invention tasks - how cognitively inspired heuristics can help (both in terms of efficiency and efficacy) in the realization of intelligent artificial systems able to reason in a human-like fashion, with results comparable to human-level performances.

Rationality and the Good Intellectual Life

Roger Pouivet[1,2,3]

[1] Université de Lorraine, 34 Cours Léopold, 54000 Nancy, France
[2] Honorary member of the Institut Universitaire de France, 1 rue Descartes, 75231, Paris Cedex 05, France
[3] Archives Henri-Poincaré - Philosophie et Recherches sur les Sciences et les Techniques (CNRS), 91 avenue de la Libération BP 454, 54001, Nancy Cedex, France

Let us distinguish two accounts of rationality. In one, rationality is a set of procedures. Rationality is then identified with reasoning. Its study can take multiple forms, logical but also psychological and sociological. The procedures can be modeled, and when formalized, they can be implemented in artifacts. We then speak of "Artificial Intelligence". In the other account, rationality constitutes the nature of a type of beings, the rational animals of which Aristotle speaks. Rationality then defines not procedures but a form of life, a way to be alive. Does such a distinction simply signal an ambiguity in the word "rationality"; or could we articulate these two accounts of rationality as procedures and as the nature of some beings? After all, the same Aristotle who invented the formal study of logical processes spoke about rationality as the nature of human beings. In this case, the study of reasoning could not only be the description of actual processes, but it presupposes and even establishes standards of what constitutes reasoning. But where do these standards come from? Even in a procedural account the question arises of why rules must be respected. Could we blame an artificial intelligence for having "thought" poorly? Is it not only in the account of rationality as nature that we could find the answer to this question? The final question on this matter would then be: Why do we have to be rational? Is it that we would have to realize our own nature as rational animals? How then is rationality related to the idea of a good intellectual human life? It will be less a question here of giving answers to these questions than of showing why they arise.

Rationality and the Good Intellectual Life

The dynamic approach of de Finetti's probability theory and its relevance in the time of AI

Giulianella Coletti ⓘ

Università degli Studi di Perugia, 1 Piazza Università, 06123 Perugia, Italy

The pivotal role science plays in society today has underscored the need to prioritize "soft" and "open" yet well-founded theories capable of managing complex real-world issues. This facilitates the construction of bridges between theories and their respective methods, thus yielding consistent hybrid models. One area where this trajectory has clearly unfolded is in the quest for the "best model" to manage uncertainty in knowledge acquisition processes and decision-making, especially when dealing with partial, imprecise, vague, and revisable information (such as perception-based data) obtained from diverse sources or conveyed in different languages, including natural language. From this perspective, the path yielding the most convincing results is one grounded in the principles and inspired by the methods of probability theory as proposed by de Finetti. In fact, adopting de Finetti's approach is not merely a "semantic" stance favoring subjectivist positions but also, and primarily, a way to leverage the "syntactic" advantages provided by an axiomatic theory that eschews the introduction of demands and constraints aimed at utilizing results obtained in other theories or pursuing the myth of the uniqueness of result in any updating procedure. To capitalize on these advantages, clarity regarding objects and axioms is necessary, and this is precisely what de Finetti's theory accomplishes, starting from the definition of events and conditional events and extending to the axioms known as the coherence of a probability or conditional probability assignment. The result is a model capable of handling uncertainty treatment based on continuous updating and reasoning under possible hypotheses.

Contents

Artificial Reasoning and Models

Abductive Reasoning with Syllogistic Forms in Large Language Models 3
  Hirohiko Abe, Risako Ando, Takanobu Morishita, Kentaro Ozeki,
  Koji Mineshima, and Mitsuhiro Okada

Data Science and Environmental Sustainability: Mastering the Information
and Knowledge 18
  Charles El Nouty and Darya Filatova

Multi-agent Simulation of Violence Emergence in Protests 35
  Julien Rosenberger, Julien Saunier, and Nicolas Sabouret

Moral Reasoning

The Social Exchange Heuristic Operation During a One-Shot Prisoner's
Dilemma Game 55
  Kaede Maeda, Hirofumi Hashimoto, and Shigehito Tanida

Development of Intention-Based Moral Judgement in Children
and Adolescents with and Without Intellectual Disability 64
  Véronique Salvano Pardieu and Valérie Pennequin

Artificial Intelligence and Cognition

Artificialization of Intelligence and Embodied Cognition 91
  Pierre Uzan

Consciousness in AI: A Return to Fetishism or Technological Progress? 110
  David Ricardo Galeano Cabral

Agent Rationality Under Partially Resolving Uncertainty and the Hurwicz
Criterion 118
  Davide Petturiti and Barbara Vantaggi

Expressing Rational Agency Through Pragmatic Meta-vocabularies 131
  Yaoli Du

The Effect of Initial Letters in Word Recognition: A Phonological
or Positional Effect ? 142
Massimo Brasdu and Baptiste Jacquet

Rationality and Dual Process

Coherence as Logical Consistency 167
Alberto Mura

Degrees of Intuition: Coherence in the New Dual Process Theory 185
Maxime Bourlier, Daniel Lassiter, and Jean Baratgin

A Tribute to Kahneman and Tversky in the Context of Mathematics
Education 196
Ulrich Hoffrage and Laura Martignon

Effect of Brief Mindfulness Training on Cognitive Rigidity 208
Léa Lachaud and Jérémy Louis

Reasoning and Special Needs

Research Trends in Information Accessibility to Web Content for Adults
with Developmental Disabilities: A Literature Review 219
Kai Seino

NAO Robot to Help People with Alzheimer's Manage the Recall
of Activities in Prospective Memory 234
Kerem Tahan and Bernard N'kaoua

Does Replacing the Experimenter with an Ignorant Student Robot Improve
the Success of Children with ASD in the False Belief Task? 250
*Marion Dubois-Sage, Yohann Mosset-Cancel, Frank Jamet,
and Jean Baratgin*

Education

Enhancing Pharmacology Education: Investigating the Efficacy of Serious
Game-Based Learning Through Virtual Simulation 291
Florian Laronze, Solène Delsuc, and Bernard N'Kaoua

Video-Based Analysis of the Mechanical Design Process in a Student
Dyadic Activity 308
Sylvain Luc Agbanglanon and Vassilis Komis

Critical Thinking and Psychorhetoric for Promoting Environmental
Awareness 323
    Laura Macchi, Laura Caravona, and Elisa Palazzi

The Effect of Screens on Children's Development: Concrete Action Taken
in Schools, Closer to Families 345
    Sabrina Reffad and Joelle Provasi

Can We Really Learn in the Metaverse? A Discussion on Learning
Through Immersive Technology 366
    Raphaël Bompy

Experimental Procedures in Cognition

Emotion-Enhanced Pain Assessment Protocol 385
    Bruna Alves, Ana Almeida, Catarina Silva, Daniela Pais,
    Rita P. Ribeiro, João Gama, José Maria Fernandes, Susana Brás,
    and Raquel Sebastião

Additional Paper From HAR-2023

Discharge of Responsibility as an Enhancer of Utilitarian Choices 403
    Maxime Bourlier, Cassandra Leroux, Hirofumi Hashimoto,
    Kaede Maeda, Hiroshi Yama, and Jean Baratgin

Author Index 411

Artificial Reasoning and Models

Abductive Reasoning with Syllogistic Forms in Large Language Models

Hirohiko Abe[1]([✉]), Risako Ando[1], Takanobu Morishita[1], Kentaro Ozeki[1,2],
Koji Mineshima[1], and Mitsuhiro Okada[1]

[1] Keio University, Tokyo, Japan
{hirohiko-abe,risakochaan,morishita}@keio.jp,
{minesima,mitsu}@abelard.flet.keio.ac.jp
[2] The University of Tokyo, Tokyo, Japan
kentaro.ozeki@gmail.com

Abstract. Research in AI using Large-Language Models (LLMs) is rapidly evolving, and the comparison of their performance with human reasoning has become a key concern. Prior studies have indicated that LLMs and humans share similar biases, such as dismissing logically valid inferences that contradict common beliefs. However, criticizing LLMs for these biases might be unfair, considering our reasoning not only involves formal deduction but also abduction, which draws tentative conclusions from limited information. Abduction can be regarded as the inverse form of syllogism in its basic structure, that is, a process of drawing a minor premise from a major premise and conclusion. This paper explores the accuracy of LLMs in abductive reasoning by converting a syllogistic dataset into one suitable for abduction. It aims to investigate whether the state-of-the-art LLMs exhibit biases in abduction and to identify potential areas for improvement, emphasizing the importance of contextualized reasoning beyond formal deduction. This investigation is vital for advancing the understanding and application of LLMs in complex reasoning tasks, offering insights into bridging the gap between machine and human cognition.

Keywords: Abduction · Deduction · Syllogism · Reasoning bias · Large Language Models

1 Introduction

Research in Large-Language Models (LLMs) is rapidly advancing, with a significant focus on comparing their performance to human reasoning. Previous studies have shown that while LLMs generally excel at reasoning tasks [6,21,32], they exhibit similar biases to humans, such as dismissing logically valid inferences that contradict common beliefs [3,8]. Although these studies often emphasize deductive reasoning, our everyday reasoning encompasses more than just deduction. Given that LLMs are developed by learning natural language used in daily contexts without specialized logical training, it would be unreasonable to criticize them for bias tendencies in deduction tasks. Since our reasoning involves not

J. Baratgin et al. (Eds.): HAR 2024, LNCS 15504, pp. 3–17, 2025.
https://doi.org/10.1007/978-3-031-84595-6_1

only formal deduction but also abduction, which draws hypotheses from limited information, it is crucial to investigate LLMs' capabilities in making abductive inferences.

Abduction is a natural form of reasoning that seeks reasons and explanation. For example, when discussing the reason for the delay of the train by asking "Why was the train late?" it is natural to trace back from the observed fact to the reason explaining it, such as "because the traffic lights failed". However, it is rare to ask in the form of a deduction, as in "The traffic lights failed, therefore the train was late". The explanatory aspect of abduction, as well as the logical consistency in deductive reasoning, is important in investigating natural explanations and realizing an explainable AI (XAI) that naturally answers *why*-questions [23]. Investigating how accurately current LLMs can perform abduction provides a theoretical basis for research into XAI.

Abduction is important in knowledge acquisition. Charles Sanders Peirce [18] regarded abduction as a process of inquiry along with deduction and induction. Abduction plays a more important role than deduction, especially when it comes to the discovery of the unknown. Evaluating LLMs' abductive reasoning abilities is essential in determining whether LLMs can gain new knowledge, particularly from limited information.

Inquiry, or the activity of acquiring knowledge, is considered one of the chief topics in recent epistemology, and norms of inquiry have been studied [15,19,20]. Given that abduction is related to inquiry in Peirce's [18] philosophy, it can be considered a form of logic that guides inquiry, for example, by providing hypotheses and guidance on what to investigate. Assessing LLMs' abductive reasoning ability is important when addressing the question of whether LLMs can be used to guide our everyday inquiry.

In this paper, we introduce a dataset to test the abductive reasoning abilities of LLMs, compare LLMs' accuracy on abductive reasoning tasks with deductive reasoning tasks, and explore whether LLMs show human-like belief biases on reasoning.[1] By LLMs, we focus on in-context learning pretrained models such as GPT [24,25] and Llama [2], rather than those requiring fine-tuning such as BERT [9]. These in-context learning models adapt to a specific task using a task description or a few examples of correct answers as input, called a *prompt*, without changing the models' parameters. We revealed that LLMs generally performed more poorly on abductive tasks compared to deductive tasks. In addition, we found that LLMs exhibit human-like belief biases in both abduction and deduction.

With regard to the comparison between deduction and abduction, it has been pointed out that in diagnostic inference, the inference from effect to cause, there is reason to believe that the deductive model is a more natural reasoning scheme for humans than the abductive model, under a probabilistic setting [29,30]. Although this paper shares interests with these trends in the comparison between deduction and abduction, our dataset specifically focuses on

---

[1] The dataset is available at https://github.com/kmineshima/abduction-syllogism-llm.

abductions that can be generated within the framework of syllogisms, especially by swapping premises and conclusions of deductive syllogisms. This approach enables a systematic comparison of abduction and deduction, thereby providing a foundation for exploring more complex forms of abductive reasoning, including causal and practical variations.

2   Background

2.1   Abductive Reasoning

Our everyday reasoning involves not only deduction but also abduction, that is, the type of reasoning that hypothetically derives new information from limited information. Abduction is believed to be ubiquitous in our ordinary life. For example, abductive reasoning is considered to be operative in cognitive process and testimonial trust. In addition, abduction is regarded as a cornerstone of scientific methodology [11, 12].

In general, abduction is a form of reasoning that leads to a hypothesis explaining an observed fact. Abduction is considered to be of two types: hypothesis selection and hypothesis generation. The former refers to the selection of the best explanation from among several hypotheses. It is also known as *Inference to the Best Explanation* (IBE). On the other hand, the latter kind involves generating a hypothesis that explains the observed fact from given observations.

Charles Sanders Peirce first introduced abduction, distinguishing it from deduction and induction. Peirce [18] understood it as "the process of forming an explanatory hypothesis" and "the only logical operation which introduces any new idea" (CP 5.171). According to Peirce, abduction is amplicative in that it adds new information besides the premises, while deduction is not.

Peirce initially organized abduction in a syllogistic framework [4]. According to Peirce, abduction is made by changing the minor premise and the conclusion in a deductively valid syllogism. The following is an example of a deductively valid syllogism, which is called a first figure syllogism.

> Major premise: All A are B
> Minor premise: C is A
> ――――――――――――――
> Conclusion: C is B

By changing the minor premise and the conclusion, we obtain the following form of abduction.

> Major premise: All A are B
> Conclusion: C is B
> ――――――――――――――
> Minor premise: C is A

Note that this form of inference is the so-called *Affirming the Consequent*, a typical instance of deductively invalid inferences. As a formal characterization of abduction, Peirce [18] says, "The surprising fact, C, is observed. But if A were true, C would be a matter of course. Hence, there is reason to suspect that A

is true" (CP 5.189). In this paper, we call the first premise of abduction Rule, the second premise Observation, and the conclusion Hypothesis. The following is a concrete example of abduction based on this terminology.

Rule: All things that were in the bag are white.
Observation: These balls are white.
_____
Hypothesis: These balls were in the bag.

In the context of AI and Natural Language Processing, there have been studies on evaluating machine learning (deep learning) models using abductive reasoning. Among others, Bhagavatula et al. [5] focuses on the form of abduction that selects the most plausible hypothesis that explains given observations. This form of abduction can be subsumed under the IBE type abduction as described above. In this paper, based on Peirce's account, we instead focus on the form of abduction that is converted from syllogism, namely, one that derives a minor premise from the major premise and conclusion of a deductively valid syllogism.

2.2 Deductive Reasoning Abilities of LLMs

With the rapid progress in research on LLMs, the importance of assessing their reasoning ability has increased and the abilities have been researched using a variety of tasks [31]. Among these, more research that compared LLMs with humans has been conducted on deductive syllogistic reasoning, which has been studied in cognitive psychology [7,16,22]. Datasets of syllogistic reasoning tasks for LLMs are recently introduced by Dong et al. [10], Guebelmann et al. [17], Aghahadi et al. [1], and Ando et al. [3]. However, they only focus on deductions and do not deal with other kinds of inferences including abduction.

Recent research on LLM's reasoning abilities with syllogisms showed that while LLMs generally perform well on syllogisms, they tend to exhibit some human-like biases [14,28]. More specifically, Dasgupta et al. [8] found that LLMs reason more accurately about believable or realistic situations in reasoning tasks including syllogism and Wason's selection task. In addition, they revealed that LLMs tend to judge inferences with believable content as valid and those with the sentences that clash our commonsense belief as invalid regardless of forms of inferences, thus failing to separate *forms* from *contents* (the content effects). Ando et al. [3] introduced a syllogism dataset called NeuBAROCO, where syllogisms are presented in both English and Japanese. They showed that LLMs exhibit reasoning biases known in the psychological studies of syllogisms, including belief biases, conversion errors, and atmosphere effects. Ozeki et al. [26] extended the NeuBAROCO dataset and conducted a more detailed evaluation of a wide range of models by implementing various reasoning tasks, including those that require translating syllogisms into logical formulas and explaining the reasoning steps.

Based on these previous findings, this paper compares the reasoning abilities of LLMs in deduction and abduction and explores whether LLMs show human-like belief biases.

Table 1. Eight patterns of abduction. R: Rule, O: Observation, H: Hypothesis. Those in yellow are correct abductions, while those in grey are incorrect. Correct abductions are those in which H explains why O is the case, given the rule R.

R: All C are B	R: All B are C	R: All C are B	R: All B are C
O: These A are B	O: These A are B	O: These A are not B	O: These A are not B
H: These A are C	H: None	H: None	H: These A are not C
R: No C are B	R: No B are C	R: No C are B	R: No B are C
O: These A are B	O: These A are B	O: These A are not B	O: These A are not B
H: None	H: None	H: These A are C	H: These A are C

Table 2. Examples abductive syllogisms labeled as *Consistent*, *Inconsistent*, and *Neutral*. The numbers in brackets show the number of each type.

Type	Example
Consistent (66)	Rule: All people that had a fun time are smiling
	Observation: These people are smiling
	Hypothesis: These people had a fun time
Inconsistent (66)	Rule: All things that are made in the sweet restaurant are spicy
	Observation: These cakes are made in the sweet restaurant
	Hypothesis: These cakes are spicy
Neutral (84)	Rule: All things that were in the bag are white
	Observation: These balls are white
	Hypothesis: These balls were in the bag

3 Datasets

We formulated abduction as an inference from Rule and Observation to Hypothesis as shown in Sect. 2.1. Rule consists of a sentence of the form *All A are B* (Universal Affirmative) or *No A are B* (Universal Negative), while Observation and Hypothesis consist of a sentence of the form *These A are B* (Particular Affirmative) or *These A are not B* (Particular Negative). We classified four patterns of abductive inference, which are shown in Table 1.

One of the essential features of abduction is that it is amplicative. Abduction contributes to the acquisition of new knowledge by drawing the conclusions whose content is beyond those contained in the premises. In this respect, abduction differs from deduction. The patterns of abduction identified above fulfil this characteristic. On the other hand, the greyed-out patterns in Table 1 do not, and they are deductively valid. Restricting the attention to syllogistic forms helps us to define what counts as correct patterns of deductive and abductive reasoning.

To construct a set of abductive inferences having these patterns, we first created a triple (A, B, C) of terms, where A is a subject term, B is an observational predicate, and C is a non-observational predicate. Observational predicates are those that can be verified through direct observation, while non-observational predicates are those that cannot. For example, *are white* is an observational predicate, while *were in the bag* is a non-observational predicate. In this study, we manually created 27 triples like (*balls, are white, were in the bag*).

By instantiating the inference patterns shown in Table 1 with these terms, we obtained 216 problems of abductive inference in total, with 108 correct patterns and 108 incorrect patterns. To annotate information about belief biases, we classified each problem by three labels, *consistent, inconsistent,* or *neutral.* The problem is *consistent* if Rule is considered to be true as inferred from our common-sense beliefs; it is *inconsistent* if Rule contradicts our common-sense beliefs; if neither holds, it is *neutral.* Table 2 shows some examples of each type.

In the same way, we constructed 216 patterns of corresponding deductive inferences, with 108 valid patterns and 108 invalid patterns. We obtained instances of deduction by changing Observation and Hypothesis in abduction. For example, from the first example of abduction in Table 2, we obtained an instance of valid deduction, *All people that had a fun time are smiling. These people are smiling. Therefore, These people had a fun time.* Table 3 shows all eight patterns of deduction in comparison with the corresponding abductions in Table 1.

Table 3. Eight patterns of deduction. P1: Major Premise, P2: Minor Premise, C: Conclusion. Those in yellow are valid deductions, while those in grey are invalid.

P1: All C are B	P1: All B are C	P1: All C are B	P1: All B are C
P2: These A are B	P2: These A are B	P2: These A are not B	P2: These A are not B
C: None	C: These A are C	C: These A are not C	C: None
P1: No C are B	P1: No B are C	P1: No C are B	P1: No B are C
P2: These A are B	P2: These A are B	P2: These A are not B	P2: These A are not B
C: These A are not C	C: These A are not C	C: None	C: None

4 Experiments

4.1 Experimental Settings and Evaluated Models

We conducted experiments on two tasks: the Abduction task and the Deduction task, using the dataset created by the method described in Sect. 3. All experiments were conducted in English. In each iteration, a single problem was provided as input along with a prompt, and the model's output was collected as an answer. The performance of the LLMs was evaluated using overall accuracy and accuracy for each problem type, providing a basic assessment of their capabilities.

In our experiments, we evaluated four state-of-the-art models with varying parameter sizes: GPT-3.5 [25], GPT-4 [24], Llama-3-8B (8 billion parameters), and Llama-3-70B (70 billion parameters) [2]. The GPT models are closed-source, and their specific details, including the exact number of parameters, are not publicly disclosed.[2] For hyperparameters of the models, we set the maximum output token length to 10 to prevent redundant responses, while keeping other hyperparameters at their default values. We employed in-context learning with prompts and did not perform any fine-tuning on the models.

4.2 Tasks

We compare two tasks in our experiments, Abduction task and Deduction task. Table 4 shows example prompts of each task.

Abduction task provides sentences for Rule and Observation and asks to choose the most plausible hypothesis. Given a hypothesis H, The answer is selected from three options: H, the negation of H, and "Neither is a good explanation," as shown in Table 4. In a similar way, the Deduction task provides two premise sentences and three options for the conclusion.

We conducted experiments on the Abduction and Deduction tasks in both zero-shot and few-shot settings [6]. For the few-shot prompts, we included eight examples using the same set of terms, corresponding to the eight patterns of abduction shown in Table 1, which were inserted between the task description and the problem. Details of the few-shot prompts can be found in Table 8 and Table 9 in the Appendix.

We have also tested alternative prompts, which are shown in Table 7 in the Appendix. However, since there was no performance improvement compared to the prompts listed in Table 4, they were not adopted.

Table 4. Example prompts for the Abduction task and Deduction task.

Input (Abduction Task)	Input (Deduction Task)
Based on Rule and Observation, from a logical perspective, select the most reasonable hypothesis that explains why Observation holds true. Choose one from the following options (1-3) and answer with the corresponding number. Note that there is a logical relationship between the Rule, Observation, and Hypothesis, where the Observation is logically derived from the Rule and Hypothesis. Rule: All things that were in the bag are white. Observation: These balls are white. Hypothesis: 1. These balls were in the bag. 2. These balls were not in the bag. 3. Neither is a good explanation. The answer is:	Select a sentence that serves as a conclusion based on the following two premises. Choose one from the following options (1-3) and answer with the corresponding number. P1: All things that were in the bag are white. P2: These balls are white. 1. These balls were in the bag. 2. These balls were not in the bag. 3. Neither. The answer is:

---

[2] The versions of the GPT models used are `gpt-3.5-turbo-0125` and `gpt-4-0613`, accessed via OpenAI's API.

Table 5. Accuracy (%) on the Abduction task ($n = 216$).

Condition	Model	Overall	Positive	Negative	Neither	Consistent	Inconsistent	Neutral
Zero-Shot	GPT-3.5	31.02	48.15	100.00	0.93	31.82	27.27	33.33
	GPT-4	41.67	80.25	92.59	0.00	46.97	34.85	42.86
	Llama3-8B	37.50	61.73	66.67	12.04	42.42	27.27	41.67
	Llama3-70B	37.04	64.20	100.00	0.93	45.45	31.82	34.52
Few-Shot	GPT-3.5	29.63	44.44	96.30	1.85	33.33	22.73	32.14
	GPT-4	28.70	65.43	22.22	2.78	31.82	19.70	33.33
	Llama-3-8B	28.70	41.98	100.00	0.93	33.33	27.27	26.19
	Llama-3-70B	75.46	90.12	81.48	62.96	74.24	72.73	78.57

Table 6. Accuracy (%) on the Deduction task ($n = 216$).

Condition	Model	Overall	Positive	Negative	Neither	Consistent	Inconsistent	Neutral
Zero-Shot	GPT-3.5	33.80	100.00	54.32	1.85	39.39	28.79	33.33
	GPT-4	72.22	100.00	100.00	44.44	74.24	68.18	73.81
	Llama-3-8B	43.52	100.00	40.74	31.48	37.88	37.88	52.38
	Llama-3-70B	53.24	100.00	80.25	21.30	60.61	40.91	57.14
Few-Shot	GPT-3.5	46.30	85.19	91.36	2.78	48.48	42.42	47.62
	GPT-4	95.83	100.00	96.30	94.44	100.00	92.42	95.24
	Llama-3-8B	49.54	100.00	98.77	0.00	50.00	50.00	48.81
	Llama-3-70B	84.72	92.59	80.25	86.11	90.91	72.73	89.29

4.3   Results

Tables 5 and 6 show the Abduction and Deduction task results. The columns labeled *Positive*, *Negative*, and *Neither* correspond to instances where the correct answer is the hypothesis H (positive form), the negation of H, and "Neither is a good explanation", respectively.

For the Abduction task in the zero-shot setting, the overall accuracy of the highest-performing model (GPT-4) was around 42%, which was slightly above the chance level. While the model achieved over 80% accuracy on problems with correct answers labeled as *Positive* or *Negative*, it performed poorly on problems where the correct answer was *Neither*. With regard to the content types, the accuracy of *Inconsistent* problems was around 10% lower than the other two types (*Consistent* and *Inconsistent*). This suggests that belief biases are also reproduced in abduction tasks.

In the few-shot setting, Llama-3-70B was the only model that showed a significant performance improvement, achieving approximately 63% accuracy on the problems whose answer type was *Neither* and around 75% overall accuracy. The overall accuracy for GPT-3.5 showed a slight improvement over the zero-shot setting, while for GPT-4, the overall accuracy was lower than in the zero-shot setting. The score for problems where the correct answer was *Neither* slightly increased for both GPT models.

For the Deduction task, the few-shot setting improved performance across all models compared to the zero-shot setting, with gains ranging from 6.02 to 31.48 points in the overall accuracy. GPT-4 was the best-performing model in both settings, achieving an overall accuracy of 72.22% in the zero-shot setting and 95.83% in the few-shot setting. However, except for Llama-3-8B, accuracy remained lower for the problems labeled *Inconsistent* compared to those labeled *Consistent* and *Neutral*.

4.4 Discussion

Are the Models Abductive Reasoners? The results on deduction tasks generally show the similar tendencies to the previous findings [3, 8, 13]. That is, LLMs' performance was quite low in the problems whose correct answer were *Neither* and the sentences in the problems contradict common sense belief. The exception is Llama-3-70B in the few-shot setting; still the accuracy for abduction (75.46%) was lower than that for deduction (84.72%).

It was anticipated that the results for the abduction task would be better than those for the deduction task, as abduction is more akin to everyday human reasoning, whereas deduction requires more reflective reasoning. However, the results surprisingly showed that the accuracy on abduction tasks was low overall. In particular, for the problems where the correct answer was *Neither*, LLMs often failed to solve them at all.

Given that abduction has been studied less than deduction, there is a possibility that while deduction cases are included more in the LLMs' training data, abduction problems are fewer. Also, considering that it is expected that sentences that perform hypothesis selection and hypothesis generation are included in natural texts, it is possible that there is difficulty in applying abduction to the syllogistic form. The observation human-like belief biases in abduction is consistent with Pereira et al. [27], which reports that belief bias is observed in abductions.

Why do the Models Tend to Mistakenly Choose *Negative*? In the Abduction task, for problems where *Neither* was the correct answer (108 problems), the distribution of GPT-4's predictions was as follows: 32 problems were answered as *Positive*, 76 as *Negative*, and 0 as *Neither*. In contrast, in the Deduction task, the distribution was as follows: 26 problems were answered as *Positive*, 34 as *Negative*, and 48 as *Neither*. Thus, in the Abduction task, there was a more pronounced tendency to incorrectly answer *Neither* problems and choose *Negative* as the correct answer.

To analyze this tendency in more detail, we calculated the rate at which a *Negative* was selected as the hypothesis when "No" or "not" appears in the Rule or Observation. In the case of GPT-4, this rate was 67.90% for the Abduction task (with the actual rate of the correct answer being *Negative* at 16.67%), while it was 70.99% for the Deduction task (with the actual rate of the correct answer being *Negative* at 50%). Thus, in both tasks, *Negative* sentences were selected at a higher rate than the actual rate of correct answers being *Negative*, with this tendency being more pronounced in the Abduction task. This tendency may be due to an effect similar to the atmosphere effects [7], where the presence of negation in the Rule or Observation leads to the selection of a hypothesis that also contains negation.

Do the Models Answer the Problems as Deduction? We examined the scores of abduction problems when labeled as if they were deduction problems. For example, an affirmation of the antecedent (e.g., *All B are C, A is B*) has no correct hypothesis for abduction, and therefore the correct answer is *Neither*. However, if it is conceived as deduction, it logically entails *A is C* and is labeled as *Positive*. When comparing the correct labels for deduction to GPT-4's predictions, the agreement rate was 51.85% for *Overall*, 100.0% for *Positive*, 93.83% for *Negative*, and 8.33% for *Neither*. In *Overall*, this was about 10 points higher than the accuracy for the original abduction labels. For example, among inferences of the form "affirmation of the antecedent", all inferences of the form *All B are C, A is B* led to the selection of the correct deductive answer, and 89% of inferences of the form *No B are C, A is B* also led to the selection of the correct deductive answer.

This suggests that LLMs are influenced by deduction when solving abduction problems. However, in general, the agreement rate does not reach the level of accuracy in the Deduction task (falling short by about 20 points for *Overall* and by about 35 points for *Neither*), suggesting that LLMs are not completely mistaking abduction problems for deduction problems.

Does the Word "Hypothesis" Mislead by Suggesting Entailment Relationship? In Natural Language Inference (NLI) tasks [33], the conclusion that entails the premise is usually called "Hypothesis". Given this fact, we investigated the possibility that the word "Hypothesis" itself does not function as a hypothesis to explain Observation, but instead it suggests an entailment or deductive relation. In particular, substituting the term "Hypothesis" with "Reason" had little effect on improving the score.

Do the Models Choose Contradictory Answers? To specify error tendencies, we investigated whether LLMs choose the answer regardless of logical consistency. We checked whether the LLMs choose the answer that contradicts the given Rule or Observation, but few such cases are observed.

5   Conclusion and Future Work

In this paper, we created a dataset to test abductive reasoning abilities of LLMs and compared LLMs' accuracy on abductive reasoning tasks with deductive reasoning tasks. The results showed that LLMs performed worse in abduction than in deduction. In addition, human-like belief biases are observed in abduction as well as in deduction.

Abduction is considered to be a more ordinary inference than deduction is. Therefore, it is expected that humans would perform better for abduction than for deduction. This expected tendency is different from the tendency of LLMs, as shown in this paper. Comparisons between LLMs and humans for the abductive reasoning tasks, as well as further investigations on the error tendencies or biases in abductive reasoning through these comparisons, are topics for future work.

We adopted Peirce's initial characterization of abduction in syllogistic framework and focused on hypothesis generation tasks. Although three options, *Positive*, *Negative*, and *Neither*, are included in each problem, the options other than the correct answer do not serve as hypothesis from the logical perspective, so the task can be seen as a simpler version of hypothesis generation, rather than a task of choosing the best one from multiple hypotheses. However, other characterizations are also possible. Abduction can be understood as *Inference to the Best Explanation*. Tasks such as selecting the best hypothesis that explains the premises from the plural candidates that are already logical explanations are expected in future work. Also, although we characterized abduction with syllogistic framework, it can be understood as a probable reasoning. Comparison of humans and LLMs by a probabilistic (Bayesian) approach to abduction is an area for future work. Furthermore, evaluating more complex types of reasoning, such as extended syllogisms and conditionals, are also left for future work.

Acknowledgements. We thank the anonymous reviewers for their helpful comments and suggestions, which have improved the paper. This work is partially supported by JST, CREST Grant Number JPMJCR2114, JST BOOST, Japan Grant Number JPMJBS2409, the KGRI Challenge Grant from the Keio University Global Research Institute, and JSPS Kakenhi Grant Numbers JP24K00004, JP21K00016, JP21H00467, JP23K20416, and JP21K18339.

A Details of Prompts

Table 7 shows examples of prompts we tested that scored lower than the prompts we finally adopted. Tables 8 and 9 show the prompts with eight exemplars (8-shot prompts) used in the few-shot setting.

Table 7. Examples of alternative prompts not adopted.

```
┌ Input (Abduction Task) ┐

Suppose the following Observation is logically
derived from Rule and Hypothesis.
Choose the most appropriate sentence for the
Hypothesis from the following options (1-3) and
answer with the corresponding number.

Rule: All things that were in the bag are white.
Hypothesis: ???
--------------------
Observation: These balls are white.

1. These balls were in the bag.
2. These balls were not in the bag.
3. Neither.

The answer is:
```

```
┌ Input (Abduction Task) ┐

You are an inquirer. You know that the following
Rule holds true in the world. Additionally,
you have recently confirmed that the following
Observation is also true.
Given this information, you want to discover the
mechanism behind why these hold true.

Based on the Rule and Observation below, select
the most plausible hypothesis from a logical
perspective that explains why the Observation is
valid.
Please respond with the corresponding number
from the numbers 1-3.

Rule: All things that were in the bag are white.
Observation: These balls are white.

Hypothesis:
1. These balls were in the bag.
2. These balls were not in the bag.
3. Neither is a good explanation.

The answer is:
```

Table 8. An example few-shot prompt for the Abduction task.

Input (Abduction Task)

Based on Rule and Observation, from a
logical perspective, select the most
reasonable hypothesis that
explains why Observation holds true.
Choose one from the following options
(1-3) and answer with the corresponding
number.
Note that there is a logical relationship
between the Rule, Observation, and
Hypothesis, where the
Observation is logically derived from the
Rule and Hypothesis.

Rule: All things that are sold at the
shop are waterproof.
Observation: These shoes are waterproof.

Hypothesis:
1. These shoes are sold at the shop.
2. These shoes are not sold at the shop.
3. Neither is a good explanation.

The answer is: 1

Rule: All things that are waterproof are
sold at the shop.
Observation: These shoes are waterproof.

Hypothesis:
1. These shoes are sold at the shop.
2. These shoes are not sold at the shop.
3. Neither is a good explanation.

The answer is: 3

Rule: All things that are sold at the
shop are waterproof.
Observation: These shoes are not
waterproof.

Hypothesis:
1. These shoes are sold at the shop.
2. These shoes are not sold at the shop.
3. Neither is a good explanation.

The answer is: 3

Rule: All things that are waterproof are
sold at the shop.
Observation: These shoes are not
waterproof.

Hypothesis:
1. These shoes are sold at the shop.
2. These shoes are not sold at the shop.
3. Neither is a good explanation.

The answer is: 2

Rule: No things that are sold at the shop
are waterproof.
Observation: These shoes are waterproof.

Hypothesis:
1. These shoes are sold at the shop.
2. These shoes are not sold at the shop.
3. Neither is a good explanation.

The answer is: 3

Rule: No things that are waterproof are
sold at the shop.
Observation: These shoes are waterproof.

Hypothesis:
1. These shoes are sold at the shop.
2. These shoes are not sold at the shop.
3. Neither is a good explanation.

The answer is: 3

Rule: No things that are sold at the shop
are waterproof.
Observation: These shoes are not
waterproof.

Hypothesis:
1. These shoes are sold at the shop.
2. These shoes are not sold at the shop.
3. Neither is a good explanation.

The answer is: 1

Rule: No things that are waterproof are
sold at the shop.
Observation: These shoes are not
waterproof.

Hypothesis:
1. These shoes are sold at the shop.
2. These shoes are not sold at the shop.
3. Neither is a good explanation.

The answer is: 1

Rule: All things that were in the bag are
white.
Observation: These balls are white.

Hypothesis:
1. These balls were in the bag.
2. These balls were not in the bag.
3. Neither is a good explanation.

The answer is:

Table 9. An example few-shot prompt for the Deduction task.

Input (Deduction Task)

Select a sentence that serves as a conclusion based on the following two premises.
Choose one from the following options (1-3) and answer with the corresponding number.

P1: All things that are sold at the shop are waterproof.
P2: These shoes are waterproof.

1. These shoes are sold at the shop.
2. These shoes are not sold at the shop.
3. Neither.

The answer is: 3

P1: All things that are waterproof are sold at the shop.
P2: These shoes are waterproof.

1. These shoes are sold at the shop.
2. These shoes are not sold at the shop.
3. Neither.

The answer is: 1

P1: All things that are sold at the shop are waterproof.
P2: These shoes are not waterproof.

1. These shoes are sold at the shop.
2. These shoes are not sold at the shop.
3. Neither.

The answer is: 2

P1: All things that are waterproof are sold at the shop.
P2: These shoes are not waterproof.

1. These shoes are sold at the shop.
2. These shoes are not sold at the shop.
3. Neither.

The answer is: 3

P1: No things that are sold at the shop are waterproof.
P2: These shoes are waterproof.

1. These shoes are sold at the shop.
2. These shoes are not sold at the shop.
3. Neither.

The answer is: 2

P1: No things that are waterproof are sold at the shop.
P2: These shoes are waterproof.

1. These shoes are sold at the shop.
2. These shoes are not sold at the shop.
3. Neither.

The answer is: 2

P1: No things that are sold at the shop are waterproof.
P2: These shoes are not waterproof.

1. These shoes are sold at the shop.
2. These shoes are not sold at the shop.
3. Neither.

The answer is: 3

P1: No things that are waterproof are sold at the shop.
P2: These shoes are not waterproof.

1. These shoes are sold at the shop.
2. These shoes are not sold at the shop.
3. Neither.

The answer is: 3

P1: All things that were in the bag are white.
P2: These balls are white.

1. These balls were in the bag.
2. These balls were not in the bag.
3. Neither.

The answer is:

References

1. Aghahadi, Z., Talebpour, A.: Avicenna: a challenge dataset for natural language generation toward commonsense syllogistic reasoning. J. Appl. Non-Classical Logics **32**(1), 55–71 (2022). https://doi.org/10.1080/11663081.2022.2041352
2. AI@Meta: Llama 3 Model Card (2024)
3. Ando, R., Morishita, T., Abe, H., Mineshima, K., Okada, M.: Evaluating large language models with NeuBAROCO: syllogistic reasoning ability and human-like biases. In: Proceedings of the 4th Natural Logic Meets Machine Learning Workshop, pp. 1–11 (2023)

4. Bellucci, F., Pietarinen, A.V.: Peirce's abduction. In: Magnani, L. (ed.) Handbook of Abductive Cognition, pp. 1–14. Springer, Cham (2022). https://doi.org/10.1007/978-3-031-10135-9_7

5. Bhagavatula, C., et al.: Abductive commonsense reasoning. In: International Conference on Learning Representations (2020)

6. Brown, T., et al.: Language models are few-shot learners. Adv. Neural. Inf. Process. Syst. **33**, 1877–1901 (2020)

7. Chater, N., Oaksford, M.: The probability heuristics model of syllogistic reasoning. Cogn. Psychol. **38**(2), 191–258 (1999). https://doi.org/10.1006/cogp.1998.0696

8. Dasgupta, I., et al.: Language models show human-like content effects on reasoning tasks. arXiv preprint arXiv:2207.07051 (2023). https://doi.org/10.48550/arXiv.2207.07051

9. Devlin, J., Chang, M.W., Lee, K., Toutanova, K.: BERT: pre-training of deep bidirectional transformers for language understanding. In: Proceedings of the 2019 Conference of the North American Chapter of the Association for Computational Linguistics, pp. 4171–4186 (2019)

10. Dong, T., Li, C., Bauckhage, C., Li, J., Wrobel, S., Cremers, A.B.: Learning syllogism with Euler neural-networks. arXiv preprint arXiv:2007.07320 (2020). https://doi.org/10.48550/arXiv.2007.07320

11. Douven, I.: Abduction. In: Zalta, E.N. (ed.) The Stanford Encyclopedia of Philosophy, Summer 2021 edn. Metaphysics Research Lab, Stanford University (2021)

12. Douven, I.: The Art of Abduction. The MIT Press (2022). https://doi.org/10.7551/mitpress/14179.001.0001

13. Eisape, T., Tessler, M., Dasgupta, I., Sha, F., van Steenkiste, S., Linzen, T.: A systematic comparison of syllogistic reasoning in humans and language models. arXiv preprint arXiv:2311.00445 (2024). https://doi.org/10.48550/arXiv.2311.00445

14. Evans, J.S.T.: Bias in Human Reasoning: Causes and Consequences. Lawrence Erlbaum Associates, Inc. (1989)

15. Friedman, J.: Zetetic epistemology. In: Reed, B., Flowerree, A.K. (eds.) Towards an Expansive Epistemology: Norms, Action, and the Social Sphere. Routledge (forthcoming)

16. Geurts, B.: Reasoning with quantifiers. Cognition **86**(3), 223–251 (2003). https://doi.org/10.1016/S0010-0277(02)00180-4

17. Gubelmann, R., Niklaus, C., Handschuh, S.: A philosophically-informed contribution to the generalization problem of neural natural language inference: Shallow heuristics, bias, and the varieties of inference. In: Proceedings of the 3rd Natural Logic Meets Machine Learning Workshop (NALOMA III), pp. 38–50 (2022)

18. Hartshorne, C., Weiss, P., Burks, A.W. (eds.): Collected Papers of Charles Sanders Peirce. Harvard University Press, Cambridge (1931–1958). Volumes 1–6 edited by Charles Hartshorne and Paul Weiss, 1931–1935; volumes 7–8 edited by Arthur W. Burks, 1958

19. Hookway, C.: Epistemology and inquiry: the primacy of practice. In: Hetherington, S. (ed.) Epistemology Futures, pp. 95–110. Oxford University Press (2006). https://doi.org/10.1093/oso/9780199273317.003.0006

20. Hookway, C.: Questions, epistemology, and inquiries. Grazer Philoso. Studien **77**(1), 1–21 (2008). https://doi.org/10.1163/18756735-90000841

21. Kojima, T., Gu, S.S., Reid, M., Matsuo, Y., Iwasawa, Y.: Large language models are zero-shot reasoners. Adv. Neural. Inf. Process. Syst. **35**, 22199–22213 (2022)

22. Manktelow, K.: Reasoning and Thinking. Psychology Press (1999)

23. Medianovskyi, K., Pietarinen, A.: On explainable AI and abductive inference. Philosophies **7**(2), 35 (2022). https://doi.org/10.3390/philosophies7020035

24. OpenAI: GPT-4 technical report. arXiv preprint arXiv:2303.08774 (2023). https://doi.org/10.48550/arXiv.2303.08774
25. Ouyang, L., et al.: Training language models to follow instructions with human feedback. Adv. Neural. Inf. Process. Syst. **35**, 27730–27744 (2022)
26. Ozeki, K., Ando, R., Morishita, T., Abe, H., Mineshima, K., Okada, M.: Exploring reasoning biases in large language models through syllogism: Insights from the NeuBAROCO dataset. In: Findings of the Association for Computational Linguistics: ACL 2024 (2024)
27. Pereira, L.M., Dietz, E.A., Hölldobler, S.: Contextual abductive reasoning with side-effects. Theory Pract. Logic Program. **14**(4–5), 633–648 (2014). https://doi.org/10.1017/S1471068414000258
28. Pohl, R.F.: Cognitive Illusions: A Handbook on Fallacies and Biases in Thinking, Judgement and Memory, 3 edn. Routledge (2022). https://doi.org/10.4324/9780203720615
29. Stilgenbauer, J.L., Baratgin, J.: Assessing the accuracy of diagnostic probability estimation: evidence for defeasible modus ponens. Int. J. Approximate Reasoning **105**, 229–240 (2019). https://doi.org/10.1016/j.ijar.2018.11.015
30. Stilgenbauer, J.L., Baratgin, J., Douven, I.: Reasoning strategies for diagnostic probability estimates in causal contexts: preference for defeasible deduction over abduction. In: Proceedings of the 4th International Workshop on Defeasible and Ampliative Reasoning (DARe-17) (2017)
31. Wang, A., et al.: Superglue: a stickier benchmark for general-purpose language understanding systems. Adv. Neural Inf. Process. Syst. **32** (2019)
32. Wei, J., et al.: Chain-of-thought prompting elicits reasoning in large language models. Adv. Neural. Inf. Process. Syst. **35**, 24824–24837 (2022)
33. Williams, A., Nangia, N., Bowman, S.: A broad-coverage challenge corpus for sentence understanding through inference. In: Proceedings of the 2018 Conference of the North American Chapter of the Association for Computational Linguistics: Human Language Technologies, Volume 1 (Long Papers), pp. 1112–1122 (2018). https://doi.org/10.18653/v1/N18-1101

Data Science and Environmental Sustainability: Mastering the Information and Knowledge

Charles El Nouty[1]([✉])[iD] and Darya Filatova[2][iD]

[1] LAGA, Université Sorbonne Paris Nord, 93430 Villetaneuse, France
`elnouty@math.univ-paris13.fr`
[2] ProductLife Groupe France, 8 av Arche, Courbevoie, France
`daria_filatova@interia.pl`

Abstract. The nowadays strategic management of an enterprise is determined by various criteria, one of which, increasingly, is the interaction with the environment. Advanced business analytics based on artificial intelligence allows organizations to achieve business goals by optimizing the use of resources while reducing the negative impact on the environment. This requires the improvement of existing and the invention of new tools for creating, sharing and consuming data, which can be compared with digital evolution. As a result of this evolution, new creators and consumers of data appear. Data, its creators and consumers, is a network of interconnected elements, part of a data ecosystem that is constantly evolving. Obviously, the ecosystem includes data that needs to be integrated from disparate sources; various types of analysis and skills to obtain information; actions in accordance with the acquired knowledge; as well as tools, applications and infrastructure for storing, processing and distributing data, depending on the purpose. Such multidimensionality makes the training of data-related specialists a rather complicated process. There is a need to constantly search for such key skills and knowledge that would make it easy to adapt to the new qualities and properties of the data ecosystem. The purpose of this work is to analyze the properties of the data ecosystem and highlight the key aspects of methodology for data science with respect to environmental sustainability of the data ecosystem. We illustrate the proposed methodology on the example of taxi demand prediction.

Keywords: Data ecosystem · Data Science · Methodology

1 Introduction

The growth of data or more precisely digital data around the world continues unabated. Among different approaches which explain this phenomenon one deals with notion of "a data ecosystem". Definitely, "an ecosystem is a system that environments and their organisms form through their interaction" [3]. Therefore,

J. Baratgin et al. (Eds.): HAR 2024, LNCS 15504, pp. 18–34, 2025.
https://doi.org/10.1007/978-3-031-84595-6_2

methods of creating data, methods of their transmission, storage, and use can be defined as certain ecosystems that have properties and laws of development comparable to living organisms. Data create data in a constant cycle: technologies create new data, which in turn create new technologies. However, can indeed any "data ecosystem" be considered as a system?

Before to answer, let us define an abstract system and name its properties. In many dictionaries, a system is defined as *a set of elements connected to each other; this set forms a certain integrity, unity.* The Fig. 1 shows some main properties of the system as integrity, emergence, structure, functionality, development [6]. The most crucial property is emergence, namely: the whole system behaves differently or at least has one "new" property which is not typical for any element of the system.

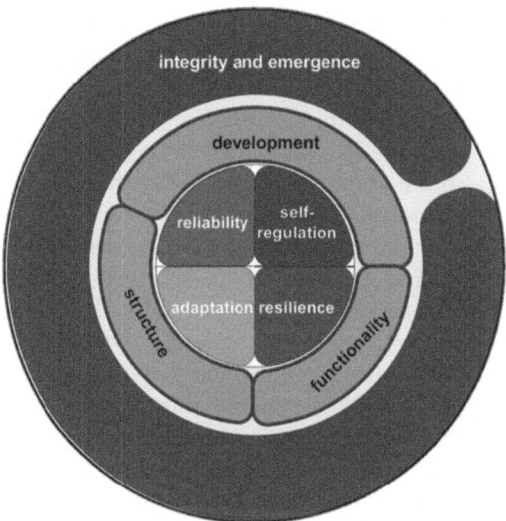

Fig. 1. The core-properties of the system

Modern data ecosystems include a whole network of interconnected, independent, and continually evolving entities. They include data that has to be integrated from disparate sources, different types of analysis and skills to recognize patterns or generate insights. Stakeholders can collaborate and act on the information they receive, as well as use variety the tools, applications, and infrastructure to store, process, and transmit data as needed. This indicates the presence of **integrity**. However, does it indicate the presence of the emergence? The insights arise as a result of data analysis carried out by the analyst. It depends on how qualified the analyst is. Understanding results of the interaction of data, analysis methods and analyst skills as new properties we can conclude on the presence of **emergence**.

Each data ecosystem has a certain structure. For example, an enterprise can accumulate vast structured, semi-structured or unstructured data about business activities in its native format. Usually, this single store of information can be considered as *"source data"*. Depending on the business task to be solved a part of data can be pulled for further analysis. To avoid the loss of information this extract is saved separately generating a kind of *"ata repository"*. The process can be repeated as many times as required. To drive valuable insights the extract is analyzed by different methodological processes available on various analytical platforms. Each process generates different transformations of extract reducing or increasing its volume. The results of these sequential modifications present *"transformed data repositories"*. Since the same data can be analyzed using different methods, the results of the analysis will correspond to models whose structure may differ significantly. This leads to the need to highlight another element of the structure of the data ecosystem - the model. Considering the chain of transformations from the data source to the model, it can be argued that the data ecosystem has some **structure** with hierarchical relationships among elements, which in turn are included to the system by certain purpose with respect to their **functionalities**. Achieving business goals is not only a manifestation of competitiveness in the face of uncertain market behavior of competitors, but also the ability to adopt new technologies that contribute to the genesis and **development** of the data ecosystem.

Under the influence of external factors, data sources are constantly updated, added, deleted and transformed. Due to the hierarchical structure of the data ecosystem, each of its elements also has dynamics. Now we can come to the conclusion that the data ecosystem is a dynamic system that responds to the influences of the external environment. To meet user's expectations the development of the data ecosystem has to be sustainable. To meet user expectations the development of the data ecosystem has to be sustainable. It has to show such properties as self-regulation, resilience, adaption, and reliability. In other words, data and related assets have to be managed to insure **availability**, **reliability**, **accessibility**, **security**, and **integrity**, **flexibility** of data sources, diverse data repositories and data models.

A data ecosystem is a complex environment of co-dependent networks and actors contributing to data collection, transfer, and use. These ecosystems can span across sectors to inform one another's practices. Key components include data intermediaries (like data trusts, exchanges, and platforms), providers, and consumers. The goal is to create, manage, and sustain data sharing across platforms and disciplines, ultimately building value through processed data.

The modern IT world understanding requires the study of the data ecosystems and related infrastructures. The purpose of this work is to analyze the properties of the data ecosystem and highlight the key aspects of the methodology for data science concerning environmental sustainability in the sense of the generalized system theory. With reference to the proposed methodology, we will illustrate the genesis of the data ecosystem and its life cycle for the real-world problem as is the New York taxi demand prediction. The choice of this example is not accidental. Data on taxis in New York is collected and published regularly on the website and is available for immediate download. We encourage readers to use the GitHub repository (hosted on https://github.com/delnouty/ On-Taxi-Demand) not only to observe the data ecosystem genesis and life circle but also to understand the results included as an illustration of the problem solution (hosted on https://github.com/delnouty/On-Taxi-Demand/blob/ main/ReportOnTaxiDemand.pdf).

The rest of the paper covers the following topics. In Sect. 2 we propose a formal definition of data science and show its relations to sustainable development. In Sect. 3 we show that a data analysis process coincides with a data ecosystem life-circle. The data ecosystem genesis and its life circle discussed in Sect. 4. The last section 5 contains conclusions about the research performed.

2 Data Science as a Scientific Field

The natural, social and formal sciences are three major branches of modern science. This classification is related to the object of study. It is obvious that the physical world is the main interest of the natural sciences as physics or biology, at the same time individuals and societies lie in the interest of social sciences as economy, sociology or psychology. From the system analysis point of view these branches study the reflective systems. The last branch, the formal sciences as mathematics or computer science, investigates formal non-reflective systems defined by axioms and sets of rules. In this context, what is data science and what does it study?

In early 60th John Tukey defined "learning from the data" or "data analysis" as-yet unrecognized science named "data science" [15]. In 2017 David Donoho presented the cross-study on data analysis development and gave an idea about data science [4]:

> "This coupling of scientific discovery and practice involves the collection, management, processing, analysis, visualization, and interpretation of vast amounts of heterogeneous data associated with a diverse array of scientific, translational, and inter-disciplinary applications".

Data science is often called as "The Fourth Paradigm" or "The Fourth Approach" [7] to scientific discovery. In this case one can conclude that there exist some systematic investigation approaches related to the object of the study, which allow by mean of description, diagnostic, prediction and prescription to correct, integrate and acquire knowledge in an accurate and reliable way. According to Chart GPT based on Large Language Models trained on Internet data (we used IBM Mistral AI LLM to get an answer) data science

> "is a multidisciplinary field that uses scientific methods, processes, algorithms, and systems to extract knowledge and insights from structured and unstructured data. It combines aspects of several fields including statistics, data analysis, machine learning, and related methods to understand and analyze actual phenomena with data".

Despite all the terms and elements are presented this definition is rather the intuitive one. It is clear that data science is a field of science that deals with the study of digital data, including their types and structures, sources and methods of their accumulation, transmission and processing with respect to development and execution environments as well as code and data management (see Fig. 2). It includes sub-disciplines such as data engineering, data mining, cloud computing, database management, machine learning, deep learning, etc. It pulls concepts and methods from other scientific fields, namely: mathematics, philosophy, economics, computer sciences. Guiding by decision making, it aims to understand the fundamental principles that govern the functioning of data ecosystems at all stages of data-life-circle. To sum up, we can give the following answer to the question of what data science is. Hence, according to us

"Data science is the scientific field that studies data ecosystems, their existence and co-existence".

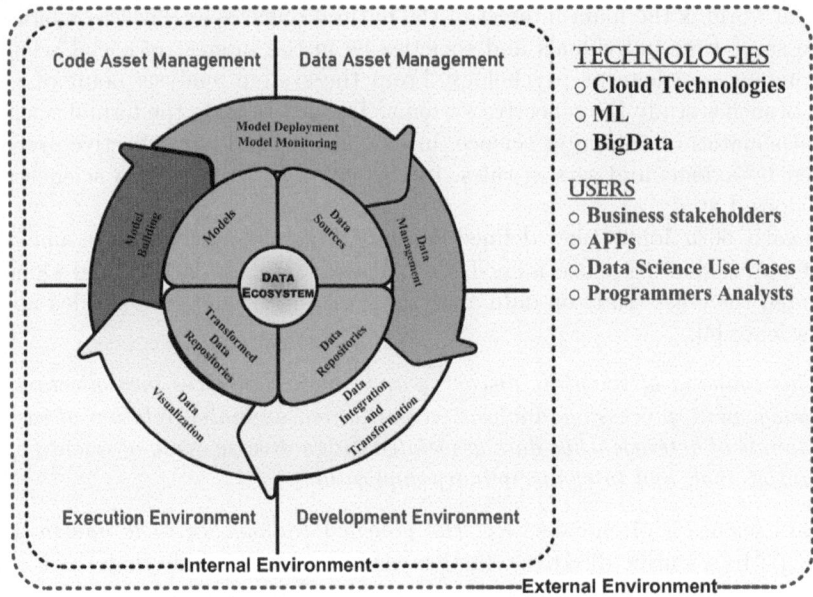

Fig. 2. The data ecosystem as the object of study

2.1   Data Science and Main Aspects of Sustainability

Nowadays science plays a critical role in understanding our world, advancing technology, and solving complex problems. The process of scientific knowledge of the world is always purposeful, and the world is dynamic. The world, which is the external environment for the process, can change under the influence of this process, i.e. it is characterized by dynamism. Therefore, it is necessary to manage the process of cognition in such a way that, as a result of interactions, the external environment is not destroyed, and its development itself is sustainable.

This kind of human activity can be understood in business context (see Fig. 3). Sustainable development of science should be considered in two aspects. Firstly, it itself must be in a state of balance, i.e. homeostasis. A data ecosystem created to solve a specific problem (value creation) must be managed, regulated and collaboratively coordinated. This requires the selection of management strategy and tactics at the modular level of the data ecosystem, providing, in general, a competitive solution to the original problem. In other words, the data ecosystem has to be sustainable. Secondly, sustainability must arise through interaction with the external environment, i.e. directly with the subject area of solving the problem. There is a bit of a duality here: data science can be used to solve the problem of sustainable development of the world around us, and its development itself must also remain sustainable.

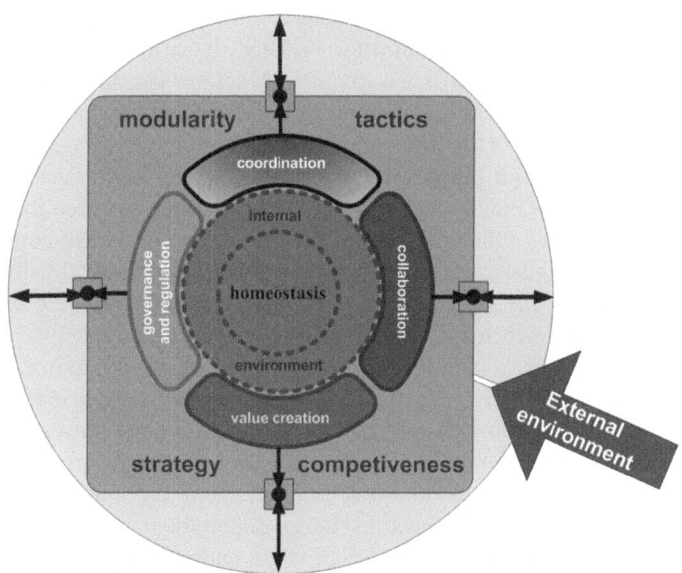

Fig. 3. The business context of data science

Sustainability is usually considered as the long-term well-being of our planet [1,10,13]. Proponents of this definition highlight the most important actions to avoid a negative mutual impact of scientific knowledge and environment, namely:

– **environment** – scientific experiments, fieldwork, and data collection can harm ecosystems, contribute to pollution, and exacerbate climate change;

- **preservation of natural resources** – depletion of natural resources (minerals, water, and fossil energy) due to reckless use hinders scientific progress;
- **ethical responsibility** – the negative impact of scientific practice due to ethical dilemmas and economic instability.

It is clear, that in this context, data science plays a critical role in advancing environmental sustainability by providing the tools, techniques, and insights needed to monitor, analyze, and address environmental challenges effectively. We can give several examples. Data-driven decision making enables the analysis of large and complex datasets related to environmental issues. By extracting insights from data, stakeholders can make informed decisions to mitigate environmental impact, promote conservation, and support sustainable development initiatives. Environmental monitoring and analysis have numerous techniques such as remote sensing, geographic information systems (GIS), and spatial analysis are instrumental in monitoring and analyzing environmental changes over time. Satellite imagery, sensor data, and other environmental datasets provide valuable information about land use, deforestation, habitat loss, air and water quality, and climate patterns, allowing scientists to track environmental trends, identify areas of concern, and develop targeted interventions. Data science enables the development of predictive models and forecasting algorithms to anticipate environmental risks and predict future outcomes. Machine learning techniques can analyze historical data to forecast trends in energy consumption, carbon emissions, sea level rise, and other environmental indicators, helping policymakers and businesses to plan and prepare for potential impacts and develop resilience strategies. Data science plays a crucial role in optimizing resource management and promoting resource efficiency. By analyzing data on energy consumption, water usage, waste generation, and supply chain operations, businesses can identify opportunities for efficiency improvements, reduce resource waste, and lower environmental footprint. Optimization algorithms can optimize processes, logistics, and infrastructure to minimize resource inputs and maximize outputs, leading to cost savings and environmental benefits. Data science supports the transition to a more sustainable energy system by analyzing data on renewable energy sources, energy generation, distribution, and consumption. By leveraging data analytics and modeling techniques, policymakers and energy planners can identify optimal locations for renewable energy deployment, optimize energy grid operations, and promote the integration of clean energy technologies. Additionally, data science facilitates the implementation of circular economy initiatives by analyzing data on product life-cycles, material flows, and supply chains. By optimizing resource use, promoting reuse, recycling, and remanufacturing, businesses can minimize waste generation, reduce environmental impact, and create more sustainable production and consumption patterns.

2.2   Data Ecosystem and Environmental Sustainability

Each data ecosystem during its life-circle and interactions with the external environments has managed in a sustainable manner. Particular attention should be paid to the following issues, namely:

- **Data Asset Management.** Implementing efficient data management practices helps reduce unnecessary data duplication, storage, and processing, leading to lower energy consumption and resource usage [8].
- **Data Life Cycle Management.** Implementing data life cycle management practices ensures that data is retained only for as long as necessary and securely disposed of when no longer needed, minimizing storage costs and environmental impact [12].
- **Data Asset Governance.** Establishing robust data governance frameworks ensures that data is managed responsibly, ethically, and in compliance with regulations. This helps prevent data breaches, mitigate risks, and build trust with stakeholders [2].
- **Optimized Infrastructure related to the internal environment.** Utilizing energy-efficient hardware, cloud-based solutions, and virtualization technologies can minimize the environmental impact of data storage and processing operations [11].
- **Data Quality and Accuracy.** Improving data quality and accuracy through data cleansing, validation, and verification reduces the need for rework and unnecessary processing, saving energy and resources [5].
- **Collaboration and Sharing.** Encouraging collaboration and data sharing within the ecosystem reduces redundancy and promotes efficient use of resources by leveraging existing data assets and infrastructure [9].
- **Continuous Improvement.** Adopting a culture of continuous improvement and innovation allows organizations to identify opportunities for optimizing data processes, reducing waste, and enhancing sustainability [14].

To conclude, by integrating sustainability principles into the design, operation, and governance of data ecosystems, organizations can minimize their environmental footprint while maximizing the value derived from data assets.

3   Data Analysis Process

The further presentation and discussion depend on the perception of the field of interest. We will concentrate our attention on a data-driven business, where data, analytics, and insights support everyday decision-making process. Which steps have to be taken in order to achieve a particular goal with respect to environmental sustainability?

Before to respond on this question, let us remind some differences between business analysis and business analytics. In business analysis a business analyst defines needs and presents recommended solutions using functions and processes. In business analytics a business intelligence analyst investigates performance and makes data-driven decisions using the systematic computational analysis of data, statistics, and reporting. In the context "from data through analytics to insights" environmentally sustainable decision-making process corresponds to the following series of actions, namely:

1. Problem assessment (understanding problem and objectives; identifying and analyzing stakeholders; defying problem statement).

2. Business requirement preparation (gather and analyze business requirements; create supporting visual model; document business requirements).
3. Solution implementation (communicate solutions; implement and iterate solutions; evaluate value created).

As we can see, the data analysis process coincide with the data ecosystem life-circle. Its sustainability is very important to internal sustainability of the data ecosystem.

4   Optimal Taxi Fleet Management

4.1   Scenario

To illustrate relations of data science and sustainable development we will use close-to-real world scenario. We will discuss the problem of optimal taxi fleet management in Manhattan and airports (JFK and LaGuardia) of New York City (NYC). We will imagine the following scenario. New York authorities have repeatedly proposed reducing the number of cars on the city's streets. This was associated with programs to improve the quality of life, as well as with the introduction of new energy sources. This time, for the owner of the Super Taxi company a partner of Yellow Taxi, whose clients live in Manhattan, this will have the following consequences. If the demand for taxis is lower than the supply, i.e. the cars will be idle waiting for clients, the company will pay a fine for every hour the taxi is idle. Obviously, in this case, the Super Taxi company will incur losses, which will negatively affect profits. The owners of the company want to instruct data specialists to analyze historical data, develop a model that predicts the "low", "medium", or "high" demand for taxis depending on the day of the week and time of day, so as to optimally calculate the required number of cars to satisfy customer demand.

4.2   Data

The choice of this case was motivated by the availability of data collected by the NYC Taxi and Limousine Commission (TLC). Trip data are published monthly on https://www.nyc.gov/site/tlc/about/tlc-trip-record-data.page and are stored in PARQUET format files (related to Yellow Taxi trip records, Green Taxi trip records, For-Hire Vehicle trip records).[1] We decided to study the problem as it was happened in 2015. This one-year-dataset related to Yellow Taxi trips stands for data source (which are in the interest of our problem). It contains 2 Crore records and leans toward the "big data" category. In this case, we extracted only some of records (10% is common practice) from the related files without losing generality and created a data repository composed from twelve CSV-files (one file for each month of the year). Next, we created a datastore and aggregate the data from the data repository and presented the aggregation by

---

[1] The data can be downloaded directly from data.cityofnewyork.us webside.

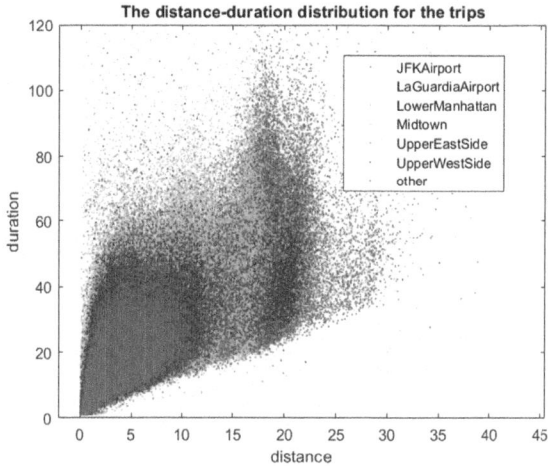

Fig. 4. The data inconsistency with respect to duration and distance of the trip

the table "TaxiData". As we suspected this data contained many inconsistency (see Fig. 4), which had to be corrected during data analysis process.

Several cleaning procedures have been completed to reduce the uncertainty related to "human-error" or "extreme-road" conditions (traffic intensity, weather conditions, etc.). It made suitable the distance-duration distribution for the regions of interest for the further analysis (see Fig. 5). Applying data restructuring and feature engineering we introduced new categorical variable related to "Taxi-Demand". It was a final step in the data repository transformation before the data modeling[2].

4.3   Demand Prediction Models

To select the independent variables which were important for demand prediction we used minimum redundancy maximum relevance (MRMR) algorithm. Using the machine learning methodology we split the dataset for training and testing. To ensure a random flip of initial data we set the random number generator seed and separate 20% of the data set for testing later on, and create the training data.

Our primary goal was to determine the class of the predictive model with the best accuracy. Since the dependent variable was the categorical one, we used a classification methodology based on supervised machine learning. To select the model type we used two alternatives: a simple decision tree and an ensemble decision tree. In both cases, 10-fold cross-validation was used to select the optimal structure of the decision tree. In the first case, the optimizable tree gave "the best model" with an accuracy 71, 9%. For the second case, optimizable ensemble

---

[2] The detailed solution can be found at https://github.com/delnouty/On-Taxi-Demand.

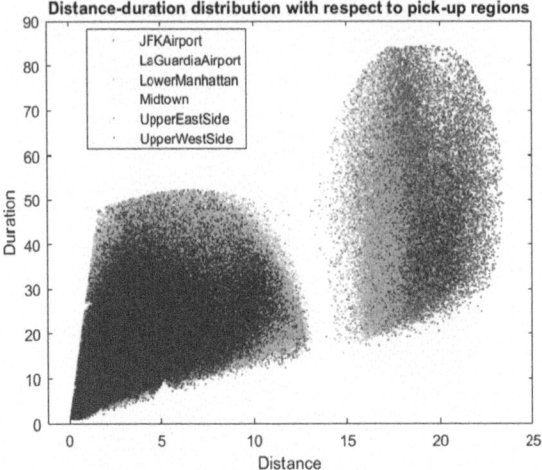

Fig. 5. The data cleaning with respect to the distance-duration distribution

gave "the best" model with 70% accuracy. Since the training was faster in the first case, we used the decision tree with parameters as indicated on Fig. 6. Its prediction accuracy for the "low" demand class was 72.0%. Thus, second goal was the introduction of some improvement of the prediction power for "low" demand class. Customizations to the cost-error was made to prioretize the accuracy of "low" demand prediction by the misclassification cost matrix. The model was trained and validated once again. It gave 73.26% accuracy. As we can conclude the precise prediction of "low" demand allows allocating the resources among regions with "medium" or "high" demand to avoid losses and penalties.

4.4   The Data Ecosystem Life Circle

We have discussed on the "taxi-demand-prediction" the genesis of the data ecosystem. It is the transformation chain – "data source" – "data repository" – "transformed data repository" – "model". These core elements have to be corrected with respect to the data ecosystem life circle. Let us discuss it (see Fig. 2).

Data Management (*collection, storage, preprocessing, integration, exploration and visualization, security and privacy, version control, governance, quality monitoring, automate data processing, backup and disaster recovery*). In our case, a dataset containing historical taxi trip records is a large one. It has to be analyzed to ensure that the data is accurate, complete, and representative of the NYC taxi system. Since the dataset is in PARQUET format, it is important to choose an appropriate storage solution: to store raw data in a data lake (e.g., Azure Data Lake Storage, Amazon S3) and use Big Data Platforms as Hadoop HDFS

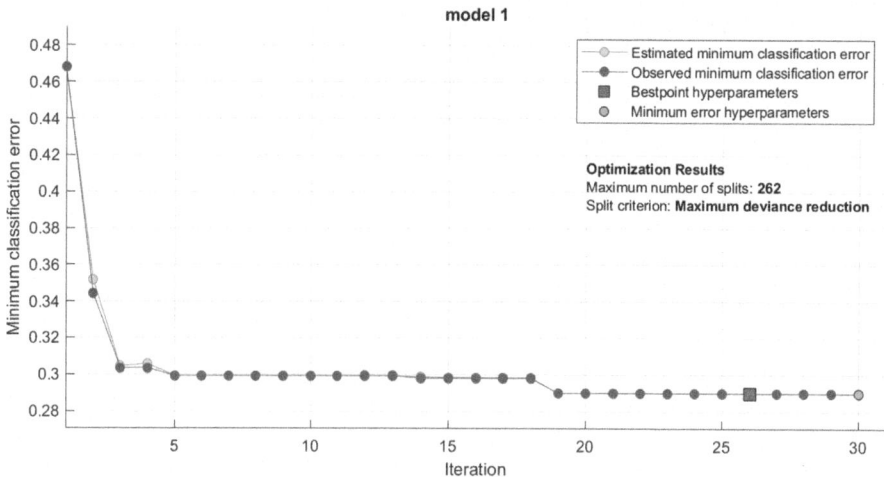

Fig. 6. The best model characteristics – fine decision tree (maximum split number - 262, split criterion - maximum deviance reduction)

or Apache Spark for distributed storage. In particular case data have to be integrated from various sources as Weather Data, Holiday Data, Geospatial Data. One has to learn data privacy regulations (e.g., GDPR and PII). Any changes in the data has to be tracked by version control system (e.g. Git). Established data governance policies concern access control to sensitive data, quality monitoring of the data, and metadata management. Data pipelines – ETL – and data streaming pipelines provide the effective automate data processing. Finally, regular back up actions as well as disaster recovery plans prevent data loss. Some software solutions can be use to manage mantioned problems – Collibra, Profisee, Google Cloud Platform, Tableau Data Management, Azure Data Factory. The effective data management ensures the reliability, accuracy, and usability of datasets, which directly impact the sustainability of the data-ecosystem.

Data Integration and Transformation. Data integration combines information from multiple sources to create a unified dataset for analysis. So, primary dataset containing records of past taxi trips can be combined with holiday and geospatial data. Since different data sources may have varying formats, structures, and quality, one has to ensure consistency across integrated data. Software tools that help achievemn seamless data integration are Microsoft SQL Server, Apache Airflow, Talend, IBM Infosphere Backstage, Oracle Data Integrator. Data transformation prepares raw data for modeling by enhancing its quality and relevance. This task can be solved by Talend Data Transformation, AWS Glue, Azure Data Factory, and Data Build Tool, Ataccama ONE, Informatica Intelligent Data Management Cloud, etc.

Data Visualization (discovering trends in data, providing perspective on the data, putting data into the correct context, telling a data story – dashboard). Visualization tools which can be used in the project are Microsoft Power BI, Tableau, Qlik Sense, Klipfolio, Looker.

Model Building. We have discussed it in details in the previous subsection.

Model Deployment and Monitoring. Deploying and monitoring a taxi demand prediction model involves several steps to ensure its effectiveness and reliability, namely:

- a user-friendly web application where users (both passengers and taxi drivers) can interact with the model. The interface should allow users to input relevant information (e.g., pickup location, drop-off location, time of day) to receive real-time predictions. One can use frameworks like Streamlit, Flask, or Django to build the web application;
- the model deployment as an API that can be accessed programmatically. One can use cloud services like AWS Lambda, Azure Functions, or Google Cloud Functions to host the API. Clients (such as mobile APPs or web applications) can make HTTP requests to this API to get predictions;
- real-time monitoring related to the deployed model's performance (prediction accuracy, response time, resource usage, alerts for any anomalies or performance degradation, data drift detection, feedback loop which allows users to rate the accuracy of predictions, to analyze how users interact with the system, etc.);
- other related questions – model versioning, security and privacy, regular maintenance, scalability, auto-scaling, load balancing.

When it comes to deploying and monitoring machine learning models, there are several powerful tools available: Qdrant, LangChain, Evidently.

4.5   Team Selection

Building a team for the taxi demand prediction project involves assembling individuals with expertise in various areas. The concept of the data ecosystem is primary generated by (see Fig. 7):

1. Data Engineers (design and implement the data pipeline to ingest and preprocess historical taxi trip data; create a robust storage infrastructure (e.g., data lake, database) for the dataset. Collaborate with data scientists to ensure data availability for modeling).
2. Data Analysts (clean and organize raw taxi trip data; analyze trends in demand patterns over time; create visualizations and dashboards to help stakeholders interpret data; present analysis results to business clients).

3. Data Scientists (develop machine learning models to predict taxi demand; incorporate features like time of day, day of the week, and etc.; evaluate model performance using appropriate metrics; work closely with data engineers to access relevant data for modeling).
4. BI Analysts (collaborate with data analysts to understand business requirements; design dashboards showing real-time taxi demand trends; monitor key performance indicators (KPIs) related to taxi service; provide actionable insights to improve fleet management and resource allocation).

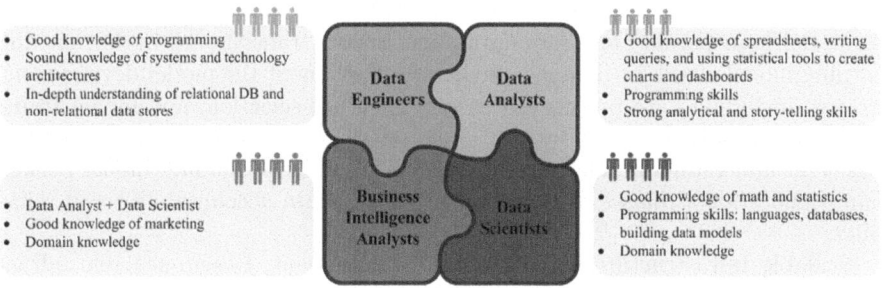

Fig. 7. A concept team

Once the concept was accepted, the other specialists can be involved to make comments on the project details and the implementation process. Among them are:

– Machine Learning Engineers (focus on implementing and optimizing machine learning models, skills: Deep learning frameworks (TensorFlow, PyTorch), model deployment);
– Software Engineers/Developers (build the web application or API for model deployment, skills: Web development (e.g., Flask, Django), API design);
– Geospatial Analysts (experts in geospatial data processing and visualization, skills: GIS tools (e.g., ArcGIS, QGIS), spatial analysis);
– Domain Experts (with knowledge of transportation systems, urban planning, and taxi operations; kills: understanding of taxi demand dynamics, traffic patterns, and city infrastructure);
– Project managers (coordinate team efforts, set deadlines, and manage resources; skills: project management, communication);
– Database Administrators (handle data storage, database design, and optimization; skills: SQL, database management systems (e.g., PostgreSQL, MySQL));
– UI/UX Designers (design the user interface for the web application, skills: User experience design, front-end development);
– Quality Assurance/Testers (ensure the application works correctly and meets requirements; skills: Testing methodologies, bug tracking).

The collaboration tools like GitHub, Slack, and Trello which facilitate communication and project management are also a part of dataecosystem of the project.

4.6   Recommendations on Sustainable Data Ecosystem

Data Governance and Privacy: Implement robust data governance practices to ensure the ethical and responsible use of data. This includes adhering to privacy regulations, anonymizing sensitive information, and obtaining appropriate consent for data usage. By respecting data privacy and security, trust is fostered with stakeholders and the community, ensuring continued access to valuable data sources.

Model Transparency and Accountability: Ensure transparency in the machine learning models used for demand prediction. Document the model development process, including data preprocessing steps, feature selection, and model training techniques. Additionally, provide explanations for model predictions to build trust and understanding among stakeholders. Regularly evaluate model performance and address biases or inaccuracies to maintain accountability and reliability.

Scalable Infrastructure and Resource Management: Design scalable infrastructure to accommodate fluctuations in data volume and computational requirements. Utilize cloud-based services or containerization platforms to efficiently manage computational resources and scale infrastructure as needed. Implement cost-effective strategies to optimize resource usage and minimize environmental impact, such as utilizing renewable energy sources or adopting energy-efficient hardware.

Continuous Improvement and Adaptation: Foster a culture of continuous improvement and adaptation within the ecosystem. Encourage collaboration and knowledge sharing among stakeholders to exchange insights and best practices. Invest in ongoing research and development to enhance predictive models, incorporate new data sources, and adapt to evolving transportation dynamics. By staying agile and responsive to changing needs, the ecosystem can remain relevant and effective over time.

Community Engagement and Stakeholder Collaboration: Engage with the community and stakeholders to solicit feedback, gather insights, and co-create solutions. Collaborate with transportation authorities, taxi companies, and local organizations to align priorities, address concerns, and ensure the relevance of the project outcomes. By fostering collaboration and inclusivity, the ecosystem can harness collective expertise and resources to drive sustainable impact.

Long-Term Planning and Investment: Develop long-term plans and strategies to sustain the ecosystem beyond the duration of the project. Secure funding and support from government agencies, private sector partners, or grant programs to sustain operations, maintain infrastructure, and support ongoing research initiatives. By establishing stable funding sources and governance structures, the ecosystem can thrive and evolve over time, delivering lasting value to stakeholders and the community.

By integrating these principles into the project's ecosystem, it can become a sustainable framework for predicting demand for Super Taxis and driving positive social, economic, and environmental outcomes in the transportation sector.

5 Conclusion

The genesis of any data ecosystem depends on the subject area, the problems that are associated with it, as well as on the knowledge and experience of specialists who solve problems at all stages of the data life cycle: from the data source to the insights that support decision making. By considering sustainable development as the subject area and creating the related data ecosystem, many environmental and social challenges related to resource management, renewable energy adoption as well as environmental monitoring and conservation can be resolved. It is clear that all of these sustainability issues involve huge amounts of data. Identification of trends and patterns leading to decision-making requires repeated transformation of the source of information, which in turn can lead to unlimited growth in the volume of information. In addition, the same problem can be solved in several ways, which also generates new information. Supporting the data ecosystem internal environment directly depends on the volume of information, and therefore can lead to unlimited resource consumption or, in other words, loss of environmental sustainability. This is why the choice of data processing methods must be carefully considered. To ensure that solving the sustainable development problem does not lead to this paradox, the data-related specialists must constantly improve their skills, assessing the sustainability of each of their solutions.

Acknowledgment. We sincerely appreciate the thorough review provided by the anonymous reviewer. Their insightful comments and constructive feedback have significantly improved the quality of our paper. We are grateful for their time and effort in helping us refine our work.

References

1. Balaprakash P., Dunn, J.B.: Overview of data science and sustainability analysis. In: Dunn, J.B., Balaprakash P. (eds.) Data Science Applied to Sustainability Analysis, pp. 1–14. Elsevier (2021)
2. Birch, K., Cochrane, D., Ward, C.: Data as asset? The measurement, governance, and valuation of digital personal data by big tech. Big Data Soc. **8**(1) (2021)
3. Chapin, F.S., Matson, P.A., Vitousek, P.M.: The ecosystem concept. In: Principles of Terrestrial Ecosystem Ecology. Springer, New York (2011)
4. Donoho, D.: 50 Years of Data Science. J. Comput. Graph. Stat. **26**(4), 745–766 (2017)
5. Eswararaj, D.: Developing a data quality framework on azure cloudâĂŕ: ensuring accuracy, completeness, and consistency. Int. J. Comput. Trends Technol. **71**(5), 62–72 (2023)

6. El-Nouty, C., Filatova, D.: The learning model for data-driven decision making of collaborating enterprises. In: Baratgin, J., Jacquet, B., Yama, H. (eds.) Human and Artificial Rationalities. HAR 2023. Lecture Notes in Computer Science, vol 14522. Springer, Cham (2024)

7. Hey, T.: The fourth paradigm – data-intensive scientific discovery. In: Kurbanoglu, S., Al, U., Erdogan, P.L., Tonta, Y., Ucak, N. (eds.) E-Science and Information Management. IMCW 2012. Communications in Computer and Information Science, vol. 317. Springer, Heidelberg (2012)

8. Judijanto, L., Uhai, S., Suri, I.: The influence of business analytics and big data on predictive maintenance and asset management. Eastasouth J. Inf. Syst. Comput. Sci. 1(03), 123–135 (2024)

9. Li, Y., Zou, H., Qin, H., Liu, B., Ji, H.: Research on government incentive and enterprise data resource sharing strategies in digital innovation ecosystems. IEEE Access 12, 25278–25295 (2024)

10. Sarker, I.H.: Data science and analytics: an overview from data-driven smart computing, decision-making and applications perspective. SN Comput. Sci. 2, 377 (2021)

11. Schleimer, A.M., Duparc, E.: Designing digital infrastructures for industrial data ecosystems –a literature review. In: Proceedings of the 18th International Conference on Wirtschaftsinformatik, Paderborn, Germany (2023)

12. Scope, N., Rasin, A., Lenard, B., Wagner, J.: Compliance and data lifecycle management in databases and backups. In: Strauss, C., Amagasa, T., Kotsis, G., Tjoa, A.M., Khalil, I. (eds.) Database and Expert Systems Applications. DEXA 2023. Lecture Notes in Computer Science, vol. 14146. Springer, Cham (2023)

13. Sultana, N., Turkina, E.: Collaboration for sustainable innovation ecosystem: the role of intermediaries. Sustainability 15, 7754 (2023)

14. Soewarno, N., Tjahjadi, B.: Eco-oriented culture and financial performance: roles of innovation strategy and eco-oriented continuous improvement in manufacturing state-owned enterprises, Indonesia. Entrep. Sustain. Issues 8(2), 341–359 (2020)

15. Tukey, J.W.: The future of data analysis. Ann. Math. Stat. 33, 1–67 (1962)

Multi-agent Simulation of Violence Emergence in Protests

Julien Rosenberger[1], Julien Saunier[2(✉)], and Nicolas Sabouret[1]

[1] Université Paris-Saclay, CNRS, LISN, 91400 Orsay, France
julien.rosenberger@student-cs.fr
[2] INSA Rouen Normandie, Univ Rouen Normandie, Université Le Havre Normandie, Normandie Univ, LITIS UR 4108, 76000 Rouen, France

Abstract. This paper presents an agent-based model to understand the emergence of aggressive mob during a protest. We study to what extent (1) protester behavior can lead to violence and (2) protest policing doctrine. The multi-agent architecture relies on the Belief Desire Intention paradigm, enriched with emotions, norms and personality. Indeed, to reproduce phenomena such as tension-building, the protester model includes emotional components that are added to Epstein's model of civil violence. In Epstein's model, aggressive behavior is triggered as a trade-off between grievance and perceived risk of arrest.

Our model is implemented on the open-source GAMA platform and applied to the study of two factors observed in real-life situations: police-protester ratio and protest configuration (police blockade of escape routes, also called kettling, or police moving in the crowd). Simulations show that the key parameters are not necessarily the police-protester ratio but rather the protest configuration.

Keywords: Crowd Simulation · Agent-based · Emotion Modelling · Protest · Violence

1 Introduction

Spurred by an increase of fuel taxes planned for 2019 that raised resentment towards economic inequalities, the Yellow Vests Protests started in France in November 2018 and gave rise to violent confrontations between police force and citizens. By December 20, 2018, injured people amounted to 1843 among civilians and 1048 among police officers [56]. One of the cause of the excessive violence is a reinterpretation by the government of the State's legitimacy to use force in order to maintain order [55]. For instance, it undermined the reversibility principle which states that the use of force should stop as soon as the problem it tried to quell down has disappeared [55]. On the one hand, without a better understanding of violence emergence during protests, the riots could become rampant since the number of political protests around the globe already skyrocketed on average of 11.5% each year between 2009 and 2019 [11]. On the other hand,

J. Baratgin et al. (Eds.): HAR 2024, LNCS 15504, pp. 35–51, 2025.
https://doi.org/10.1007/978-3-031-84595-6_3

from statistical analysis of campaigns originating from 1900 to 2006, Stephan and Chenoweth [52] questioned the necessity of using force through two main results: violence to repress a movement statistically does not impact the outcomes of a campaign; and nonviolent resistances are almost twice as likely to reach their goals than violent ones.

In this context, research has been conducted on protest management. Riot forecasting is investigated in order to understand radicalization processes [1] or shed light on the circumstances that are more likely to precede the riot such as recent repressions and the protest organization [33]. At the level of the event itself, norms and practical guidelines are examined for police force to keep the protest peaceful [44]. Those guidelines strongly rely on an understanding of the causal mechanisms turning a protest into a riot. To do so, a first approach is to empirically analyze what occurs during peaceful protests and before riots, and to relate it to theoretical research. Along that line, Anne Nassauer grounds her work on many different kinds of documents like videos and reports from the police, the court and researchers [45]. Another approach is to use simulations. It allows to explore theories as well as formulate new hypotheses [20]. It also shows advantages in substituting to real experiments when experimenting is too expensive, impracticable or unethical.

The goal of this article is to follow-up on this question: can we understand how violence emerges during protests? To answer it, we pursue a computational approach called multi-agent modeling where the endeavour is to reproduce violence emergence within a crowd through the formulation of individual rules for the protesters and police officers.

The article is organized as follows. Section 2 studies the theories in social sciences that endeavour to tackle this issue along with computational approaches. Section 3 introduces the agent model we propose to study the emergence and dynamics of violence in protests. Section 4 examines the obtained experimental results. Finally, Sect. 5 summarizes the conclusions.

2   Related Work

According to the literature [13], a distinction can be made between passive crowds, called audiences, and active crowds, called mobs. Those latter crowds can be further categorized into four subgroups: aggressive, escaping, acquisitive, or expressive [13]. We call aggressive mob or riot a crowd trying to physically hurt others or damage objects [21]. We will first explain insights coming from social sciences on the appearance of aggressive behavior before presenting computational models that try to tackle the same issue, focusing on the emotional factors that should be taken into account.

2.1   Social Sciences' Approach

As we intend to better understand violence emergence within crowds, the model introduced in this article is grounded on studies in social sciences. We can distinguish these related works on violence emergence between the ones focusing on the individual and the others focusing on the group.

At the individual level, the General Aggression Model (GAM) [2] bridges the gap between different theories like excitation transfer theory [57,58] and social learning theory [4] to highlight the main factors involved in aggression behavior. It separates dispositional causes, such as personality, from situational causes, which are composed of the environment the subject evolves in and of the subject's current mental states on the current event. Although it may have less predictive power than domain-specific theories for certain behaviors [21], the model gives a general toolbox of the factors that underlies individual aggression behavior.

At the scale of the crowd, the individual may indulge in violence after weighting up the relative cost of this behavior [34]. It explains the influence of recent repressions on crowd behavior found by Ives and Lewis [34] as well as the influence of forbidding nonviolent protests: if the peaceful protest is as hazardous as the riot, it translates into a low relative cost of violence and protesters are more likely to become aggressive. Although the less organized the protest, the more likely it may turn into a riot [34], multimedia data analysis reveals that riots cannot only be due to a violent group infiltrating the crowd [45]. Riots mostly emerge after a first nonviolent phase of tension-building in parallel with an increase of fear [16,45]. This first phase lasting between one and three hours makes way for a riot that is ignited once a side loses the advantage: the police-protester lines breaks up, some are outnumbered by the opposing group or even fall down on the ground [45].

The multi-agent model undertaken in this article consequently formalizes the individual decision-making processes and the course of tension-building leading to aggressive behavior. This model has the objective of enlightening the influence of violent subgroups over the rest of the crowd thanks to the study of the spread of aggressive behavior.

2.2   Computational Approach

Several computational models are already designed to simulate aggressive behavior in different situations [28]. Some are used to reproduce the behavior of aggressive car drivers [19,31,46], or for linking biological factors such as testosterone, adrenalin and blood sugar with situational factors with the aim of understanding the behavior of people affected by psychological disorders [8].

The most referenced model of civil violence is the threshold-based model of Epstein [24]. This model is abstract in the sense that police officers and citizens are set on a fictive environment in a grid, and that the condition for an individual to turn from passive to active is based on abstract concepts. A citizen agent engages in a rebellion by computing its relative cost of turning in, which compares its grievance with its net risk of being active. The grievance is fixed at initialization by a combination of the perceived hardship and the legitimacy of the government. Subsequent models succeed in clarifying some of its facets. In the scenario of worker protest, Kim and Hanneman [36] expressed the grievance in terms of a comparison between the agent's wage and the perceived mean wage around it. To clarify spatial phenomena, Davies, Fry and Wilson's model [17] reckoned with the selection of the assembling site for each citizen agent.

Nonetheless, such models do not consider what happens during the protest itself. To cope with this shortcoming, Torrens and McDaniel [54] designed geographic functionalities for agents within crowds. Their model takes into account collisions avoidance, social steering, way-finding and object obstruction within the vision field by means of ray-casting. Those functionalities can easily be combined with other models, either physical or psychological. The authors used Epstein's work [24] with the exact same condition ruling behavior shifts, and biased the movement of rebels and police officers with a set of weights. Six scenarios were investigated including a walled space which entailed lower time periods for riot peak and duration compared to a square-type setting. One of the limitations of this article was the possible roles of the agents. Lemos [39] extended Epstein's model's with by differentiating "active" from "violent" protesters and initializing protester agents into one of three sub-types defined according to their propensity for violence.

In this article, we focus on the transitions between one state to another without considering any sub-type within protesters. We assume to achieve the same outcomes since the agents' initial grievances against the government, inhibition thresholds for violence and personalities already delimit those sub-types. Rather than visual obstruction because of objects, we take a closer look at the impact of emotions on perception and decision.

2.3   Emotions and Violence

Emotions are empirically found in protest visual recordings [45] and can better reflect the decision-making process of protesters before engaging in violence. However, scarce are the micro-level models of riot events taking emotions into account.

Emotions have significant influences on judgement and decision-making [42] and were noticed during protests on visual data and reports [45]. Extensive research has identified anger as a measure of desire to fight [38] or desire for revenge [48]. Anger was also found to reduce risk aversion [40] and to more easily blame [29] or patronize others [7]. Besides, emotions can arise a specific phenomenon within groups called emotional contagion, which is the influence on an agent's emotions of the perception of the neighboring agents' own emotions [5]. This phenomenon can be relevant to explain social mechanisms [9].

Interestingly, the emotion detected during protests is mostly fear [45] which contradicts the traditional view that the action tendency for fear is avoidance [27]. A middle ground can be found in excitation transfer theory where residues of excitation created by fearful events can transform into a misattributed anger [58]. The attribution of intent to the agent who causes negative emotions may be key for aggressive behavior analysis or modeling because hostile attribution was robustly related to aggressive behavior [18]. Additionally, other negative emotions than those reported and detected on images may be involved. Research in fight-or-flight mechanisms [14] already draws links between aggression and negative emotions, like anxiety [37] and frustration [26]. Anxiety, which is related to uncertainty about a future threat [30], will be subsumed under the risk assessment of the given situation in our model; and frustration, associated to uncertainty about the causes of the events that created them [51], seems unrelated to the studied situations at the microscopic level. Protesters have only one source of threat: the police force. At the macroscopic level, however, frustration can fuel the resentment felt against the government and motivate aggression.

We agree with the EROS (Enhance Realism Of Simulation) paradigm, which states designing social models from psychological theories improves realism and validity of results [35], and argue for emotions as a main component for aggressive behavior emergence within protests. Therefore, we endeavour to reproduce emotion dynamics and their impacts on decision-making in our model.

3   Model

3.1   Overview

The purpose of our model is to understand how aggressive behavior can emerge in a decision-making process by modeling how situations create emotions and how emotions impact the participants behaviour.

Protesters, police officers, buildings, walls and damageable items are the only physical entities among agents. Added to those entities, teams and arrest teams are abstract agents representing collectives of police officers to better coordinate them and reflect the hierarchical structure of police force.

The police officers' and the protesters' decision process rely on the BDI paradigm [12] to explain and reproduce their actions as a result of interactions between the agents' beliefs, desires and intentions. Structured around this paradigm and the OCC model of emotions [47], the protester's emotions are modeled using the BEN (Behavior with Emotions and Norms) architecture [10]. The emotion decay of the BEN architecture is adapted so that an emotion only lasts during the event that sparked it as per the definition that "emotions are short-lived psychological-physiological phenomena that represent efficient modes of adaptation to changing environmental demands" [41].

In this model, we denote an emotion $Em_{emotionName}$ with $emotionName$ the name of the emotion (fear, sadness, anger...) and a belief $B_{beliefName}$ with $beliefName$ the name of the belief. An agent a at time t has an emotion base $Em(a,t)$, a belief base $B(a,t)$ and a current intention $I(a,t)$.

3.2   Damageable Items

Damageable items are inanimate physical objects such as street lamps or trash cans. They have a specific attribute R standing for the current resistance before being broken or not being of interest for the protesters anymore. Once the resistance falls below zero, the item disappears from the simulation. Depending on the scenario, it can reappear after a period of time.

3.3 Damageable Items

Damageable items are inanimate physical objects such as street lamps or trash cans. They have a specific attribute R standing for the current resistance before being broken or not being of interest for the protesters anymore. Once the resistance falls below zero, the item disappears from the simulation. Depending on the scenario, it can reappear after a period of time.

3.4 Police Force

Officer agents detect violent offenders at perception distance at 360° around them. When a rioter disappears from the perception field, the officer immediately forgets about the rioter, emulating strategies of hiding among the crowd or fleeing. Officers share the offenders they detected with their team.

Police officers are always affiliated to a team agent that sends them their specific locations within the formation. Those locations can be set before simulation, or adapted by the team during the simulation to barricade the road or position police officers regularly on lines and columns. Officers are assumed to obey their affiliated teams' orders and adapt their desires' priorities accordingly.

The team is the one dispatching the members to apprehend a violent offender by creating a specific arrest team agent for them. A team dispatches members for an arrest when violent offenders are recorded around and the current number of members is sufficiently high. The target of arrest is the violent offender who is the closest to the center of the team and who is not targeted by another arrest team. The dispatched members are the ones closest to the arrest target.

A member of an arrest team tracks the target of this arrest team and sends contributions whenever it is close enough to the target. When the contributions are higher than the initial resistance $R_{arrestInit}$ of the target, the arrest team ends the process and disappears, which lets former members go back to their team positions. An arrested protester disappears to represent its removal from the protest.

When a police officer involved in an arrest strays too far from its assigned position within the original team, he leaves the arrest team and goes back to his position in the team's formation. This is useful when enforcing line formation across streets or maintaining the police officers united.

Police officers can be targeted by rioters but never disappear from the simulation.

3.5 Protesters

Protester agents are characterized by an energy value and a state which can take three distinct values: "peaceful", "violent" or "retreat".

In **peaceful** state, the agent participates in the protest. To implement group behavior of a protest march, peaceful agents follow the three flocking rules of boids [49], which are the rules of separation, cohesion and imitation. The equations for those are taken from Rochefort et al. [50].

In **attack** state, the agent considers its mere presence not enough to express its demands and its violent desire translates into the intention $Attack(v)$ with v another agent, object or police officer. The agent a_0 in this state can attack an object at time t only if its current energy $E(a, t)$ is positive. The victim v is chosen by maximizing a utility function U defined as:

$$U : \mathcal{A}_{Item} \cup \mathcal{A}_{Officer} \times \mathbb{R} \to \mathbb{R}$$

$$a, t \mapsto \frac{log_2(2 + card(\{a' : \boldsymbol{I}(a', t) = Attack(a)\}))}{d(\lambda(a, t), \lambda(i, t))^2}$$

with λ the function giving the location of an agent at a given time and d a distance function. $\mathcal{A}_{Item}$ and $\mathcal{A}_{Officer}$ are the respective sets for damageable item and police officer agents. The log_2 function was chosen to saturate the influence of the number of agents on the attractiveness of a target. At some point, if many offenders are already on a target, the violent agent would not bring a significant help to the rest of the group either and will account for this into its decision.

In **retreat** state, the agent is scared of police officers and flees from them as long as its current energy is positive. This energy emulates exhaustion from running or vandalizing, and is never recharged.

The priorities of protesters' desires evolve with their perceptions, beliefs and emotions. The main attributes ruling state transitions of any given protester agent a among the set of agents $\mathcal{A}$ at a time t are the intensity $G(a) \in [0, 1]$ of its complaints against the government, $R(a) \in [0, 1]$ summarizing its preference to stay safe in uncertain situations, the threshold $T_v \in [0, 1]$ inhibiting its violent behavior, a probability of arrest $P(a, t)$ identical to Epstein's model [24] and $\mathcal{L}_b^t(a) \in [-1, 1]$ the attitude of the agent towards another agent b from which is derived $\mathcal{L}_p^t(a)$ the aggregation of its attitude towards police officers. This attitude partly encodes the hostility perceived from another agent. The transitions are handled by the subsequent Algorithm 1 executed during each agent's time step where $\forall x \in \mathbb{R}$, $f(x) := 1 - 0.5 \times |x| \times \mathbb{1}_{\mathbb{R}^-}(x)$ function completes the Epstein's model [24] by involving emotions in the decision.

This algorithm can be explained as follows. A peaceful protester agent becomes violent if its energy is not depleted and its grievance is greater than its net risk of becoming violent of an inhibition threshold T_v that can be lowered because of its hatred for police officers. It can also starts fleeing if one of its fear is confirmed. From a "violent" state, the protester agent can start fleeing if one of its fear is confirmed and it does not remember fleeing. It can also go back to the "peaceful" state in case its situation has become significantly more risky. The protester agent trying to retreat can deem running as a lost cause and start to struggle against officers. The agent gives up when it feels surrounded or its energy is not sufficient to run anymore. Potentially without energy, it will not attack officers or items but may be a motivation for neighboring protesters to rebel. Naturally, if its confirmed fear is reassessed and disappears, the agent calms down.

Algorithm 1. Protester a_0's Desires Adaptation

---

switch $state(a_0)$ **do**

    case *peaceful*

        if $G(a_0) - R(a_0) \times P(a_0, t) > T_v \times f(\overline{\mathcal{L}_p^t(a_0)})$ and $E(a, t) > 0$ **then**

            $state(a_0) \leftarrow violent$

        else if $\exists \boldsymbol{Em}_{fearConfirmed} \in \boldsymbol{Em}(a_0)$ **then**

            $state(a_0) \leftarrow retreat$

        end if

    case *violent*

        if $\exists \boldsymbol{Em}_{fearConfirmed} \in \boldsymbol{Em}(a_0)$ and $\nexists \boldsymbol{B}_{triedRetreat} \in \boldsymbol{B}(a_0)$ **then**

            $state(a_0) \leftarrow retreat$

        else if $\exists \boldsymbol{B}_{arrestOtherAround} \in \boldsymbol{B}(a_0)$ and

            $G(a_0) - R(a_0) \times P(a_0, t) < 0.5 \times T_v \times f(\overline{\mathcal{L}_p^t(a_0)})$ **then**

            $state(a_0) \leftarrow peaceful$

        end if

    case *retreat*

        if $[\exists \boldsymbol{Em}_{fearConfirmed} \in \boldsymbol{Em}(a_0)$ and $E(a, t) < 0]$ or $\exists \boldsymbol{B}_{surrounded} \in \boldsymbol{B}(a_0)$

then

            $state(a_0) \leftarrow violent$

        else if $\nexists \boldsymbol{Em}_{fearConfirmed} \in \boldsymbol{Em}(a_0)$ **then**

            $state(a_0) \leftarrow peaceful$

        end if

The advantage of using the emotion $\boldsymbol{Em}_{fearConfirmed}$ is being a proxy for the various events that lead to that emotion. Within the BEN architecture, emotion is derived from desires, ideals and beliefs. The desires that generate fear in our model are the desire of being safe, while the desire and ideal of justice are the ones creating anger. Those two emotions of interest are caused by police officers on protesters. Table 1 introduces triggering events caused by the police force and the emotions they raise within the protester.

The emotions depend on the current protester's state since appraisal theory argues that an event is partly perceived with respect to the current goals and needs of the individual [23]. This is why a fleeing protester agent in our model construes every introduced event as a confirmation of its belief of not being safe. This accentuates its fear and extend its state of panic (see Table 1). Furthermore, feeling or being arrested while being peaceful turns the agent both fearful, to not be arrested, and angry, against the arrest perceived as unjust.

Spatial incursion is detected when a police officer is closer than 2.1 m to the agent, which is the upper bound of the near mode in social distances [32]. The protester feels outnumbered when there are three times more officers than protesters around them, which is the amount of officers usually required for an arrest [45]. The agent feels arrested if an arrest team goes in its direction, and knows it is arrested when it detects an arrest team with itself as target. It feels surrounded when three out of its four cardinal direction are obstructed by police officers or walls. Additionally, the protester is subjected to emotional contagion [5] and follows the rules set in the BEN architecture [10] with the exception of

Table 1. Protester agent emotions raised by events depending on the agent's state when interacting with the police force.

Event	State "peaceful"	State "violent"	State "retreat"
spatial incursion	fear	fear	fearConfirmed
outnumbered	fear	fear	fearConfirmed
feeling arrested	fearConfirmed, anger	fear	fearConfirmed
being arrested	fearConfirmed, anger	fearConfirmed	fearConfirmed
surrounded	fear	fear	fearConfirmed

the fear confirmed emotion that translates into fear when perceived in neighbors in the "violent" state. The reason is that the "violent" protester agent confirms its fears only when an arrest team is directly against it.

From those emotions is estimated the police liking value of protesters. It is inspired from the degree of liking introduced in BEN model [10] and adapted to dismiss the formulation of emotion intensity:

$$\forall (a, b, t) \in \mathcal{A}^2 \times \mathbb{R}^+, \ \mathcal{L}_b^{t+h}(a) - \mathcal{L}_b^t(a) = k_L(1 - N(a))(nP_{b,a} - nN_{b,a})(1 - |\mathcal{L}_b^t(a)|)h$$

with $h > 0$ the time step of the simulation. $k_L \in \mathbb{R}^+$ is a parameter that adapts the speed of liking variations while $N(a)$ is the agent's neuroticism coming from BEN architecture [10]. For all a and b in $\mathcal{A}$, $nP_{b,a}$ is the number of positive emotions caused by b felt by a; similarly, $nN_{b,a}$ is the number of positive emotions caused by b felt by a. Positive emotions include joy and hope; negative emotions sadness and fear [10]. Then, $\overline{\mathcal{L}_p^t(a)}$ is the average liking for the police officers met by a.

It is important to notice that anger in BEN architecture [10] results from reproach and sadness, and is thereby counted in the negative emotions. Furthermore, this liking value finds a middle ground between the empirical findings of Nassauer [45] supporting fear is a crucial emotion to understand shifts to aggressiveness, and theories about emotions advocating anger as the motivation for revenge, even through violence [38,48]. This middle ground revolves around the formalisation of hostility attribution which motivates aggression behavior [18].

4 Simulations

4.1 Overview of the Simulations

The two scenarios we have retained for our simulations are a protest taking place in a large public square patrolled by a single police team and a protest contained within a street because police officers are blocking both ends of the street using the kettling tactic. Those scenarios were selected because they represent the two opposites of observed situations in western protests' configurations [15,22].

Kettling is also a controversial tactic because it may infringe on the citizen's freedom of movement or of protesting. The press release from the French Conseil d'Etat of June 10, 2021 forbid the kettling tactic in the absence of specific conditions, which the release does not specify [6]. Thus, it is of interest to provide new tools to analyze the necessary conditions for kettling management. Finally, those scenarios were selected to reproduce the variety of behaviors that arise during protests. For instance, the kettling setting may spur anxiety and turn people violent more easily than in the public square setting.

4.2   Implementation

The simulation platform chosen is the open-source GAMA platform [53]. The main benefit of using an already existing platform is to clearly delineate the model we formulated from its implementation. Moreover, the GAMA platform comes with its own language that is easy to understand for non computer scientists, a lot of services such as loading Geographic Information System (GIS), and an already existing implementation of the BEN architecture we use [10].

In our implementation, every dynamic agents, which are of species protester, police officer, team or arrest team, are activated at every step. The implementation is done in an asynchronous fashion. To limit simulation artifacts, the order of agents' execution is shuffled at every step. One step represents 1 s in the simulation. One simulation is equivalent to 3 h of protest.

The world in the kettling setting comes with a main road of 50 m over 20 m. The first one is toric and its two side roads on each side are 10 m wide. The public square spreads over a 80 m by 80 m area in a toric world. Table 2 summarizes those settings.

Table 2. Scenarios' parameters.

	Square	Kettling
Dimensions	80 m × 80 m	50 m × 20 m
Toric	True	False
Number of protesters	200	200
Number of police officers	10/20/30/40	10/20/30/40

Regarding parameters calibration, we chose to set the vision distance to 7 m for both protesters and police officers [43]. For protesters, the boid's rule of separation has a span of 0.5 m [32]. The distance required to arrest a protester is the same. While protesters have $R_{arrestInit}$ set so that they can be arrested in 5 s by three officers, damageable items can be destroyed in 1 min by three violent protesters and reappear after 5 min. Thanks to this setting, items draw rioters' interest regularly. The energy consumption from attacking or fleeing is the same, and the protester is initially endowed with enough energy to flee for

5 min. Then, we chose k_L equal to 8.5×10^{-4} so that the social liking reaches the minimum -1 when an agent causes another two negative emotions for an hour and a half. Finally, the inhibition threshold T_v is set to 1.0 for every protester which allows violence only through a drop in police social liking, whereas the grievance G, the risk aversion R and the neuroticism N are heterogeneous and sampled uniformly over $[0, 1]$ at initialization.

4.3 Results and Analysis

For each of the 8 different configurations, varying the protest configuration as well as the Police Officer-Protester Ratio (POPR), 5 simulations are executed. The means of the results of those simulations are introduced in Table 3 while their standard errors are in Table 4.

Table 3. Means of the results obtained for different police officer-protester ratios (POPR) over 5 simulations. Measures are Time Before Breaking The First Item (TBB1), Time Before First Aggression on Police (TBA1), Number of Arrests (NbA), Maximum Ratio of Violent Protesters (MaxVP), Time of Maximum Ratio of Violent Protesters (TVP), Maximum Ratio of Fleeing Protesters (MaxFP), Time of Maximum Ratio of Fleeing Protesters (TFP), Time First Retreat (TR1), Min Mean PSL (MPSL)

Config.	Square				Kettling			
POPR (%)	5	10	15	20	5	10	15	20
TBB1	02:00:06	02:10:30	02:04:53	01:19:14	00:01:57	00:01:47	00:04:29	00:36:49
TBA1	02:03:42	01:49:49	01:16:47	01:33:02	01:49:39	01:26:23	01:50:51	01:45:32
NbA	1.2	2	2	1.8	14.8	12.4	17.2	18.4
MaxVP (%)	0.50	0.50	0.70	0.50	23.42	22.04	24.61	26.48
TVP	01:52:09	01:30:26	01:07:07	01:30:12	00:23:15	00:19:28	00:16:56	00:48:43
MaxFP (%)	1.90	2.40	3.00	2.40	35.58	35.03	33.75	34.57
TFP	01:34:39	01:01:05	00:48:10	00:52:10	00:02:20	00:02:27	00:03:12	00:03:15
TR1	02:00:43	00:43:16	00:29:51	00:47:47	01:48:09	01:24:23	01:48:03	01:43:05
MPSL	−0.007	−0.009	−0.010	−0.008	−0.294	−0.277	−0.222	−0.196

In a basic situation with 5% of police officers compared to protesters in a square, the protest never turns violent. Increasing the number of police officers, thereby increasing the number of interactions between police officers and protesters and potentially heightening the tension between the two groups, we notice that the square scenario never reaches a riot. Only a few protesters become violent, and are thereafter quickly handled by the police force. Indeed, the Maximum Ratio of Violent Protesters (MaxVP) is on average below 0.70% and the Number of Arrests (NbA) is below 2. The Maximum Ratio of Fleeing Protesters (MaxFP) consequently remains low and its mean never reaches higher than 3.00%. This peaceful situation is due to the Mean Police Social Liking (MPSL)

Table 4. Standard errors of the results obtained for different police officer-protester ratios (POPR) over 5 simulations. Measures are Time Before Breaking The First Item (TBB1), Time Before First Aggression on Police (TBA1), Number of Arrests (NbA), Maximum Ratio of Violent Protesters (MaxVP), Time of Maximum Ratio of Violent Protesters (TVP), Maximum Ratio of Fleeing Protesters (MaxFP), Time of Maximum Ratio of Fleeing Protesters (TFP), Time First Retreat (TR1), Min Mean PSL (MPSL)

Config.	Square				Kettling			
POPR (%)	5	10	15	20	5	10	15	20
TBB1	00:59:03	01:09:40	01:16:11	00:40:52	00:00:14	00:00:07	00:02:13	01:15:07
TBA1	00:41:19	00:55:22	01:00:27	00:53:16	01:36:21	01:14:06	01:34:42	01:30:09
NbA	0.8	0.7	1.2	0.8	20.3	17.34	23.6	25.4
MaxVP (%)	0.00	0.00	0.44	0.00	30.70	28.82	31.89	35.12
TVP	00:33:54	00:40:26	00:25:51	00:29:26	00:36:32	00:28:03	00:16:09	01:06:53
MaxFP (%)	1.24	0.42	1.17	0.41	48.74	47.96	46.22	47.35
TFP	01:00:17	00:32:39	00:20:18	00:13:08	00:03:12	00:03:24	00:04:23	00:04:29
TR1	00:42:41	00:19:23	00:19:01	00:11:31	01:38:25	01:16:49	01:38:32	01:33:30
MPSL	0.000	0.000	0.000	0.000	0.142	0.130	0.112	0.118

staying close to zero. Furthermore, this situation is stable across the simulations with low standard errors in Table 4. Especially, the MPSL is null because the police officers are always wandering across the plaza.

On the reverse side of the spectrum, kettling does result in aggressive behaviors among the crowd. The MaxVP stands at 23.42% on average with 5% of POPR. On the one hand, this value remains approximately constant when the number of officers increases. On the other hand, the Time of Maximum Ratio of Violent Protesters (TVP) decreases before attaining one hour and seven minutes when the POPR is at 20%. An explanation to this result is that the number of officers plays a role in dissuading protesters for indulging in violence. For lower values of POPR, the tension created by interacting with police officers exceeds the dissuasion power.

What is prominent in the kettling scenario is the significant variations in its results. Further inquiries show that one out of two simulations becomes violent with around 50% of protesters turning violent until the end of the simulation, no matter the POPR value. This happens because of the presence of arrests. In our simulations, when police officers start an arrest, peaceful protesters around will get scared and start to panic. They will therefore enter the "retreat" state and spread their fears with others more quickly.

The reason it does not occur in every simulation is that the individual factors inhibiting aggressive behavior (the grievance G and the risk aversion R) are randomly sampled and may let appear more or less violence-prone protester agents. These agents are qualified as violence-prone since even a high inhibition $T_v \times f(\overline{\mathcal{L}_p}(.))$ enables these agents to turn violent. Events, such as the spatial incursion or feeling outnumbered, will be deemed as a sufficient reason to turn aggressive. In future work, it may be of interest to study in more detail the circumstances leading to this phenomenon and understand why it does not occur in the previous square scenario.

5   Conclusion

This article has introduced a first emotional multi-agent model to reproduce violence emergence within the protest. Starting from the BEN architecture [10], the model adapted the former degree of liking to embody the attitude of protesters towards police officers, including the tension-building observed in Nassauer's data analysis [45]. By using emotions in its formulation, the model can be a base for future extensions.

The model was implemented in the GAMA platform [53] and was tested on two features: the protest configuration and the ratio of police officers versus protesters. The protest configurations are a square patrolled by a single police team and officers employing a kettling tactic to block the street. While the first scenario presented a good situation for peacekeeping, violence sometimes appeared in the second. In the future, the reasons for this emergence could be further investigated by running and studying more configurations. Also, the kettling scenario demonstrated a change in dynamics when the number of police officers increased sufficiently. It could be interesting to characterize when the number of police officers becomes more dissuasive than a source of conflict, or the reverse.

New insights can ensue from drawing further attention to the macroscopic phenomena involved in this work. For instance, the existence and properties of the resulting macroscopic equilibrium states of the current study may be explained by the macroscopic model of Filatova and Baratgin [25] who studied a phenomenon of hesitation between two opposites, like fight and flight, but for political preferences.

Finally, this model focused only on psychological facets. Yet, Torrens and McDaniel [54] argued for the importance of realistic physical behavior on agents within the crowd. These physical properties could further allow new perceptions for our model. For instance, feeling tightly packed could be a new source of tension for the agent, or heat, rising because of people movements, could be reckoned with since it has also been shown to make people more aggressive [3].

Acknowledgments. The authors would like to thank Pierre Wieser for lending the computational resources needed for the simulations of this article.

References

1. Adam-Troian, J., Çelebi, E., Mahfud, Y.: "Return of the repressed": exposure to police violence increases protest and self-sacrifice intentions for the yellow vests. Group Process. Intergroup Relat. **23**(8), 1171–1186 (2020)

2. Allen, J.J., Anderson, C.A., Bushman, B.J.: The general aggression model. Curr. Opin. Psychol. **19**, 75–80 (2018). https://doi.org/10.1016/j.copsyc.2017.03.034, https://linkinghub.elsevier.com/retrieve/pii/S2352250X17300830
3. Anderson, C.A.: Heat and violence. Curr. Dir. Psychol. Sci. **10**(1), 33–38 (2001). https://doi.org/10.1111/1467-8721.00109
4. Bandura, A.: The social learning theory of aggression. In: The War System, pp. 141–156. Routledge (1980)
5. Barsade, S.G.: The ripple effect: emotional contagion and its influence on group behavior. Adm. Sci. Q. **47**(4), 644–675 (2002). https://doi.org/10.2307/3094912, http://journals.sagepub.com/doi/10.2307/3094912
6. Benlolo Carabot, M., Domingo, L.: Report for the conseil d'État n°444849. Recueil Lebon (2021). https://www.conseil-etat.fr/fr/arianeweb/CE/decision/2021-06-10/444849
7. Bodenhausen, G.V., Sheppard, L.A., Kramer, G.P.: Negative affect and social judgment: the differential impact of anger and sadness. Eur. J. Soc. Psychol. **24**(1), 45–62 (1994)
8. Bosse, T., Gerritsen, C., Treur, J.: Towards integration of biological, psychological and social aspects in agent-based simulation of violent offenders. SIMULATION **85**(10), 635–660 (2009). https://doi.org/10.1177/0037549709103407, http://journals.sagepub.com/doi/10.1177/0037549709103407
9. Bosse, T., Hoogendoorn, M., Klein, M.C.A., Treur, J., van der Wal, C.N., van Wissen, A.: Modelling collective decision making in groups and crowds: integrating social contagion and interacting emotions, beliefs and intentions. Auton. Agent. Multi-Agent Syst. **27**(1), 52–84 (2013). https://doi.org/10.1007/s10458-012-9201-1
10. Bourgais, M., Taillandier, P., Vercouter, L.: BEN: an architecture for the behavior of social agents. J. Artif. Soc. Soc. Simul. **23**(4), 12 (2020). https://doi.org/10.18564/jasss.4437, http://jasss.soc.surrey.ac.uk/23/4/12.html
11. Brannen, S., Haig, C., Schmidt, K.: The age of mass protests: understanding an escalating global trend. Technical report, Center for Strategic and International Studies (2020)
12. Bratman, M.: Intention, Plans, and Practical Reason. Harvard University Press, Cambridge (1987). https://doi.org/10.2307/2185304
13. Brown, R.: Mass phenomena. Handb. Soc. Psychol. **2**, 833–877 (1954)
14. Cannon, W.B.: Bodily Changes in Pain, Hunger, Fear and Rage: An Account of Recent Researches into the Function of Emotional Excitement, 2nd edn. Appleton-Century-Crofts, New York (1925)
15. Collectif: Entre facebook et le rond-point, ≪ la double originalité du mouvement des "gilets jaunes" ≫. Le Monde (2019)
16. Collins, R.: Violence: A Micro-Sociological Theory. Princeton University Press, Princeton (2008)
17. Davies, T.P., Fry, H.M., Wilson, A.G., Bishop, S.R.: A mathematical model of the London riots and their policing. Sci. Rep. **3**(1), 1303 (2013). https://doi.org/10.1038/srep01303, https://www.nature.com/articles/srep01303
18. De Castro, B.O., Veerman, J.W., Koops, W., Bosch, J.D., Monshouwer, H.J.: Hostile attribution of intent and aggressive behavior: a meta-analysis. Child Dev. **73**(3), 916–934 (2002)
19. Deffenbacher, J.L., Deffenbacher, D.M., Lynch, R.S., Richards, T.L.: Anger, aggression, and risky behavior: a comparison of high and low anger drivers. Behav. Res. Ther. **41**(6), 701–718 (2003). https://doi.org/10.1016/S0005-7967(02)00046-3, https://linkinghub.elsevier.com/retrieve/pii/S0005796702000463

20. Dennett, D.C.: Two contrasts: folk craft vs folk science and belief vs opinion. In: Greenwood, J.D. (ed.) The Future of Folk Psychology, pp. 135–148. Cambridge University Press (1991)
21. DeWall, C.N., Anderson, C.A., Bushman, B.J.: The general aggression model: theoretical extensions to violence. Psychol. Violence **1**(3), 245–258 (2011). https://doi.org/10.1037/a0023842, http://doi.apa.org/getdoi.cfm?doi=10.1037/a0023842
22. Dodd, V., Lewis, P.: Kettling of G20 protesters by police was illegal, high court rules. The Guardian (2011). https://www.theguardian.com/uk/2011/apr/14/kettling-g20-protesters-police-illegal
23. Ellsworth, P.C., Scherer, K.R.: Appraisal processes in emotion. In: Handbook of Affective Sciences. Oxford University Press (2002). https://doi.org/10.1093/oso/9780195126013.003.0029
24. Epstein, J.M.: Modeling civil violence: an agent-based computational approach. Proc. Natl. Acad. Sci. **99**(suppl_3), 7243–7250 (2002). https://doi.org/10.1073/pnas.092080199, https://pnas.org/doi/full/10.1073/pnas.092080199
25. Filatova, D., Baratgin, J.: Multi-agent social choice model and some related questions. In: 2018 11th International Conference on Human System Interaction (HSI), pp. 425–431 (2018). https://doi.org/10.1109/HSI.2018.8431333
26. Fite, P.J., Raine, A., Stouthamer-Loeber, M., Loeber, R., Pardini, D.A.: Reactive and proactive aggression in adolescent males: examining differential outcomes 10 years later in early adulthood. Crim. Justice Behav. **37**(2), 141–157 (2010)
27. Frijda, N.H.: Emotion, cognitive structure, and action tendency. Cogn. Emot. **1**(2), 115–143 (1987). https://doi.org/10.1080/02699938708408043
28. Ghasem-Aghaee, N., Khalesi, B., Kazemifard, M., Ören, T.I.: Anger and aggressive behavior in agent simulation. In: Proceedings of the 2009 Summer Computer Simulation Conference, pp. 267–274. Citeseer (2009)
29. Goldberg, J.H., Lerner, J.S., Tetlock, P.E.: Rage and reason: the psychology of the intuitive prosecutor. Eur. J. Soc. Psychol. **29**(5–6), 781–795 (1999)
30. Grupe, D.W., Nitschke, J.B.: Uncertainty and anticipation in anxiety: an integrated neurobiological and psychological perspective. Nat. Rev. Neurosci. **14**(7), 488–501 (2013)
31. Habtemichael, F.G., De Picado Santos, L.: Crash risk evaluation of aggressive driving on motorways: microscopic traffic simulation approach. Transp. Res. Part F: Traffic Psychol. Behav. **23**, 101–112 (2014). https://doi.org/10.1016/j.trf.2013.12.022, https://linkinghub.elsevier.com/retrieve/pii/S1369847813001514
32. Hall, E.: Les distances chez l'homme. La dimension cachée, Le Seuil, Paris, pp. 143–160 (1971)
33. Ives, B., Lewis, J.S.: From rallies to riots: why some protests become violent. J. Conflict Resolut. **64**(5), 958–986 (2020)
34. Ives, B., Lewis, J.S.: From rallies to riots: why some protests become violent. J. Conflict Resolut. **64**(5), 958–986 (2020). https://doi.org/10.1177/0022002719887491, http://journals.sagepub.com/doi/10.1177/0022002719887491
35. Jager, W.: Enhancing the realism of simulation (EROS): on implementing and developing psychological theory in social simulation. Jasss-J. Artif. Soc. Soc. Simul. **20**(3) (2017). https://doi.org/10.18564/jasss.3522
36. Kim, J.W., Hanneman, R.: A computational model of worker protest. J. Artif. Soc. Soc. Simul. **14**(3), 1 (2011)
37. Kunimatsu, M.M., Marsee, M.A.: Examining the presence of anxiety in aggressive individuals: the illuminating role of fight-or-flight mechanisms. In: Child & Youth Care Forum, vol. 41, pp. 247–258. Springer (2012)

38. Lazarus, R.S.: Cognition and motivation in emotion. Am. Psychol. **46**(4), 352 (1991)
39. Lemos, C.M., Coelho, H., Lopes, R.J.: ProtestLab: a computational laboratory for studying street protests. In: Nemiche, M., Essaaidi, M. (eds.) Advances in Complex Societal, Environmental and Engineered Systems. NSC, vol. 18, pp. 3–29. Springer, Cham (2017). https://doi.org/10.1007/978-3-319-46164-9_1
40. Lerner, J.S., Keltner, D.: Fear, anger, and risk. J. Pers. Soc. Psychol. **81**(1), 146 (2001)
41. Levenson, R.W.: Human emotion: a functional view. Nat. Emot.: Fundam. Quest. **1**, 123–126 (1994)
42. Loewenstein, G., Lerner, J.S., et al.: The role of affect in decision making. Handb. Affect. Sci. **619**(642), 3 (2003)
43. Moussaïd, M., Helbing, D., Theraulaz, G.: How simple rules determine pedestrian behavior and crowd disasters. Proc. Natl. Acad. Sci. **108**(17), 6884–6888 (2011). https://doi.org/10.1073/pnas.1016507108
44. Nassauer, A.: Effective crowd policing: empirical insights on avoiding protest violence. Policing: Int. J. Police Strategies Manage. **38**(1), 3–23 (2015)
45. Nassauer, A.: From peaceful marches to violent clashes: a micro-situational analysis. Soc. Mov. Stud. **15**(5), 515–530 (2016). https://doi.org/10.1080/14742837.2016.1150161, http://www.tandfonline.com/doi/full/10.1080/14742837.2016.1150161
46. Neubauer, J., Wood, E.: Accounting for the variation of driver aggression in the simulation of conventional and advanced vehicles. Technical report, National Renewable Energy Lab.(NREL), Golden, CO, USA (2013)
47. Ortony, A., Clore, G.L., Collins, A.: The Cognitive Structure of Emotions. Cambridge University Press (1988)
48. Petersen, R., Zukerman, S.: Anger, violence, and political science. In: International Handbook of Anger: Constituent and Concomitant Biological, Psychological, and Social Processes, pp. 561–581. Springer (2009)
49. Reynolds, C.W.: Flocks, herds, and schools: a distributed behavioral model, pp. 273–282. Association for Computing Machinery, New York (1998). https://doi.org/10.1145/280811.281008
50. Rochefort, Y., Piet-Lahanier, H., Bertrand, S., Beauvois, D., Dumur, D.: Guidance of flocks of vehicles using virtual signposts. IFAC Proc. Vol. **44**(1), 5999–6004 (2011)
51. Smith, C.A., Ellsworth, P.C.: Patterns of cognitive appraisal in emotion. J. Pers. Soc. Psychol. **48**(4), 813 (1985)
52. Stephan, M.J., Chenoweth, E.: Why civil resistance works: the strategic logic of nonviolent conflict. Int. Secur. **33**(1), 7–44 (2008)
53. Taillandier, P., et al.: Building, composing and experimenting complex spatial models with the GAMA platform. GeoInformatica **23**(2), 299–322 (2019). https://doi.org/10.1007/s10707-018-00339-6
54. Torrens, P.M., McDaniel, A.W.: Modeling geographic behavior in riotous crowds. Ann. Assoc. Am. Geogr. **103**(1), 20–46 (2013). https://doi.org/10.1080/00045608.2012.685047, http://www.tandfonline.com/doi/abs/10.1080/00045608.2012.685047
55. Trouillard, P.: Repressing the protests through law, police and discourse: the example of the yellow vests' movement in France. J. Contemp. Eur. Stud. **30**(3), 506–520 (2022). https://doi.org/10.1080/14782804.2021.1915257, https://www.tandfonline.com/doi/full/10.1080/14782804.2021.1915257

56. Vecchio, M.: Gilets jaunes et lycéens: 2891 blessés depuis le début du mouvement. BFM TV (2018). https://www.bfmtv.com/police-justice/gilets-jaunes-et-lyceens-2891-blesses-depuis-le-debut-du-mouvement_AN-201812200115.html

57. Zillmann, D.: Excitation transfer theory. In: Donsbach, W. (ed.) The International Encyclopedia of Communication, 1 edn. Wiley (2008). https://doi.org/10.1002/9781405186407.wbiece049, https://onlinelibrary.wiley.com/doi/10.1002/9781405186407.wbiece049

58. Zillmann, D., Bryant, J.: Effect of residual excitation on the emotional response to provocation and delayed aggressive behavior. J. Pers. Soc. Psychol. **30**(6), 782–791 (1974). https://doi.org/10.1037/h0037541, https://doi.apa.org/doi/10.1037/h0037541

Moral Reasoning

The Social Exchange Heuristic Operation During a One-Shot Prisoner's Dilemma Game

Kaede Maeda[1]([✉]) [ID], Hirofumi Hashimoto[2] [ID], and Shigehito Tanida[3]

[1] Rikkyo University, 1-2-26 Kitano, Niiza 352-8558, Saitama, Japan
k.maeda8221@gmail.com
[2] Osaka Metropolitan University, 3-3-138 Sugimoto, Sumiyoshi-ku, Osaka-shi 558-8585, Osaka, Japan
[3] Taisho University, 3-20-1 Nishisugamo, Toshima-ku, Tokyo 170-0001, Japan

Abstract. This study aimed to analyze the decision-making process during a one-shot prisoner's dilemma game and test the social exchange heuristic hypothesis. To this end, we used an eye-tracking measure to examine whether the cooperator subjectively transformed the one-shot prisoner's dilemma game into the assurance game. According to the group heuristic model and the social exchange hypothesis, cooperation in a one-shot prisoner's dilemma game is more pronounced toward in-group members, and the mechanism is thought to arise from the subjective transformation; furthermore, cooperation in a one-shot prisoner's dilemma game, especially toward in-group members, should be based on quick decision making. To test these hypotheses, we carefully analyzed the decision-making process of participants during one-shot prisoner's dilemma games. Our results suggest that decision time is faster for those who cooperate with in-group members than for those who do not cooperate. In addition, cooperators, regardless of who they cooperate with, generally pay more attention to the outcome of mutual cooperation than non-cooperators. These findings support the social exchange heuristic hypothesis.

Keywords: cooperation · heuristic · prisoner's dilemma game · eye-tracking · intuitive cooperation model · social exchange heuristic

1 Introduction

Recent empirical studies that apply the dual-process theory [1–4] have suggested that individuals' decision-making may differ based on the use of intuitive versus deliberative processes [5]. For instance, the intuitive cooperation model [6] suggests that individuals who exhibit high levels of cooperativeness tend to reach decisions faster in economic games, such as the public goods or prisoner's dilemma game (PDG). This model exhibits a certain degree of consistency with the social exchange heuristic hypothesis proposed by Kiyonari et al. [7]. Specifically, Kiyonari et al. have argued that people perceive the PDG as an *assurance game* because of an intuitive cognitive bias in processing information about social exchange. In the assurance game, there is no dominant choice. Defection results in an individually better outcome if the partner is also a defector. If

© The Author(s), under exclusive license to Springer Nature Switzerland AG 2025
J. Baratgin et al. (Eds.): HAR 2024, LNCS 15504, pp. 55–63, 2025.
https://doi.org/10.1007/978-3-031-84595-6_4

the partner cooperates, however, cooperation produces an individually better outcome. With the subjective transformation, people intuitively perceive most mixed-motive incentive structures as ones in which mutual cooperation is personally more desirable–that is, produces personally better outcomes–than defection, provided that the partner also cooperates. It therefore can be reasonably assumed that individuals process information in an intuitive and fast manner aimed at mutual cooperation in economic games [8]. Based on the above, it is hypothesized that the time allowed for decision making will influence cooperative behavior in a one-shot PDG. The primary aim of the present study was to test the hypothesis that cooperation during a one-shot PDG would be based on quick (rather than slow) decision-making (Hypothesis 1). Theoretical literature regarding whether the decision times of cooperators and defectors differ under time pressure lacks clarity. Nevertheless, the finding that individuals who make quick decisions are more likely to cooperate than those who make slow decisions, coupled with the finding that people cooperate more under time pressure (as demonstrated in studies by Rand et al. [6]) provides sufficient reasons to assume that time pressure can affect individuals to employ cooperative decision-making.

The secondary aim of this study was to examine the potential effect of group membership on fast cooperative decision-making. Shared group membership can be a key factor that affects decisions in an economic game. A classic study of social psychology [9] consistently showed that people tended to act more cooperatively with their own group (in-group) members than with non-group (out-group) members. The conventional wisdom regarding in-group favoritism is based on social identity theory, as explained by Tajfel and Turner [10, 11]. However, the present study examined an alternative theory, the group heuristic model (GHM), as proposed by Yamagishi et al. [8, 12–14]. In proposing the GHM, Yamagishi et al. argued that individuals utilize a set of beliefs and decision rules among members belonging to the same community in a cooperative manner by default. Theoretical studies in mathematical biology [15] explain the evolution of cooperation through indirect reciprocity. This can be achieved through group- [16, 17] or spatial-structured indirect reciprocity. Therefore, humans' cooperation can be sustained within groups of individuals who behave in an indirectly reciprocal manner. Indirect reciprocity can be defined as a strategy employed to act cooperatively toward individuals who have a reputation for being altruistic toward their own kind. In terms of reputation within a group, it can be argued that the GHM is applicable to an explanation of cooperation via indirect reciprocity; the group heuristic makes individuals minimize the risk of exclusion from a reputation-based, closed, and tight relationship. According to this GHM, we predicted that the intuitive and fast cooperation observed in previous studies is a reflection of a default decision-making strategy in social exchange with in-group members. Specifically, combining the intuitive cooperation model and GHM, we hypothesized that in-group favoritism would be conspicuously shown through an intuitive process (Hypothesis 2). To test this hypothesis, we employed a minimal group paradigm, a traditional method in social psychological studies, and constrained decision time in one-shot anonymous interactions with in- and out-group members using PDG.

In summary, this study hypothesized that cooperation during a one-shot PDG would be based on quick (as opposed to slow) decision-making. Furthermore, this type of cooperation would be limited where the interaction partner is an in-group member; only in-group favoritism would be intuitive. In other words, in-group favoritism in a minimal group would be more salient with an intuitive process. Maeda and Hashimoto [18] demonstrated that cooperative behavior during a one-shot PDG under time pressure was limited to in-group members. However, only a few studies have examined this issue. Moreover, the process of intuitive decision-making in the context of the shared group membership remains unclear. To test the aforementioned hypotheses and address the aforementioned issues, the present study employed the minimal group paradigm, constrained the decision-making time in the PDG to 15 s, and set three conditions as within-participant factors: in-group members, out-group members, and unspecified partners in a one-shot PDG. Furthermore, eye movements of the participants were examined in the analysis. Specifically, we examined eye movements in accordance with the research of Maeda, Hashimoto, and Tanida [19], which examined eye movement during the decision-making process in a one-shot PDG to ascertain whether subjective transformation of the PDG in a biased way occurred in an exploratory manner. During a one-shot PDG, in which it is economically rational to defect rather than cooperate, many players choose to cooperate. As a potential explanation regarding this type of cooperation, Kiyonari et al. argued that people subjectively transform the PDG in a biased way. They proposed that people have a cognitive bias in their information processing of social exchange, where they perceive PDG-like situations as an assurance game. This subjective transformation leads individuals to perceive most mixed-motive incentive structures as those in which mutual cooperation is personally more desirable—or produces better outcomes personally—than defection insofar as the partner also cooperates. The inclination to regard mutual cooperation as a favorable outcome in PDG scenarios is evidenced by the self-rating results. In particular, the previous study repeatedly demonstrated that the outcome of mutual cooperation, defined as both players cooperating, is the most desirable. Moreover, the use of an eye-tracking device revealed that cooperators exhibited a greater focus on the outcome of mutual cooperation in the one-shot PDG. Thus, the present study also analyzed the decision-making process in the one-shot PDG and examined whether the cooperator subjectively transforms the one-shot PDG into the assurance game.

2   Methods

2.1   Participants

A total of 27 undergraduate students (mean age $= 20.04$ years) voluntarily participated in this study. The participants were told that they would receive the money determined by their actual decisions in the study. Specifically, as will be described below, participants were required to make a total of three decisions in this experiment. However, they were informed that they would receive a reward in accordance with the outcome of one of the decisions, which was randomly selected.

2.2   Experimental Design Subsection Sample

In accordance with the minimal group paradigm, the participants were arbitrarily divided into two groups. This was achieved by presenting nine pairs of paintings by renowned artists Paul Klee and Wassily Kandinsky and asking the participants to indicate which painting they preferred. Based on the results of this preference task, the participants were divided into either the Klee or Kandinsky group. After dividing the participants into two groups, one-shot PDGs were conducted on a monitor screen in the laboratory with in-group members, out-group members, and those whose group affiliation was unknown. The experimental condition of the partner's group membership was set as a within-participant factor, and the participants repeated the PDG a total of three times with different partners (i.e., in-group, out-group, and unknown partners). The counterbalance was applied exclusively with regard to the partner group membership in the initial one-shot PDG (that is, whether the partner was an in-group or out-group member). In the third PDG, all participants engaged in a one-shot PDG with an unknown partner. To clarify the economic incentive, the financial rewards of the three times were determined by lottery, and the rewards were paid as determined by the participant's decision.

2.3 Experimental Procedures

Upon arrival at the laboratory, participants were escorted individually to a soundproof room. Once seated, participants first completed a calibration task in which they tracked their eye movements following a white dot on the screen. After completing this calibration task, participants were asked to follow on-screen instructions regarding the PDG at their own pace after confirming that their eye movements could be accurately recorded. The participants were informed of the rules of PDG. Subsequently, they were asked whether they would give their partner 500 yen ($\leq$\$4). In the explanations, the words "cooperation" and "defection" were not used. Participants were encouraged to respond within 15 s. This was done owing to the study's focus on the intuitive and fast aspect of cooperation.

2.4 Apparatus

The PDG instructions and pay-off matrix were displayed on a 23.0-inch monitor (ThinkVision T23i-20). A screen-based eye-tracking device, Tobii pro fusion (Tobii Technology, Inc), with a sampling rate of 250 Hz, was used to record participants' eye movement. The video data of eye movement was continuously recorded from the calibration period to the end of the experiment.

2.5 Area of Interest Settings

In PDG, two players are presented with two options: either cooperate by giving 500 yen or defect by keeping 500 yen. This yields four possible outcomes. The analysis focused on the decision time at which the participants' eyes were fixed on each of the four (and eight) possible outcomes in the pay-off matrix, which was divided into the following four (or eight) divisions: CC, CD, DC, and DD, which were combinations of acronyms, with CC representing the result of mutual cooperation and DD signifying the result of mutual defection. The area of interest (AOI) represented the combined rate of gaze at each pay-off matrix (e.g., CC area) and the rate of gaze at the corresponding option (e.g., L and K buttons). For example, the gaze rate of the AOI for CC was calculated by dividing the gaze time for CC and the corresponding option by the total gaze time. The gaze time to the corresponding option was doubled for convenience. The eight divisions, rather than the four that were previously employed, were based on the distinction between the gazing time directed toward one's own gain and that of the partner's for each area and the AOI. The gazing rate was calculated in the same manner.

3 Results

3.1 Cooperation Rate and Decision-Making Time

The cooperation rate of participants in the one-shot PDG was 59.26% when the partner was an in-group member, 55.66% when the partner was an out-group member, and 29.63% for those whose group membership was unknown. Figure 1 shows the cooperation for each round, considering the counterbalance. As illustrated, the rate of cooperation decreased with the round. The findings indicate that participants in this study, regardless of the affiliation of their partner, initially exhibit an intuitive tendency to cooperate. However, this inclination diminishes with repetition. The results suggest that participants may have interpreted PDG in a biased manner (subjective transformation to AG), but upon repeating the game, they may have developed an understanding of it that is not contingent on social exchange. Furthermore, our analysis revealed noteworthy findings regarding decision-making time. An examination of the initial decision-making time when interacting with an in-group member revealed a statistically significant difference between cooperating and defecting individuals ($t = 2.42, p = .03$). This indicates that the decision-making time of cooperating individuals was shorter. However, the difference was not significant for decision-making time when dealing with out-group members ($t = 0.38, p = .71$).

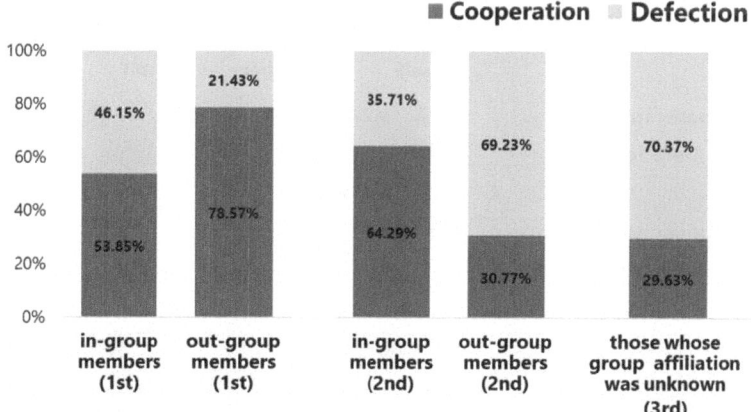

Fig. 1. The rate of cooperation of the participants in a total of three decision making sessions.

3.2 Gazing Patterns for the PDG Matrix

A two-factor analysis of variance was conducted to examine the relationship between the rate of attention for the outcome of mutual cooperation (CC) (also the dependent variable) and two independent variables: the partner's group membership (in-group, out-group, or unknown) and participants' decision (cooperation or non-cooperation). The analysis revealed a significant main effect of decision-making [F (1, 23) = 9.14, p = .006]. The same analysis was conducted for the percentage of attention for own only non-cooperation (DC) and mutual non-cooperation (DD), respectively. The main effect of decision-making was significant [F (1,23) $\geq$ 5.98, $p \leq$ .023]. The details are shown in Figs. 2 and 3. Overall, the results were consistent with the social exchange heuristic, indicating that cooperators are more likely to attend to outcomes of mutual cooperation and less likely to attend to cells in which they themselves take a non-cooperative action.

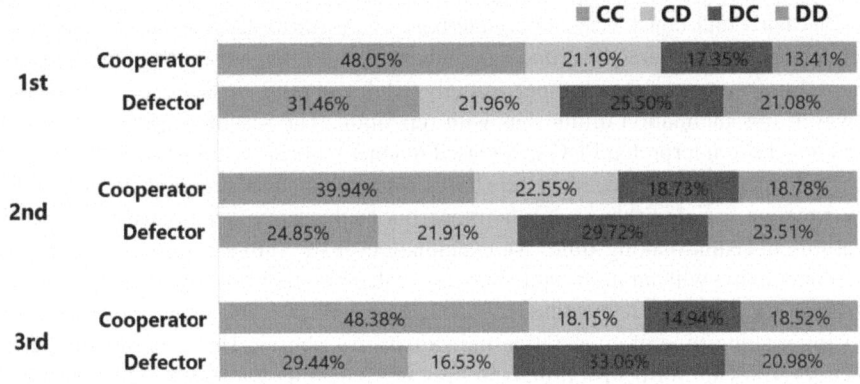

Fig. 2. Percentage of eye gaze directed toward the four pay-off matrices per condition and participants' decisions.

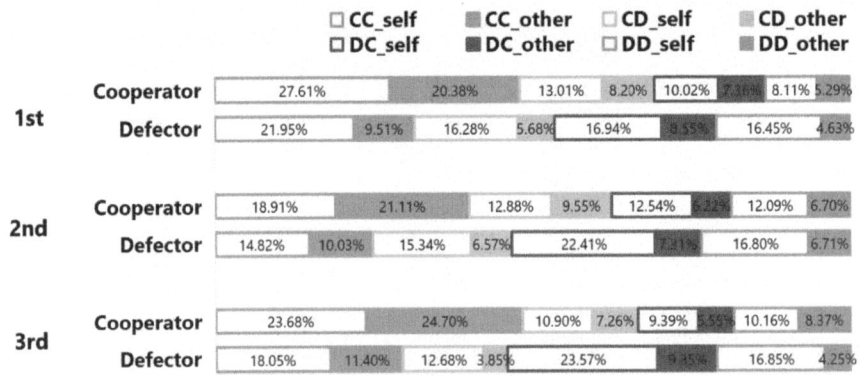

Fig. 3. Eye gaze percentage directed toward the eight areas (with distinction between the participants' own gains and those of their partners) per participants' decisions.

4   Discussion

The results did not support our hypotheses. Nevertheless, the results indicated a potential direction for further research. First, the results demonstrated the absence of group differences in intuitive cooperative behavior in the one-shot anonymous PDG. This indicates that our study could not confirm the robustness of the finding reported by Maeda and Hashimoto [18], which suggested that group differences occur in intuitive cooperation. It is challenging to provide a definitive interpretation, but it is worth noting that there were differences in the experimental procedures between Maeda and Hashimoto [18] and our study. These include whether the specific amount of gain in the one-shot PDG was presented before decision-making. In most previous studies, including that by Maeda and Hashimoto [18], the amount of money in the matrix in one-shot PDG was revealed (or suggested) at the instructional stage. For instance, in the study by Maeda and Hashimoto [18], the participants were informed that they would each receive 1,000 yen if they chose to cooperate with each other before the decision was made. In contrast, the present study did not provide specific amounts of money obtained as part of the explanation. Instead, these amounts were replaced by question marks. This may indicate that a high cooperation rate in the first session was based on an intuitive decision process because it was only at the time of decision-making that one was able to understand their own and others' interests and make decisions based on them. The observed decline in cooperation rates between the second and third sessions can be attributed to participants' comprehension of the specific amount of gain in the PDG, leading to decisions being made without the involvement of intuition. It would be beneficial to examine whether the observed pattern in Fig. 1 will be replicated in future studies.

Second, an examination of the degree of attention paid to the PDG matrix during decision-making indicated that those inclined toward cooperation demonstrated a higher level of focus on the CC domain and a reduced interest in the possibility of non-cooperation, which aligns with the social exchange heuristic hypothesis. It is important to acknowledge that the interpretation of these results was not solely dependent on this analysis. Nevertheless, this eye gaze pattern is consistent with the findings of Maeda, Hashimoto, and Tanida [19]. It would be beneficial for future studies to examine more carefully the differences in decision-making modes between cooperators and non-cooperators. Furthermore, the findings regarding decision-making time when interacting with an in-group member demonstrated that cooperators took less time making decisions than did non-cooperators. These findings have been repeatedly confirmed in our studies and it is necessary to consider the implications of what these results suggest.

It is important to note that this study is subject to a number of limitations. The most significant of these is the relatively small sample size. Further rigorous replicational studies will be needed to confirm the reliability and generalizability of the findings. Additionally, the use of within-participant factors made interpretation of the results difficult. For example, it was not possible to determine whether the low cooperation rate for the third decision in this study was due to the participants' lack of knowledge regarding the other participant's group membership or whether it was a consequence of repeated decision-making. It would be preferable to conduct further studies using between-participant factors. Despite these limitations, it should be noted that the results of this study provide significant directions for future research.

Acknowledgments. This study was supported by Grants-in-Aid 21K02992, 21KK0042, and 24K16798 from the Japan Society for the Promotion of Science.

Disclosure of Interests. The authors declare that the research was conducted in the absence of any commercial or financial relationships that could be construed as a potential conflict of interest.

References

1. Evans, J.S.B., Stanovich, K.E.: Dual-process theories of higher cognition: advancing the debate. Perspect. Psychol. Sci. **8**, 223–241 (2013). https://doi.org/10.1177/1745691612460685
2. Evans, J.S.B.T.: Dual-processing accounts of reasoning, judgment, and social cognition. Ann. Rev. Psychol. **59**, 255–278 (2008). https://doi.org/10.1146/annurev.psych.59.103006.093629
3. Kahneman, D.: Thinking, fast and slow. Farrar, Straus & Giroux (2011)
4. Petty, R.E., Cacioppo, J.T.: The elaboration likelihood model of persuasion. In: Berkowitz, L. (ed.) Advances in Experimental Social Psychology, vol. 19, pp. 123–205. Academic Press (1986). https://doi.org/10.1016/S0065-2601(08)60214-2
5. Capraro, V.: The dual-process approach to human sociality: meta-analytic evidence for a theory of internalized heuristics for self-preservation. J. Pers. Soc. Psychol. **126**, 719 (2024). https://doi.org/10.1037/pspa0000375
6. Rand, D.G., Greene, J.D., Nowak, M.A.: Spontaneous giving and calculated greed. Nature **489**, 427–430 (2012). https://doi.org/10.1038/nature11467
7. Kiyonari, T., Tanida, S., Yamagishi, T.: Social exchange and reciprocity: confusion or a heuristic? Evol. Hum. Behav. **21**, 411–427 (2000). https://doi.org/10.1016/S1090-5138(00)00055-6
8. Yamagishi, T., Jin, N., Kiyonari, T.: Bounded generalized reciprocity. Adv. Group Process. **16**, 161–197 (1999)
9. Tajfel, H., Billig, M.G., Bundy, R.P., Flament, C.: Social categorization and intergroup behaviour. Eur. J. Soc. Psychol. **1**, 149–178 (1971). https://doi.org/10.1002/ejsp.2420010202
10. Tajfel, H., Turner, J.C.: An integrative theory of intergroup conflict. In: Austin, W.G., Worchel, S. (eds.) The Psychology of Intergroup Relations, pp. 33–47. Nelson-Hall, Monterey (1979)
11. Tajfel, H., Turner, J.C.: The social identity theory of intergroup behavior. In: Austin, W.G., Worchel, S. (eds.) Psychology of Intergroup Behavior, pp. 7–24. Nelson Hall, Chicago (1986)
12. Yamagishi, T., Mifune, N., Liu, J.H.: Exchanges of group-based favours: ingroup bias in the prisoner's dilemma game with minimal groups in Japan and New Zealand. Asian J. Soc. Psychol. **11**, 196–207 (2008). https://doi.org/10.1111/j.1467-839X.2008.00258.x
13. Yamagishi, T., Kiyonari, T.: The group as the container of generalized reciprocity. Soc. Psychol. Q. **63**, 116–132 (2000). https://doi.org/10.2307/2695887
14. Yamagishi, T.: The social exchange heuristic: a psychological mechanism that makes a system of generalized exchange self-sustaining. In: Radford, M., Ohnuma, S., Yamagishi, T. (eds.) Cultural and Ecological Foundations of the Mind, pp. 11–37. Hokkaido University Press, Sapporo (2007)
15. Nowak, M.A., Sigmund, K.: Evolution of indirect reciprocity. Nature **437**, 1291–1298 (2005). https://doi.org/10.1038/nature04131
16. Masuda, N.: Ingroup favoritism and intergroup cooperation under indirect reciprocity based on group reputation. J. Theor. Biol. **311**, 8–18 (2012). https://doi.org/10.1016/j.jtbi.2012.07.002
17. Nax, H.H., Perc, M., Szolnoki, A., Helbing, D.: Stability of cooperation under image scoring in group interactions. Sci. Rep. **5**, 12145 (2015). https://doi.org/10.1038/srep12145

18. Maeda, K., Hashimoto, H.: Time pressure and in-group favoritism in a minimal group paradigm. Front. Psychol. **11**, 603117 (2020). https://doi.org/10.3389/fpsyg.2020.603117
19. Maeda, K., Hashimoto, H., Tanida, S.: Cooperators pay more attention to the outcome of mutual cooperation in the one-shot prisoner's dilemma game: empirical evidence from an eye-tracking study. Lett. Evol. Behav. Sci. **14**, 8–12 (2023). https://doi.org/10.5178/lebs.2023.101

Development of Intention-Based Moral Judgement in Children and Adolescents with and Without Intellectual Disability

Véronique Salvano Pardieu[1,2(✉)] and Valérie Pennequin[2]

[1] Laboratoire CHArt-UP8 RNSR 200515259U, 2 Rue de la Liberté, 93200 Saint-Denis, France
Veronique.pardieu@wanadoo.fr
[2] Université de Tours EA2114 PAVEA, 3 rue des Tanneurs, 37000 Tours, France

Abstract. In this study using a method based on Anderson's functional theory of cognition, judgement of blame was studied with 124 participants divided into four groups; 31 children and 31 adolescents with mild intellectual disability (ID) and 31 children and 31 teenagers typically developing (TD). Participants had to judge social interactions between two characters in four situations: "intentional harm" (with bad intent-with bad consequence), "attempted harm" (with bad intent-without consequence), "accidental harm" (without bad intent-with bad consequence) and "no harm" (without bad intent-without bad consequence) within three different levels of aggressiveness (low, medium and high). Results showed that children, and primarily ID children, focus their judgement on the consequence but not on the intention. In contrast, TD teenagers judge the action according to the intention. TD children and teenagers with ID were able to judge according to the intention rather than according to the consequence, but not systematically, showing a developing stage. Only children with ID had difficulties in classifying deliberate action according to the level of aggressiveness: low, medium and high.

This result suggests the implication of two systems: one evaluating the consequence of the actions that is not altered by intellectual disability; and the other evaluating the agent's intention, based on Theory of Mind, which has not yet been attained in the young TD children and which is delayed in ID participants. This latter system appears to be correlated with intellectual development and age and confirms previous result showing that judgement based on intent depends on ToM.

Keywords: Moral judgment · deontic reasoning · perspective taking and theory of mind · Intellectual disability and development · level of aggressiveness

1 Introduction

1.1 Anderson's Theory of Cognition

Moral judgment is the ability to judge what is considered as "good" or "bad" according to the values of a social group. Piaget (1932) and later Kohlberg (1964) studied moral judgment using semi directive interviews. This method is mainly based on verbal explanations following long fictive stories or dilemmas. A more recent approach: Anderson's

© The Author(s), under exclusive license to Springer Nature Switzerland AG 2025
J. Baratgin et al. (Eds.): HAR 2024, LNCS 15504, pp. 64–87, 2025.
https://doi.org/10.1007/978-3-031-84595-6_5

Theory of Cognition (1996, 2008) measures judgment on an intensity scale by weighting factors such as Intent and Consequence. This method which does not need verbalization avoids criticism by using short vignettes describing daily life situations instead of fictive dilemmas and semi directive interview usually used in experiments on moral judgment (Hommers and Lee 2010, Hommers et al. 2012). Moreover, the intent and the consequence factors are clearly described, which makes the short vignettes easy to memorize. Finally, this method make comparison easier between groups since it remains regardless of the age and cognitive development of the participants (Anderson 1996, 2008).

Anderson defines moral judgment and especially the judgment of blame, as a decision-making mechanism defined by two factors: intention of the actor and consequence of the action. In social interactions, a person judges a situation by taking into account the intent to cause harm and the consequences of the action.

Thus, the level of importance or "weight" given to each factor, "intent" and "consequence", and the two modalities: with and without, lead to a cognitive structure of judgement called: "moral algebra". If we use two factors such as "bad intent" and "bad consequence or bad outcome" and two modalities such as "with" and "without", we obtain four possibilities: 1- with bad intent - with bad outcome ("intentional harm"), 2- with bad intent- without bad outcome ("attempted harm"), 3- without bad intent- with bad consequence ("accidental harm") and 4- without bad intent- without bad consequence ("no harm"). The weight given to the factors "Intent" and "Consequence" is evaluated globally for each situation. Each participant has to choose on the intensity scale, the level of blame they want to give to each combination of Intent and Consequence (i.e. each situation). There are three possibilities: in the first case: only one factor either "intent" or "consequence" is taken into account, this mono-factorial algebra is not often observed. In the second case: both factors intent and consequence are taken into account but evaluated independently. No interaction between the two factors is observed. This additive moral algebra is mainly observed with children (Przygotzki and Mullet 1997). Finally, in the third possibility, both factors "intent" and "consequence" are taken into account but the weight of one factor is modulated according to the other and an interaction is observed. This multiplicative moral algebra is mainly observed with adolescents and adults.

Previous research (Leon 1984; Surber 1982; Przygotzki and Mullet 1997) has shown that over their life span, most people use the same rule of judgment by weighting both factors, although the weight given to each factor, Intent and Consequence, varies with age. Children usually give more importance to the consequence than the intent while adolescents and adults give more importance to the intent than the consequence. (Przygotzki and Mullet 1997; Salvano-Pardieu et al. 2016). In addition, numerous studies on children's moral judgment confirm this result: children give more weight to consequences than to intentions (Killen Mulvey, Richardson, Jampol, & Woodward 2011; Rogé and Mullet 2011; Cushman et al. 2013, Salvano-Pardieu et al. 2020).

1.1.1 Structures of Moral Judgment

When a person blames an action, she needs to know the social rules to apply to a violation detection rule. The person has to determine if the rule was violated, if the action was deliberate or accidental and if the consequence is serious. Therefore, two components seem essential in the judgment of blame: the knowledge of the group's rules to which the person belongs and the ability to understand the thought of the other i.e. the actor and the victim, to determine how much blame the actor deserves. The first component: the knowledge of the rule and what is allowed or not in our social group is based on "deontic reasoning" (Manktelow and Over 1991, Manktelow et al. 1995; Manktelow 1999; 2012). This cognitive ability is observed at about 2–3 years of age, when children have learned social rules.

In addition, determining whether the action has been done deliberately or not, requires the ability to understand the intent of the actor i.e. and his perspective, this is the perspective-taking ability. This concept is based on the Theory of Mind (ToM). ToM is the ability to understand our own thoughts, beliefs and emotions, as well as those of others.

It is also the ability to attribute beliefs, emotions and intentions to others, and to recognize these beliefs, emotions and intentions as different from one's own (Premack and Woodruff 1978). This ability allows people to understand and explain the behaviour of others in social interactions. ToM develops around 4–5 years old and continues to develop via social interactions (Hughes and Leekam 2004) and during adolescence and early adulthood (Gweon et al. 2012; Miller 2012; Valle et al. 2015).

The first stage of the theory of mind called "first order Theory of Mind" is the ability to attribute belief, thought and emotion to another person. This step develops from 3 to 8 years old (y.o.).

Later, around 8 y.o. and during adolescence the second order ToM develops. Children and adolescents are able to attribute thought to one person about another person. For example: Mary thinks that Peter thinks that… Wellman and Liu (2004) used a scale measuring the different levels of first order ToM. This scale evaluates the level of ToM with children from 3 to 8 y.o, by asking them 7 questions with an increasing difficulty: the understanding of diverse desires, diverse believes, knowledge access, understanding of false knowledge and understanding emotion and hidden emotion. Salvano-Pardieu and Pennequin (2024) evaluated the level of ToM with children aged 4 to 7 y.o. with Wellman & Liu' scale and compared the level of ToM with the ability to take into account the intent in moral judgement when judging actions based on intent of the character and consequence of his action. These authors showed that children with the lowest level of ToM judged social interactions based on the consequence more than the intention. Conversely, children with the highest level of ToM were the most able to judge social interactions based on the character's intention. This effect increases with age and probably with cognitive development. Therefore, the comparison between children and adolescents typically developing (TD) and children and adolescents intellectually disabled (ID) who are cognitively delayed will confirm the role of ToM in cognitive development and in the ability to consider intention in moral judgment.

1.2 Dual System's Models: "Action Evaluation System" vs "Intent Evaluation System"

In previous studies, Fontaine et al. (2004); Salvano-Pardieu et al. (2016), suggest that two components, Deontic Reasoning and Perspective-Taking, are involved in moral judgment. According to these authors, judging social interactions requires the ability to take into account the intent of the actor, which depends on Perspective-Taking, and the ability to evaluate the outcome of the action, (if the action is allowed and what are the consequences) which depends on Deontic Reasoning. The moral algebra of people, i.e. the way they combine and weight Intent and Consequence in social situations, reflects, the activation of Deontic Reasoning and Perspective-Taking. Cushman (2008; 2013) and Buon et al. (2016) explained moral judgment within the framework of a dual system: the first system "Action Evaluation System" evaluates the action, and therefore its consequences, and the second system "Intent Evaluation System" evaluates the intention of the actor. The latter system would be supported by ToM (Buon et al. 2016). According to these authors, these two systems can be "congruent", when the outputs of the two systems are identical or not congruent, when the two systems act in opposition, and each system leads to a different output. Table 1 presents the four situations of moral algebra: with bad intent - with bad consequence, "intentional harm", without bad intent - without bad consequence "no harm", without bad intent - with bad consequence, "accidental harm" and with bad intent – without bad consequence, "attempted harm". While Buon et al. (2016) developed a model of moral judgment based on three components: Emotion, Theory of Mind, and inhibitory Control (ETIC), Fontaine et al. (2004); Salvano-Pardieu et al. (2016); Salvano-Pardieu and Pennequin (2024) assume the implication in moral judgment of two main components: deontic reasoning and perspective-taking relying on ToM.

As shown in Table 1, in "intentional harm" and "no harm" the Action Evaluation System (AES) and the Intent Evaluation System (IES) are congruent and since each system leads to the same output, the same result could be obtained using both systems or one of them such as the "action evaluation system", referring to the consequences and based on deontological reasoning. In contrast, when the outputs of the two systems are not congruent such as in "accidental harm" and in "attempted harm", the judgment resulting from these situations is more complex and cannot be explained only with deontic reasoning. The activation of the IES referring to the Intent and based on ToM is necessary to render the judgment. In "accidental harm", the IES is weakly activated (there is no intention to harm) while the AES is activated (consequence is negative). In this case, the two systems are in opposition and ToM will allow consideration that the action is accidental and not blameworthy. According to Buon et al. (2016) the inhibitory control will prevent the AES to blame harshly since there is no bad intention. The opposite is observed in the "attempted harm" situation. In this case, the IES is activated (intention to harm is present) while the AES is not activated (there is no consequence). In this situation, ToM ability allows one to blame the negative intent on the actor and the potential bad consequence for the victim, leading to blaming even though there is no actual consequence. ToM is therefore crucial to evaluate the agent's intention, determine whether the action was perpetrated deliberately or not, and apportion blame accordingly.

In evaluating moral algebra, we can determine how participants weight the Intent and Consequence factors and we could measure the activation of the two components of the dual system: ToM and Deontic Reasoning. Therefore, if ToM is not yet developed as in young children, or if it is delayed as in ID children and teenagers, the blame would not be apportioned according to the intent of the actor, and the IES would not be activated

Table 1. The four situations of Moral Algebra: "with bad intent – with bad consequence", "without bad intent - without bad consequence", "without bad intent - with bad consequence" and "with bad intent - without bad consequence" and the two components: Deontic Reasoning and ToM involved in Action Evaluation System (AES) and in Intention Evaluation System (IES).

Both systems are Congruent					
Intentional harm	bad intent- bad consequence			Output of the components of the system	Moral judgement
A.E.S. : Action Evaluation System	Action's effect (consequence) : Harm	**Deontic reasoning** (he caused harm; it is forbidden and wrong).	**Blame**		
I.E.S. : Intent Evaluation System	Intention : to cause harm	**Perspective Taking based on ToM** (he wanted to cause harm and make someone suffer, it is bad).	**Blame**	**Blame**	
No harm	No bad intent- no bad consequence			Output of the components of the system	Moral judgement
A.E.S.	Action's effect (consequence) : No Harm	**Deontic reasoning** (he did nothing forbidden or wrong).	No Blame		
I.E.S.	Intention : not to cause harm	**Perspective Taking based on ToM** (he did not want to cause harm).	No Blame	No Blame	

Both systems are Not Congruent					
Accidental harm	No bad intent- bad consequence			Output of the component of the system	Moral judgement
A.E.S.	Action's effect (consequence): Harm	**Deontic reasoning** (he caused harm; it is forbidden, it is bad).	**Blame**	**Blame**	
I.E.S.	Intention: not to cause harm	**Perspective Taking based on ToM** (he did not want to cause harm).	No Blame	No Blame	
Attempted harm	bad intent- no bad consequence			Output of the component of the system	Moral judgement
A.E.S.	Action's effect (consequence): No harm	**Deontic reasoning** (he did nothing forbidden, or wrong; it is not bad).	No Blame	No Blame	
I.E.S.	Intention: to cause harm	**Perspective Taking based on ToM** (he wanted to cause harm and make someone suffer. It is bad).	**Blame**	**Blame**	

or only weakly. Rather, the blame would be apportioned according to the consequence, and the AES would be activated.

1.3   Intellectual Disability

Intellectual disability (ID) is characterized by a low intelligence quotient score (IQ score below 70) and impairments of general cognitive abilities (memory, language, reasoning …) and social abilities (empathy, social judgment…). Most ID individuals have difficulties in social interaction and adaptive behaviours (DSM 5 2013). Intellectual disability is identified as mild, moderate, or severe, but mild intellectual disability is most frequent. Although few studies have focused on moral judgment and judgment of blame of people with ID, results up until now have highlighted that ID individuals' moral development could be slower than typical individual's moral development (Yirmiya et al. 1998; Langdon et al. 2010). In addition, van Vugt et al. (2011); Oubrahim et al. (2019), have shown that children and adolescents with ID present more social behaviour difficulties, than typically developing (TD) children and adolescents. Crick and Dodge (1996) explain this difficulty with a cognitive bias in the attribution of intention, leading to ambiguous and incorrect interpretations of social situations. This cognitive bias could be due to a deficit in the ToM. Most studies on moral judgment have shown that children judge social interactions primarily on the basis of the outcome of the action (Killen et al. 2011; Rogé and Mullet 2011; Salvano-Pardieu, et al. 2020). Killen et al. (2011) reported that children could heed the actor's intention, and not blame accidental action, because they understood that the actor did not want harm neither a bad outcome. However, children with ToM skills less developed still punish accidental action (Salvano-Pardieu and Pennequin 2024). In addition, several studies (Hughes and Leekam 2004; Willaye and Magerotte 2008) support the idea that the understanding of the intention seems to depend on cognitive and intellectual development and reflects ToM ability.

1.4   The Present Study

In this study, we wanted to compare two levels of age (children vs adolescents) and two levels of cognitive and intellectual development (Typical Development (TD) vs Intellectual Disability (ID)) to assess the link between intellectual development and the ability to consider intent in social interaction and see how the Theory of Mind evolves with cognitive development. We hypothesize that the weight given to the factors "Intent" and "Consequence" in moral algebra evolves not only according to age, but also according to intellectual development.

  When the output of the "action evaluation system" and "intent evaluation system" are congruent, no difference would be observed with age and intellectual development. This could be explained by the congruence between the two components involved in moral judgment: deontic reasoning and Perspective Taking based on ToM. For all participants, the "intentional harm" situation should lead to a comparable level of blame, while the "no harm" situation should lead, for all, to no blame or very low blame, as expected in the model (see Table 1). When the outputs of the dual system, AES and IES, are not congruent such as in "accidental harm" and in "attempted harm" situations, a difference is expected between TD children and adolescents as well as between ID and TD participants. Referring to previous research, (Killen et al. 2011; Rogé and Mullet 2011; Salvano-Pardieu, et al. 2020, Salvano-Pardieu and Pennequin 2024), we expect that TD children will judge social interactions using the AES based on deontic reasoning more than the IES based on Perspective-Taking. Therefore, a difference should be observed between children and adolescents typically developing. TD children should blame more harshly "accidental harm" than "attempted harm" while TD adolescents should attribute more blame to "attempted harm" than "accidental harm". Indeed, TD children should judge on the basis of the output rather than on the basis of intention, since Perspective-Taking ability is still developing (Peterson et al. 2005; 2012; de Villiers and de Villiers 2014).

In addition, in mild ID children and adolescents, the level of ToM development is lower than in the TD children and adolescents of the same age (Yirmiya, et al. 1998; Thirion-Marissiaux and Nader-Grosbois 2008; Smogorzewska, et al. 2018). Therefore, our hypothesis is that the ability to evaluate Intent, more than the ability to evaluate Consequence, should be affected by intellectual disability. Indeed, if judging an action on the basis of the Intent requires ToM that depends on cognitive and intellectual development, then participants with ID, delayed in their cognitive development, should also be delayed in their ToM abilities, and would not blame the Intent but rather the outcome of the action. For this reason, ID participants, and especially ID children would be even more focused on the consequence than TD children in situations where AES and IES outputs are not congruent, i.e., in Accidental harm and in Attempted harm. Furthermore, since ToM develops with age and social experience (Hughes and Leekam 2004), ID adolescents, older than children and presumably more socially experienced, should take the perspective of the other and, therefore, the Intention into account more than ID children, but less than TD adolescents.

Finally, our last hypothesis is that adolescents, and particularly TD adolescents, should rank the different levels of aggressiveness of the actions more accurately than children and even more than ID children, as they have experienced more social interactions and are not delayed in their ToM development. For this reason, we varied in this experiment the level of aggressiveness (low, medium, or high) of the action, in addition to the four moral algebra situations.

2 Material and Method

This study compares the moral algebra of children and adolescents with mild intellectual disabilities to those who are typically developing. The judgment of blame of the participants was measured with the factors: intention, consequence, and aggressiveness of the action. The levels of aggressiveness were chosen according to daily life situations, which children and adolescents might have witnessed or heard of in their social environment. Material, method and the 12 short stories used in this experiment were the same as those used by Salvano-Pardieu et al. (2016).

2.1 Participants

There were 124 participants divided into two groups.

Group 1 comprised of 62 participants: 31 children (11 girls - 20 boys, mean age 9.5 years, SD 0.9, range 8–10.9 years) and 31 adolescents (14 girls - 17 boys, mean age 12.8 years, SD 1.2, range 11–15 years) with mild intellectual disability. Children and adolescents of this group were enrolled for the study from special education classrooms for pupils with cognitive impairment, in the vicinity of Tours (France) and in La Chaux de Fonds in the canton of Neuchatel (Switzerland). All participants (50 French and 12 Swiss), came from comparable socio-economic backgrounds and comparable cultures; the 12 Swiss participants were all French native speakers, living near the border with France. All participants with intellectual disability (ID) were tested with the WISC IV. The IQ scores were below the average, in the mild deficiency range (55–70).

Group 2 comprised of 62 participants: 31 children and 31 adolescents typically developing (TD). The children's age (12 girls - 19 boys) matched with the age of the children with ID (mean age 9.3 years; SD 1.0; range 7.6–10.9 years). The age of the adolescents (13 girls - 18 boys) matched with the age of the adolescents with ID (mean age 12.8 years, SD 1.1; range 11–15 years). TD participants were volunteer pupils from primary and middle schools in Tours and surrounding areas in France. All participants, assessed by the school doctor, had typical cognitive development and no disease affecting brain function. All the participants came from similar socio-economic backgrounds. All the children enrolled in this experiment had basic reading skills and were able to read. However, some children needed help to read the text, while all the adolescents were able to read alone. All the participants were French native speakers and gave their consent with the written agreement of their parents to participate in this experiment. This agreement was based on the confidentiality and the anonymity of the data. All the participants were volunteers receiving no payment. All the participants took part in this experiment at school with the agreement of their teachers and the Head of the school.

2.2 Materials

The materials consisted of twelve short stories describing, in a few short lines, simple and concrete social interactions that participants had already experienced in their own life or would have heard about. These social interactions involved two characters, either children or adolescents. Each story contained the following information: (a) the bad intent of the character (with or without); (b) the bad outcome of the action (with or without); (c) the level of aggressiveness of the action (low, medium, or high). The 12 stories presented to the participants were obtained by crossing orthogonally these three factors: Intent of the actor, outcome or consequences of the action and level of aggressiveness of the action. Four situations: "intentional harm" (with bad intent - with bad consequence), "attempted harm" (with bad intent-without bad consequence), "accidental harm" (without bad intent-with bad consequence) and "no harm" (without bad intent- without bad consequence) were used for each independent level of aggressiveness (Push, Punch and Knife).

The low level/Push story described a push and a fall with the consequence of a knee injury. The fall was caused by an accidental push during a game with friends which resulted in a knee injury, (accidental harm) or in no injury (no harm), or by a deliberate push from a girl to another girl she dislikes. In this case, the victim is injured: the fall resulting in a knee injury, (intentional harm), or the victim is not injured (attempted harm).

Example of the Vignette "Intentional Harm" with the Push Theme and the Response Scale

Marie and Fanny play together in the playground. Claire comes to join them, but Fanny does not like Claire and is not happy. Fanny pushes Claire. Claire falls to the ground and hurts her knee.

How much blame do you assign to Fanny?

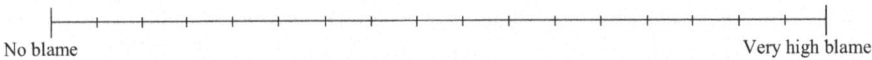

No blame   Very high blame

The medium level of aggression/Punch story described a punch which resulted in a broken nose. In the accidental situations, two friends are playing together, opposite each other, with a punch bag, and one of them misses the punch bag and punches his friend's face. His friend has a broken nose (accidental harm) or he punches into the air near the head of his friend who is not injured (no harm). In the two deliberate situations, a boy who dislikes another boy punches him in his face, breaking his nose (intentional harm) or tries to punch his face but the other boy moves his head and escapes the punch. He is not injured (attempted harm).

Example of the Vignette "Attempted Harm" with the Punch Theme.

Philippe does not like Franck. During break-time, Philippe in a bad mood argues with Franck. He tries to punch him in the face, but Franck turns his head at the last moment. Franck is not injured.

How much blame do you assign to Philippe?

The high level of aggression/Knife story described a stabbing in which the consequence is a serious injury to the leg. In this stab situation, two friends are playing by throwing a knife. The aim of this game is to throw the blade of the knife in the ground, as far as possible. One of them throws the knife clumsily and stabs his friend's leg that is seriously injured (accidental harm) or throws the knife clumsily, near his friend's leg who is not injured (no harm). In the deliberate actions, the aggressor dislikes another boy and insults him. The victim replies and the aggressor who is very angry, takes a knife out of his pocket and stabs the victim's leg who is seriously injured (intentional harm), or the aggressor tries to stab the victim who manages to run away and is not injured (attempted harm).

Example of the Vignette "Accidental Harm" with the Knife Theme

Theo and Timothy are good friends. One day after school they meet at Timothy's house to play in his garden. They have fun throwing a knife as far as they can. The aim of this game is to throw the blade of the knife in the ground, as far as possible. At one point, Theo throws his knife badly, the blade hits Timothy's leg. Timothy's leg is seriously injured.

How much blame do you assign to Theo?

Example of the Vignette "no Harm" with the Knife Theme

Dave and Simon are good friends. One day after school they meet at Dave's house to play in his garden. They have fun throwing a knife as far as they can. The aim of this game is to throw the blade of the knife in the ground, as far as possible. At one point, Dave throws his knife badly, the blade almost hitting Simon's leg. Fortunately, Simon's is not injured.

How much blame do you assign to Dave?

In the vignettes, only the minimum information was given, to avoid a description being too long that could prevent children and participants with ID memorizing crucial information. For this reason, the age of the protagonists was not specified in the vignettes, but it was told to the participants that both characters were pupils of the same age. In the vignettes, characters are friends when the action occurs accidentally, and the aggressor dislikes his victim when the action occurs deliberately. This was made explicit in the text, to avoid any possible ambiguity of the intention of the character, and to help children and participants with ID to understand "accidental" and "attempted" harm situations. We didn't display situations where the two protagonists were friends, but one hurt the other on purpose, or were enemies but one accidentally hurt the other. We assumed that it will be too difficult to understand for children and especially ID children. In addition, the accidental or deliberate actions were clearly specified in the vignettes. Under each scenario a response scale of 16 cm was presented with: "no blame" as the left anchor and "very severe blame" as the right anchor. Participants had to tick a notch wherever they wanted between "0" and "16" on the response scale to record the intensity of their blame. A 16 intervals colour scale from green to yellow, orange and red was presented to the children. This scale was topped with five smileys. Dark and pale green, smileys near "0", means the blame was very low (the green smileys smile). Yellow and orange smileys in the middle part of the scale means the blame was medium (smileys do not smile) and the red smiley near "16" means the blame was high or very high (the red smiley is unhappy). The experimenter explained to the children how to use the scale with the five coloured smileys. In addition, he asked them to give some examples of wrong behaviours at home or at school and then to say how much these behaviours deserved punishment using the "smiley" scale. Children were asked to respond on the intensity of the blame they decide to give but not on the type of punishment. Once children understood how to use the scale, the familiarization phase, and then the experiment started.

2.3 Procedure

Children and adolescents participated in this experiment during school time, in small groups, with an experimenter. During the familiarization phase, participants were provided with two shorts stories and the scale below each story. The experimenter explained to each participant that they had to read a story and decide the level of blame they want to give to the character of the story by ticking a notch on the scale. Participants could compare their answer for the two stories and ask questions. Then, in the experimental phase, each participant was provided with a booklet containing 12 stories. Each booklet presented the stories in a different order, to avoid order effects. Participants judged the character of each story by ticking a notch on the intensity scale below each story. It was not possible to compare the level of blame given in the different stories nor to change a previous answer. No time limit was given to the participants to complete the task. The distance between the tick on the scale and the left anchor "0" was measured in cm and this numerical value was recorded. The participants did not answer "0" or "16" system-atically except for four typically developing adolescents who were excluded from the sample. The answer "0" for each question, or "16" for each question does not take into account the difference between the questions, which is of no interest for the analysis of the results. The duration of the experiment varied from 20 to 45 min depending on the age and the intellectual development of the participants. A small number of children and adolescents who did not take part in the experiment were asked to read the scenarios to test clarity and understanding. Finally, the youngest children and ID children had to give their answer by putting a cross on a scale of 16 color levels ranging from green to yellow, orange and red. Dark green meant no blame, light green meant weak blame, yellow meant medium blame, orange meant strong blame, and red meant very strong blame.

The experimenter explained to the children how to use this scale. It also explained to children the intensity of blame based on their experiences at school or at home when they were scolded or punished for doing something wrong. It was specified to each child that he had to answer only on the intensity of the blame he wanted to give. The experimenter ensured that the child understood how to use the scale by asking him to use this scale with two stories presented during the familiarization phase and asking him to justify his answer, before starting the experiment.

3 Results

3.1 Global Result

A 2 Group (TD or ID) x 2 Age (adolescents or children) x 3 Level of Aggressiveness of the action (Story): (Push, Punch or Knife) x 2 Intent: (with or without) x 2 Consequences: (with or without) ANOVA was conducted on the entire sample of participants. In this "repeated measures" ANOVA, Group and Age was a between-subjects factor, but Level of Aggressiveness of the action (Story), Intent and Consequence were within-subjects factors. The effect size of each of these factors was estimated with a partial η^2.

3.1.1 Main Effect of: Group, Age, Intent, Consequence and Level of Aggressiveness

Each of these factors was found to be statistically significant. The ID group blames the different scenarios more severely (8.23) than the TD group (7.36), [F (1, 120) = 6.73, p < .01], and children blame the scenarios more severely (8.17) than adolescents (7.43), [F (1, 120) = 4.81, p < 0.05]. On average, actions with negative intent are blamed more severely (M = 10.07) than accidental actions (M = 5.53), [F (1, 120) = 361.22, p < 0.0001], and actions with a 'bad' consequence (M = 10.15) are more severely blamed than actions without consequences (M = 5.44), [F (1, 120) = 457.61, p < 0.0001]. Finally, the severity of blame increased significantly with the three levels of aggressiveness, "Push" (M = 4.93), "Punch" (M = 7.72) and "Knife" (M = 10.74), [F (1, 120) = 291.98, p < 0.001].

3.1.2 Effect of the Interactions Between: Group x Intent, Group x Consequence, Group x Level of Aggressiveness

While ID and TD participants blame to the same extent actions with bad intent (M = 10.09 (ID); 10.06 (TD)), a difference is observed when the actions are accidental. ID participants blame much more severely accidental actions (M = 6.38) than the TD group (M = 4.66). [F (1, 120) = 12.57, p < 0.001]. When the actions have no consequence, both groups blame at the same level (5.58 ID; 5.32 TD). In contrast, when the actions have a negative consequence, the ID group, on average, blame the actions more severely (10.89) than the TD group (9.40), [F (1, 120) = 7.93, p < 0.01].

Although both groups attribute the weakest blame to the Push story and the highest to the Knife story, this difference of severity between the three levels of aggressiveness is much stronger with the TD group (3.63; 7.20; 11.24) than with the ID group (6.22; 8.24; 10.24). [F (2, 240) = 27.92, p < .00001]. The planned comparison analysis between TD and ID groups and the three levels of aggressiveness reveals that the ID group blames the "Push" and "Punch" stories more severely than the TD group F (1, 120) = 37.61; p < 0.01 (Push story) and F (1, 120) = 5.70; p < 0.05 (Punch story). In contrast, they rate the "Knife" story less severely than the TD group F (1, 120) = 4.78; p < 0.05. This result shows that the TD group discriminates the aggressiveness of the action more accurately than the ID group does.

3.1.3 Effects of the Interactions Between: Age x Intent, Age x Consequence, Age x Aggressiveness of the Action

On average, children and adolescents blame with the same severity (10.16 (children) and 9.98 (adolescents)) deliberate actions, while accidental actions are blamed more severely by children (6.18) than adolescents (4.87). [F (1, 120) = 5.7, p < 0.05].

While the two groups blame at the same level actions without consequences [5.59 (children); 5.31 (adolescents)], when the actions lead to an adverse outcome, children blame more severely the action (10.76) than the adolescents (9.54), [F (1, 120) = 4.44, p < 0.05].

Adolescents blame more severely than children the difference between the three levels of aggressiveness. This difference between both groups is statistically significant, [F (2, 240) = 5.96, p < 0.01].

A planned comparison shows that no difference is observed between children and adolescents when they judge the "Knife story", [10.63 (children); 10.85 (adolescents)], F (1, 120) = 0.22 NS. By contrast, a significant difference is observed when they judge the Push story: [5.57 (children); 4.31 (adolescents)], F (1, 120) = 8.70; p < 0.01; and the Punch story; [8.32 (children); 7.12 (adolescents)], F (1,120) = 7.75; p < 0.01. These results show that adolescents discriminate more precisely the three level of aggressiveness of the action than children do.

3.1.4 Effect of the Interactions Between: Group x Intent x Aggressiveness of the Action; Age x Intent x Aggressiveness of the Action; Group x Age x Intent x Aggressiveness of the Action

The significant interaction: **group x intent x aggressiveness of the action** [F (2, 240) = 3.87, p < 0.05], shows that ID participants do not blame the intent as much as the TD participants do. Indeed, the difference between actions perpetrated with bad intent and those without, in each story, is much less in the ID group (Push: 3.89; Punch: 3.85 and Knife: 3.34) than in the TD group (Push: 4.32; Punch: 6.56 and Knife: 5.31).

In addition, the significant interaction: **age x intent x aggressiveness of the action** [F (2, 240) = 4.70, p < 0.01], shows that for each scenario, the group of adolescents blames the intent harsher than the group of children. Indeed, the difference between actions perpetrated deliberately and those made accidentally is higher in the group of adolescents (Push: 4.06; Punch: 5.74 and Knife: 5.56) than in the group of children (Push: 4.16; Punch: 4.67 and Knife: 3.10).

Finally, the significant interaction: **Group x Age x Intent x Aggressiveness of the action** [F (2, 240) = 7.50, p < .001] shows that the four groups: ID Children, ID adolescents, TD children and TD adolescents do not combine intent and aggressiveness of the action in the same way.

As shown in Fig. 1, adolescents of the TD group are the most "intentional", they blame actions according to the presence of a negative intent and the aggressiveness of the action. In contrast, they reduce the level of blame when actions are accidental. For this group (Fig. 1A) the difference of blame between actions with and without bad intent is the highest compared with the other groups. Unlike TD adolescents, adolescents with ID give less importance to the intent (Fig. 1B), although they blame according to the aggressiveness of the action. The pattern of responses of TD children (Fig. 1A) is very close to the pattern of response of the ID adolescents. Like ID adolescents, TD children give less importance to the intention, than to the outcome of the action. Finally, children with ID (Fig. 1B) are the less "intentional". The weakest difference is observed in the level of blame between actions with and without negative intent. In addition, the hierarchy observed in the three other groups between the three levels of aggressiveness of the action is not observed with the ID children. When the actions are accidental, ID children take into account the level of aggressiveness and blame more severely the "Knife" story (9.17) than "Punch" story (6.67). By contrast, when the actions are perpetrated deliberately, they blame to the same degree the "Punch" story (11.01) and "Knife" story (10.71). A Bonferroni post hoc analysis revealed that this weak difference of blame between these two stories was not statistically significant.

In addition, unlike the other three groups, when the consequence of the action is a stab in the leg, ID children blame this action with negative intent (10.71) or without (9.17) with the same severity. A Bonferroni post hoc analysis showed that this weak difference was not statistically significant.

3.2 Specific Results

In order to analyze more specifically the judgment of the four groups: ID children, ID adolescents, TD children and TD adolescents, we compared their moral algebra to understand how they combine Intent and Consequence for each level of aggression.

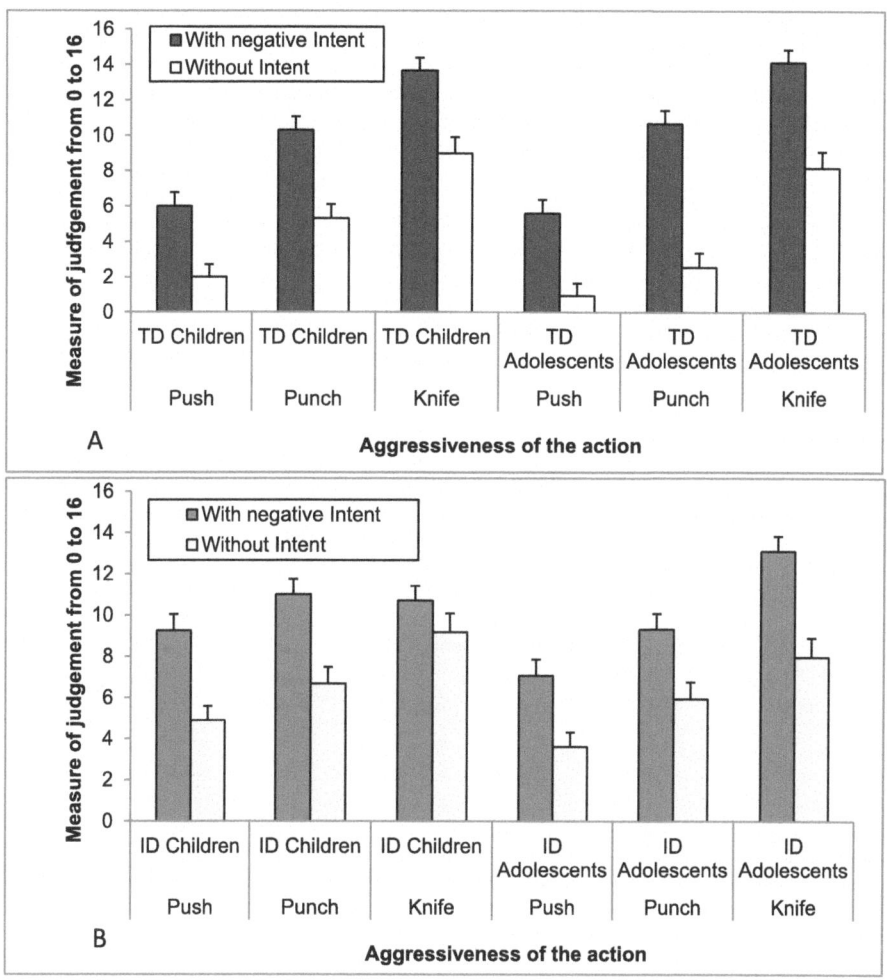

Fig. 1. Mean blame judgments and standard error of TD children and TD adolescents (A), ID children and ID adolescents (B), for the three levels of action aggressiveness (Push, Punch, Knife), regarding actions done deliberately, with bad intention or accidentally without bad intention.

3.2.1 Low Level of Aggressiveness: "Push Story"

In the "Push story" the moral algebra of ID and TD groups is different. While ID children and ID adolescents present an additive moral algebra, TD children and adolescents present a multiplicative moral algebra. As shown in Fig. 2A, in the two ID groups, no interaction is observed between intent and consequence ([F $(1, 30) = 0.80$, NS] for the ID children and [F $(1, 30) = 0.47$, NS] for the ID adolescents). Children with ID give the same weight to the consequences: $\eta^2_p = 0.55$ and to the intent: $\eta^2_p = 0.56$, while adolescents with ID judge the action according to the intent ($\eta^2_p = 0.60$) slightly more than the consequence ($\eta^2_p = 0.50$). On average, as shown in Fig. 2B, children and adolescents with ID blame to the same extent "accidental harm" and "attempted harm".

By contrast, as displayed in Fig. 2A, the moral algebra of the typically developing children and adolescents is multiplicative. In the group of TD children, the interaction between intent and consequence is statistically significant: $F (1, 30) = 3.81$, $p < .05$. TD Children give slightly more weight to the consequences ($\eta^2_p = 0.58$) than to the intent ($\eta^2_p = 0.51$). In the group of TD adolescents the interaction between the factors intent and consequence is also statistically significant $F (1, 30) = 8.43$, $p < .01$, but TD adolescents give even more weight to the intent factor, ($\eta^2_p = 0.72$) than to the consequences, ($\eta^2_p = 0.61$). As shown in Fig. 2B, TD children and adolescents blame leniently, and to a similar degree, accidental actions with a bad consequence, since the

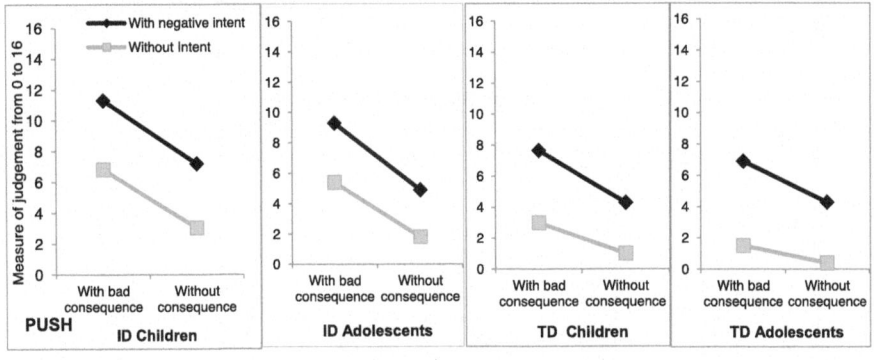

A

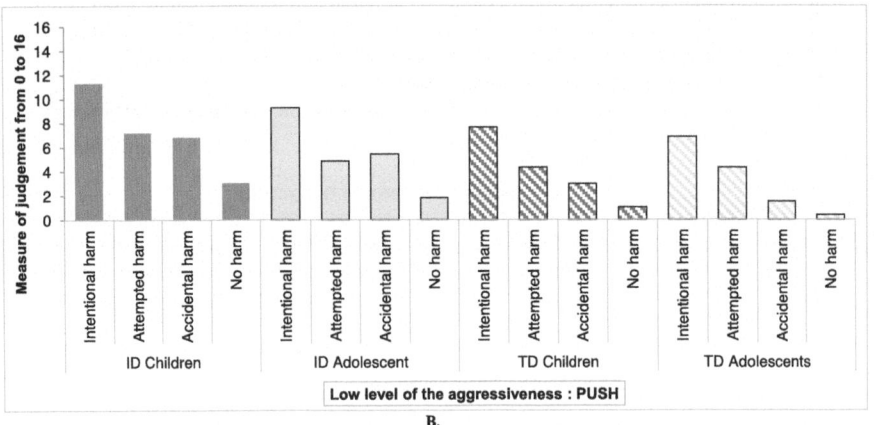

B.

Fig. 2. (a) Moral algebra: interaction effects for judgement of blame for the low level of aggressiveness: Push story, showing actions with and without intent and with and without consequences for all four groups: ID children, ID adolescents, TD children, and TD adolescents. The moral algebra of the TD children and TD adolescents is multiplicative, for the two other groups moral algebra is additive. (b) Moral algebra of the four groups: ID children, ID adolescents, TD children, and TD adolescents, for the low level of aggressiveness: Push story, and for the four situations of judgement: "**intentional harm**": negative intent- bad consequences, "**attempted harm**": negative intent- no consequences, "**accidental harm**": no intent- bad consequences and "**no harm**": no intent- no consequences.

consequence is not serious, but they blame more severely "attempted harm" and even more "intentional harm".

3.2.2 Medium Level of Aggressiveness: Punch Story

When the level of aggressiveness increases, the moral algebra of the TD group varies. As shown in Fig. 3A, TD Children show, like children and adolescents with ID, an additive moral algebra. Only the adolescents of the TD group present an algebra that tends to be multiplicative and close to being significant: $F (1, 30) = 3.28$, $p < .10$. While the TD adolescents give more weight to the intent ($\eta^2_p = 0.87$) than to the consequences ($\eta^2_p =$

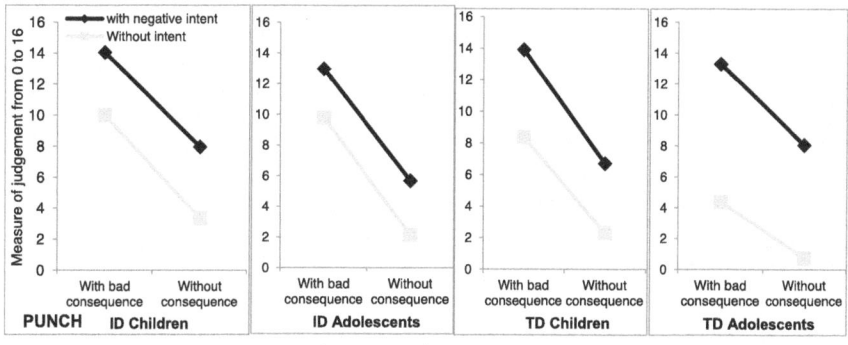

A

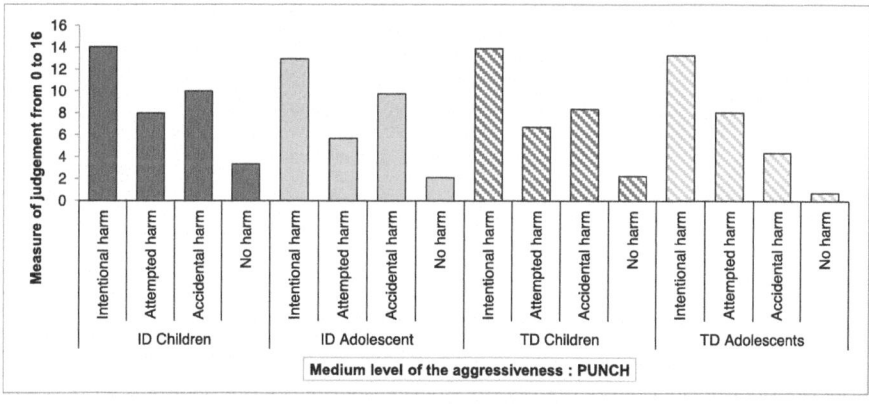

B.

Fig. 3. (a) Moral algebra: interaction effects for judgement of blame for the medium level of aggressiveness: Punch story, showing actions with and without intent and with and without consequences for all four groups: ID children, ID adolescents, TD children, and TD adolescents. Only the moral algebra of TD adolescents is multiplicative, for the three other groups moral algebra is additive. (b) Moral algebra of the four groups: ID children, ID adolescents, TD children, and TD adolescents, for the medium level of aggressiveness: Punch story, and for the four situations of judgement: "**intentional harm**": negative intent- bad consequences, "**attempted harm**": negative intent- no consequences, "**accidental harm**": no intent- bad consequences and "**no harm**": no intent- no consequences.

0.74), the other three groups consider the consequences ($\eta^2_p = 0.75$ for the TD children; $\eta^2_p = 0.72$ and $\eta^2_p = 0.76$ for the children and adolescents with ID, respectively) more important than the intent ($\eta^2_p = 0.62$, TD children, $\eta^2_p = 0.53$ ID children; and $\eta^2_p = 0.42$ ID adolescents). As shown in Fig. 3B, they consider more blameworthy an "accidental harm" resulting in a broken nose than an "attempted harm" that reveals intent to harm and could have resulted in a broken nose. Only TD adolescents blame

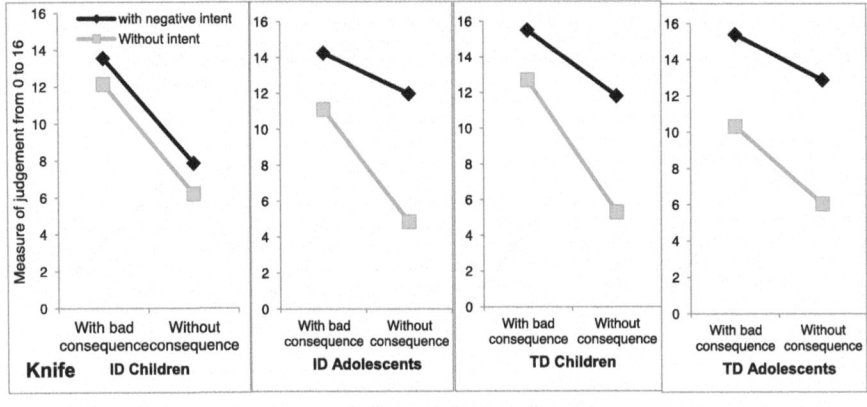

A

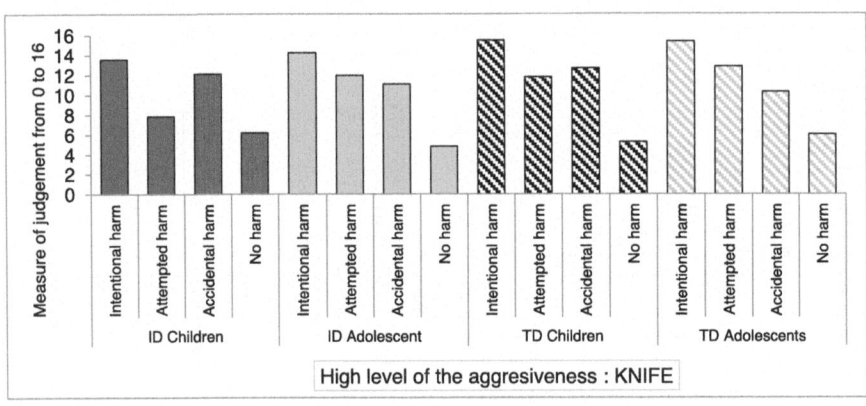

B.

Fig. 4. (a) Moral algebra: interaction effects for judgement of blame for the high level of aggressiveness: Knife story, showing actions with and without intent and with and without consequences for all four groups: ID children, ID adolescents, TD children, and TD adolescents. Only the moral algebra of ID children is additive, for the three other groups moral algebra is. Multiplicative. (b) Moral algebra of the four groups: ID children, ID adolescents, TD children, and TD adolescents, for the high level of aggressiveness: Knife story, and for the four situations of judgement: "**intentional harm**": negative intent- bad consequences, "**attempted harm**": negative intent- no consequences, "**accidental harm**": no intent- bad consequences and "**no harm**": no intent- no consequences.

more harshly attempted harm than accidental harm revealing their ability to consider intention even if the consequence of accidental harm is serious.

3.2.3   High Level of Aggressiveness: Knife Story

Finally, as shown in Fig. 4A, when the aggressiveness of the action is as serious as a stab in the leg, apart from the children with ID, who still show an additive moral algebra, all the other groups present a multiplicative moral algebra ($F (1, 30) = 9.14$ $p < .01$ (ID adolescents); $F (1, 30) = 8.99$ $p < .01$ (TD children) and $F (1, 30) = 4.53$ $p < .05$ (TD adolescents). The multiplicative algebra for this story is the opposite of the multiplicative algebra observed with the low level of aggressiveness (Fig. 2A). While TD adolescents give more weight to the intent $\eta^2_p = 0.78$ than to the consequences $\eta^2_p = 0.63$. The opposite is observed with the three other groups and even more with the ID children who are more focused on the consequences ($\eta^2_p = 0.52$ ID children; $\eta^2_p = 0.78$ TD children and $\eta^2_p = 0.66$ ID adolescents), rather than on the intent ($\eta^2_p = 0.14$: ID children; $\eta^2_p = 0.51$: TD children and $\eta^2_p = 0.61$: ID adolescents). In the Knife story, the consequences are so severe that the actions are severely blamed even in the "no harm" situation. As shown in Fig. 4B, adolescents with and without ID blame more harshly the intentional action in "attempted harm", than the serious consequence in "accidental harm". This result, although observed in both groups, is much stronger in the TD group. In contrast, children with and without ID blame the actions according to the consequences and blame more harshly accidental actions with bad consequence than deliberate bad action without consequence.

4   Discussion

The difference in moral algebra observed between children and adolescents with and without ID is congruent with the assumption that moral judgment is supported by a dual system, one evaluating the outcome of the action (AES) and the other the intention of the actor (IES). The Action Evaluation System (AES) would be based on deontic reasoning while the Intention Evaluation System (IES) would be based on Perspective-Taking of others and therefore on the Theory of Mind. The ID participant's moral algebra, based more on AES than on IES, suggests, in congruence with previous research, that the ability to take the perspective of the other is adversely affected by intellectual disability.

4.1   The Impact of Age and Intellectual Development on the Four Situations of Judgment

Participants with ID do not blame the character's intention as much as their TD counterpart but blame the outcome of the action much more than the TD participants. Furthermore, TD children, blame the character's intention less than adolescents and the action's outcome more than adolescents. TD Adolescents show with all stories greater effect sizes for intent, while children with ID show with all stories a greater effect size for the consequences.

This result supports the assumption that taking the perspective of the other is not as developed in ID participants as in TD participants and not as developed in TD children as in TD adolescents. This result could be explained by the delay of ToM in ID participants as well as its immaturity in TD children. Indeed, Salvano-Pardieu and Pennequin (2024), using the same vignettes "Push" and "Punch" stories that those used in this experiment, confirmed in TD children aged 4 to 7 years, the link between the ability to judge social interactions based on the character's intention and the level of ToM measured with the scale of Welman and Liu (2004). These authors showed that children with the lowest level of ToM were those who judged the consequence of the action more than the character's intention. In contrast, children with the highest level of ToM were those who judged social interactions most based on the character's intention. According to these authors this link depends also on the age of participants, on average, the youngest children aged 4 years, had a lower level of ToM than the oldest aged 7 years and were also those judging on the basis of the consequence of the action, rather than on the basis of the intention of the character. Children, around 9 years old in our sample, did not blame the character's intention as much as adolescents did, which supports the assumption that ToM ability is still in development at 9 years of age. This result is congruent with previous research reporting that ToM is less developed in children than in adolescents (Wellman and Liu 2004; Miller 2012; Peterson et al. 2012; de Villiers and de Villiers 2014, Valle et al. 2015).

Even though the ToM does emerge around 4 years old (Wimmer and Perner 1983), it is not mature enough at 9 years old to allow children to first consider intent, when making a judgment of blame.

Furthermore, using a moral judgment task depicting social interactions between two characters acting either with bad intention or accidentally, Oubrahim et al. (2019) reported that the most aggressive ID participants were also those who blamed harder accidental actions leading to a negative outcome, rather than actions perpetrated with negative intent, but without bad consequences. This result was mainly observed within ID children who had the most social difficulties.

This misunderstanding of character's intention in social interactions can be explained by the low level of ToM observed in Intellectual Disability (Hughes and Leekam 2004) but also with the low IQ (Willaye and Magerotte 2008). Additionally, some studies measuring ToM with the Wellman and Liu scale showed that children and adolescents with mild ID, had a lower level of ToM development than their TD children and adolescents counterparts (Thirion-Marissiaux and Nader-Grosbois 2008; Smogorzewska et al. 2018). Hughes and Leekam (2004) and Willaye and Magerotte (2008) showed that the lower the level of ToM, the higher the sanction for accidental actions. This result highlights the link between understanding the actor's intention in social relationships and ToM abilities.

4.2 Explanation of Our Results According to the Dual System AES and IES

The Congruence of the Two Systems

Moral algebra of the four groups (ID and TD children, ID and TD adolescents) in "intentional harm" and "no harm" situations shows a similar pattern of responses. All the groups blame the "Intentional harm" situation the most and the "No harm" situation the least. This result already observed comparing moral judgement of adults with and without Alzheimer disease (Fontaine et al. 2004) and adolescents with and without Asperger's syndrome (Salvano-Pardieu et al. 2016), could be explained by the congruence of the output of the dual system. In the ID group and in TD children, the low level of activation of the "intent evaluation system" relying on Perspective-Taking and ToM ability would be compensated by the activation of the "action evaluation system" which relies on deontic reasoning. For this reason, similar moral algebras are observed in the four groups. This result seems to confirm that deontic reasoning is still preserved in mild ID individuals and could be enough to blame the "intentional harm" situation.

The Incongruence of the Two Systems
At the opposite, in the "accidental harm" scenarios our results showed a strong difference between the four groups. Unlike TD adolescents who blame more harshly intention of the actor than the consequence of the action, the three other groups (TD children, ID children, and ID adolescents) blame the outcome of the action more harshly than the intention of the character. This result could be explained by the non-congruent outputs of the dual system of judgement. In the "accidental harm" situation, AES and IES are in opposition, the blame in this situation means that the consequence, but not the intention, is considered and therefore the AES is primarily activated. Since ID participants and TD children attribute greater blame to the "accidental harm" situation, it means that the IES is weakly activated. This confirms that perspective-Taking of the other et therefore ToM ability is not developed in TD children and ID participants, as much as in TD adolescents. It also suggests that this component of moral judgment is affected by age and intellectual development, unlike deontic reasoning. Intellectual disability affects moral development especially in ToM. Furthermore, although TD children and ID group blame the "accidental harm" situation, focusing on the outcome, ID children blame the consequence harsher than the other groups. This result suggests that ID children have the most difficulty taking the perspective of the other and their intent. This difficulty could be due to the immaturity of ToM. This immaturity would lead to impairment in the social adaptation that would explain the aggressive behaviour of these children. This assumption is congruent with a recent result (Oubrahim, et al. 2019) reporting that ID children have less prosocial and more aggressive behaviours than ID adolescents. Finally, unlike TD adolescents, TD children and ID groups blame leniently the aggressor when the deliberate action does not lead to a bad consequence. They do not blame the bad intent since no consequence is observed. Therefore, one could consider that they use mainly deontic reasoning to blame the aggressor in this situation. This result seems to confirm that the outcome of the action, and therefore the AES based on deontic reasoning is easier to activate, than the IES based on Perspective-Taking and ToM. Indeed, previous studies (Knobe 2005; Leslie et al. 2006; Petit and Knobe 2009; Killen et al. 2011), have shown that when judging social interactions, people tend to evaluate the outcome of the action to infer the intention of the actor. According to Killen et al. (2011), people tend to think that an action is perpetrated deliberately when the consequence is bad, while they are inclined to think the intention is not bad when the action is not followed by a negative consequence.

4.3 The Level of Aggressiveness

In this study, we assumed that adolescents, and particularly TD adolescents, would rank more accurately the different levels of aggressiveness (Push, Punch and Knife story), than children, and especially ID children, who have experienced fewer social interactions and are delayed on ToM. Our result confirms this assumption, ID children have difficulties ranking, as precisely as the other groups, the three levels of aggressiveness. ID adolescents and the TD group rank the three stories (Push, Punch and Knife) from the less serious "Push" to the more serious "Knife" story. Indeed, they blame a push leading to a fall and a bruised knee less harshly than a punch in the face leading to a broken nose, which is less reproached than a stab causing a serious injury to the leg. This difference in the blame between these three levels of aggressiveness is highlighted in TD adolescents and confirms previous results already observed with TD adolescents (Salvano-Pardieu et al. 2016). By contrast, even though ID children consider the outcome of the "Punch" story more serious than the output of the "Push" story, they blame almost to the same degree the "Punch" and the "Knife" story, especially for deliberate actions. In addition, they blame more harshly the "Push" story than TD children and the two groups of adolescents. This result suggests that children with ID have difficulty understanding the severity of consequences for the victim and confirms that the theory of mind involved in the ability to understand the other's point of view is impaired. Finally, understanding aggressiveness of social interactions depends not only on cognitive development but also on age and social experiences. Indeed, TD adolescents have experienced more social relations, than TD children, and ID adolescents more than ID children. This lack of experience, and therefore the lack of training in Perspective Taking, could explain why ID children blame nearly to the same level a deliberate punch leading to a broken nose and a deliberate stab leading to a wounded leg.

Finally, social and cultural factor could explain these differences. Indeed, Barrett et al. (2016) pointed out, there are social differences in the extent to which intentionality is taken into account in moral judgments. In our experiment the participants come all from a middle class, but it could be a difference between French and Swiss participants. So, it could be very interesting, in a further experiment, to analyse the difference in taking intention into account depending on the social and cultural level of the participants.

5 Conclusion

This study seems to confirm the role of a dual system in the judgment of blame: the action evaluation system (AES) involved in the judgment of the action's output which is based on deontic reasoning and refers to the knowledge of social rules, and the intention evaluation system (IES), involved in the judgment of the actor's intention, and based on the ToM. Deontic Reasoning on one hand and ToM on the other hand do not follow the same development. Deontic reasoning appears earlier in moral reasoning and would not be affected by intellectual disability. In contrast, ToM ability seems to develop later than deontic reasoning and seems to depend on intellectual development. Indeed, the ToM development seems to be delayed in participants with ID since they present difficulties to take into account the intent of the actor. In addition, the ability to understand and consider the perspective of the others would improve with the experience of social interactions and thereby allow evaluating the level of seriousness of an aggression. For this reason, children with less social experiences and particularly ID children have additional difficulties differentiating accurately the various levels of aggression.

These findings reported in this paper help explain the differences observed in moral judgments between children and adolescents with and without intellectual disability. The late development of ToM ability involved in the Intention Evaluation System leads ID individuals to inaccurately assess the intent in social interactions.

References

American Psychiatric Association: Diagnostic and Statistical Manual of Mental Disorders, 5th edn. American Psychiatric Association, Washington, DC (2013)

Anderson, N.H.: A Functional Theory of Cognition. Erlbaum, Hillsdale (1996)

Anderson, N.H.: Unified Social Cognition. Psychology Press, New York (2008)

Barrett, H.C., et al.: Small-scale societies exhibit fundamental variation in the role of intentions in moral judgment. Proc. Natl. Acad. Sci. **113**(17), 4688–4693 (2016)

Buon, M., Seara-Cardoso, A., Viding, E.: Why (and How) should we study the interplay between emotional arousal, theory of mind, and inhibitory control to understand moral cognition? Psychon. Bull. Rev. **23**(6), 1660–1680 (2016). https://doi.org/10.3758/s13423-016-1042-5

Crick, N.R., Dodge, K.A.: Social information-processing mechanisms in reactive and proactive aggression. Child Dev. **67**(3), 993–1002 (1996)

Cushman, F.: Crime and punishment: distinguishing the roles of casual and intentional analysis in moral judgment. Cognition **108**(2), 353–380 (2008)

Cushman, F.: Action, outcome and value: a dual-system framework for morality. Pers. Soc. Psychol. Rev. **17**(3), 273–292 (2013)

Cushman, F., Sheketoff, R., Wharton, S., Carey, S.: The development of intent-based moral judgment. Cognition **127**(1), 6–21 (2013)

Fontaine, R., Salvano-Pardieu, V., Renoux, P., Pulford, B.: Judgment of blame in Alzheimer's disease sufferers. Aging Neuropsychol. Cogn. **11**(4), 379–394 (2004)

Gweon, H., Dodell-Feder, D., Bedny, M., Saxe, R.: Theory of mind performance in children correlates with functional specialization of a brain region for thinking about thoughts. Child Dev. **83**(6), 1853–1868 (2012)

Hommers, W., Lee, W.Y.: Unifying Kohlberg with information integration: The moral algebra of recompense and of Kohlbergian moral informers. Psicológica **31**, 689–706 (2010)

Hommers, W., Lewand, M., Ehrmann, D.: Testing the moral algebra of two Kohlbergian informers. Psicológica **33**(3), 515–532 (2012)

Hughes, C., Leekam, S.: What are the links between theory of mind and social relations? Review Reflections and new directions for studies of typical and atypical development. Soc. Dev. **13**, 590–619 (2004)

Killen, M., Mulvey, K.L., Richardson, C., Jampol, N., Woodward, A.: The accidental transgressor: morally-relevant theory of mind. Cognition **119**, 197–215 (2011)

Knobe, J.: Theory of mind and moral cognition: exploring the connections. Trends Cogn. Sci. **9**, 357–359 (2005)

Langdon, P.E., Clare, I.C.H., Murphy, G.H.: Developing an understanding of the literature relating to the moral development of people with intellectual disabilities. Dev. Rev. **30**, 273–293 (2010)

Leon, M.: Rules mothers and sons use to integrate intent and damage information in their moral judgments. Child Dev. **55**, 2106–2113 (1984)

Leslie, A., Knobe, J., Cohen, A.: Acting intentionally and the side effect: theory of mind and moral judgment. Psychol. Sci. **17**, 421–427 (2006)

Manktelow, K.: Thinking and Reasoning: An Introduction to the psychology of Reason, Judgment and Decision Making. Psychology Press, New York (2012)

Manktelow, K.: Reasoning and thinking. Psychology Press, Hove (1999)

Manktelow, K.I., Sutherland, E.J., Over, D.E.: Probabilistic factors in deontic reasoning. Thinking Reason. 1(3), 201–219 (1995)

Manktelow, K.I., Over, D.E.: Social roles and utilities in reasoning with deontic conditionals. Cognition 39, 85–105 (1991)

Miller, S.A.: Theory of Mind: Beyond the Preschool Years. Psychology Press, New York (2012)

Oubrahim, L., Combalbert, N., Salvano-Pardieu, V.: Moral judgment and aggressiveness in children and adolescents with intellectual disability. J. Intellect. Disabil. Offending Behav. 10(2), 21–33 (2019)

Petit, D., Knobe, J.: The pervasive impact of moral judgment. Mind Lang. 24, 586–604 (2009)

Peterson, C.C., Wellman, H.M., Liu, D.: Steps in theory-of-mind development for children with deafness and autism. Child Dev. 2005(76), 502–517 (2005)

Peterson, C.C., Wellman, H.M., Slaughter, V.: The mind behind the message: advancing theory-of-mind scales for typically developing children, and those with deafness, autism, or Asperger syndrome. Child Dev. 83(2), 469–485 (2012)

Premack, D., Woodruff, G.: Does the chimpanzee have a theory of mind? Behav. Brain Sci. 1(4), 515–528 (1978)

Przygotzki, N., Mullet, E.: Moral judgment and aging. Revue Européenne de Psychologie Appliquée 47, 15–21 (1997)

Rogé, B., Mullet, E.: Moral judgment among children, adolescents and adults with autism. Autism 15, 702–712 (2011)

Salvano-Pardieu, V., et al.: Judgment of blame in teenagers with Asperger's syndrome. Think. Reason. 22, 251–273 (2016)

Salvano-Pardieu, V., Olivrie, M., Pennequin, V., Pulford, B.: Does role playing improve moral reasoning's structures in young children? In: Yama, H., Pardieu, V. (eds.) Adapting Human Thinking and Moral Reasoning in Contemporary Society. IGI Global (2020)

Salvano-Pardieu, V., Pennequin, V.: Relationship between theory of mind and judgement based on intention in 4–7 Y.O. Children. In: Baratgin, J., Jacquet, B., Yama, H. (eds.) Human and Artificial Rationalities. Lecture Notes in Computer Science, vol. 14522, pp. 64–85. Springer, Cham (2024). https://doi.org/10.1007/978-3-031-55245-8_5

Smogorzewska, J., Szumski, G., Grygiel, P.: Same or different? Theory of mind among children with and without disabilities. PLoSONE 13(10) (2018)

Surber, C.F.: Separable effects of motives, consequences, and presentation order on children's judgments. Dev. Psychol. 18, 257–266 (1982)

Thirion-Marissiaux, A.F., Nader-Grosbois, N.: Theory of Mind "beliefs", developmental characteristics and social understanding in children and adolescents with intellectual disabilities. Res. Dev. Disabil. 29, 547–566 (2008)

Valle, A., Massaro, D., Castelli, I., Marchetti, A.: Theory of mind development in adolescence and early adulthood: the growing complexity of recursive thinking ability. Eur. J. Psychol. 11(1), 112–124 (2015)

van Vugt, E., Asscher, J., Stams, G.J., Hendriks, J., Bijleveld, C., van der Laan, P.: Moral judgment of young sex offenders with and without intellectual disabilities. Res. Dev. Disabil. 32(6), 2841–2846 (2011)

de Villiers, J.G., de Villiers, P.A.: The role of language in theory of mind development. Top. Lang. Disord. 34(4), 313–328 (2014)

Wechsler, D.: WISC-IV. Echelle d'intelligence de Wechsler pour enfants – Quatrième édition. ECPA, Paris (2005)

Wellman, H.M., Liu, D.: Scaling of theory-of-mind tasks. Child Dev. 75(2), 523–541 (2004)

Willaye, E., Magerotte, G.: Évaluation et intervention auprès des comportements défis. Déficience intellectuelle et/ou autisme. De Boeck, Bruxelles (2008)

Wimmer, H., Perner, J.: Beliefs about beliefs: Representation and constraining function of wrong beliefs in young children's understanding of deception. Cognition **13**(1), 103–128 (1983)

Yirmiya, N., Erel, O., Shaked, M., Solomonica-Levi, D.: Meta-analyses comparing theory of mind abilities of individuals with autism, individuals with mental retardation, and normally developing individuals. Psychol. Bull. **124**, 283–307 (1998)

Artificial Intelligence and Cognition

Artificialization of Intelligence and Embodied Cognition

Pierre Uzan[1,2](✉)

[1] Catholic University of Paris, Paris, France
uzanpier@gmail.com
[2] CHArt Laboratory, Paris, France

Abstract. The project of artificializing human intelligence currently faces the difficulty of simulating generalist intelligence. "Generalist" intelligence, whose essential properties are autonomy, versatility and fluidity, requires almost instantaneous access to an immense quantity of relevant information about the situations encountered, information that can range from simple factual data to understanding their emotional and social context. The difficulty in creating an intelligent machine in this general sense finds its origin in the computational paradigm of cognition, which assimilates the cognitive subject to a sophisticated calculator endowed with cognitive abilities specified by algorithms, and whose task would be to represent, using symbols or global states, a pre-existing world with equally well predefined properties. I will show that this limitation could be overcome by rebasing the project of artificialization of intelligence on the embodied approach to cognition explored by Francisco Varela, an approach which takes into account the essential role of the body and its insertion in the world.

Keywords: Artificial intelligence · Embodied cognition · Situated robots

1 Introduction

There can be no doubt about the growing success of artificial intelligence in performing highly specialized tasks. These specialized tasks range from the development of high-performance gaming software to the detection of computer viruses and bank fraud, or, in the medical field, the recognition of the nature and severity of tumors. However, the project to artificialize intelligence is currently stumbling over the difficulty of creating machines that simulate an essential feature of human intelligence, namely its "generality". As highlighted by Daniel Andler [1], human intelligence is "general" in the sense that, in addition to make people autonomous in their decisions and actions, it presupposes great versatility, enabling them to accomplish an indefinite variety of tasks that may, moreover, be totally unforeseen, and fluidity, enabling them to rapidly switch from one task to another, according to the demands of the moment. This requires almost instantaneous access to an immense quantity of relevant information about the situations they encounter, ranging from simple perceptive data to representations of the environment and the understanding of the emotional and social context of these lived situations. General intelligence is thus a situational intelligence, based on our perception of the current situation, and depends essentially on our insertion into the world.

© The Author(s), under exclusive license to Springer Nature Switzerland AG 2025
J. Baratgin et al. (Eds.): HAR 2024, LNCS 15504, pp. 91–109, 2025.
https://doi.org/10.1007/978-3-031-84595-6_6

This difficulty in realizing an "intelligent" machine in this general sense has its origins in the computational paradigm of the mind, based on the disembodied symbolic and connectionist approaches to cognition. The computational paradigm of the mind likens the cognitive subject to a sophisticated calculator whose operation is specified once and for all, namely, to perform a given cognitive task independently of any possible interaction with the environment. Its cognitive abilities, also predetermined by its algorithms, have the sole task of representing, by means of symbols or global states, a pre-existing world, which is thus supposedly endowed with determined properties. As a result, it cannot simulate the properties of autonomy, versatility and fluidity of general intelligence, enabling the individual to interact optimally with his or her environment.

This article shows that the difficulties of the computational model of AI can be overcome by rebasing the project of artificializing intelligence on the embodied approach to cognition, which considers the essential role of the body and its insertion into the world. Section 2 will briefly present the computational paradigm of artificial intelligence and the limitations it encounters. Section 3 will present the embodied approach to cognition and its phenomenological roots, an approach introduced by Chilean neurobiologist and philosopher Francisco Varela in the 1980s. Section 3.2 will focus in particular on the scientific study of embodied cognition, citing significant experimental results showing the essential links between cognition and corporeality. Section 4 will detail and comment on the very promising achievements of embodied AI, developed in close connection with robotics.

2 The Computational Model of AI

2.1 About the Artificial Intelligence Project

Can machines reason, simulate subjective experience or the understanding of words and sentences? Can it interact with the world, adapting its responses and actions to its biological, cultural and social environment? Is it possible to build such "intelligent" machines, capable of fully simulating all aspects of human intelligence? These questions, which does not a priori presuppose adherence to the computational metaphor of the mind as an information-processing system [2], merely extend and clarify the question posed by Turing in his seminal article *Computing Machinery and Intelligence* [3], published in the journal *Mind*: "Can machines think?".

Turing's innovative idea, which gave decisive impetus to the development of artificial intelligence, was to define a machine's intelligence not by identifying it with the human capacity to feel, understand or create, but as its ability to *simulate* all these. Turing made it clear in a BBC interview [4] that, while we can attribute the ability to think to the machine, the latter "thinks like a machine: it merely simulates human behavior and intelligence."

Defining a machine's intelligence in terms of its ability to simulate human behavior thus avoids all the false trials against Turing and, more generally, against the project of artificializing intelligence. These critics attribute to Turing and to this project the claim to manufacture artificial beings capable of experiencing human emotions, of understanding and creating *as a human does* [5–8]. However, *simulating* human intelligence with a machine does not require to endow that machine with human intentionality, conscious-ness or soul! As an example of such a false trial, let's mention Searle's Chinese Room argument [6], intended to show that a machine that manages to manipulate Chinese characters well enough to deceive a Chinese interrogator, without of course having his culture and knowledge of the Chinese language, "does not understand" the meaning of the questions it is asked and the answers it gives. But this argument, like those denounc-ing the impossibility of "authentic" machine creativity [7] or subjective feeling [8], is in no way a criticism of the modern project to artificialize intelligence, whose aim is to simulate human intelligence, not to produce or reproduce it artificially. These "argu-ments", which are not really arguments at all, are in fact, at most, intended to criticize the mythical project - part of what is known today as strong AI - of endowing an artificial being with *human* sensitivity, creativity and consciousness, as exemplified, for example, by the myth of Pygmalion and Galatea (Ovid, 43 BC), in which Pygmalion, in love with his statue, wished to make it alive, sensitive and conscious [9].

Indeed, Turing's behavioral definition of machine intelligence made it a quantifi-able, measurable quantity (notably through the Turing's test), and is thus at the origin of the lightning development of AI that we are witnessing. The project to simulate human thought with a machine then developed according to two approaches to cognition, com-plementary rather than opposed. The symbolic approach (often referred to as "Good Old Fashion Artificial Intelligence"), exemplified by the Turing machine, equates thought with the manipulation of symbols according to rules. The connectionist or emergentist approach is based on the idea that mental states, thought and consciousness are "emer-gent properties" of the brain, leading to the design of artificial neural networks whose operation simulates that of the human brain. While these two approaches have given rise to a first, rapid development of the project to artificialize intelligence, they nevertheless fall within the computational model of cognition and intelligence, the limits of which we will show.

2.2 The Symbolic Approach

According to the symbolic approach to cognition, thought can be conceived in terms of the manipulation of symbols (numbers, words, ideograms, etc...). One of its most important precursors is Leibniz [10], whose project was to develop a "*lingua charac-teristica universalis*" that would enable reasoning to be carried out in a mechanical, purely calculatory way. The Leibnizian idea that "thought is fundamentally a calculus" was taken up by Boole [11], who introduced the propositional calculus in his *Fonde-ments des opérations de l'esprit*. In his *Begriffschrift* [12], Frege also took up Leibniz's unfinished project and sought a "formal language of pure thought built on the model of arithmetic, in which the content of thought can be expressed", which led him to introduce the predicate calculus.

The symbolic approach takes as its paradigm the Turing machine, which processes information sequentially according to a specified algorithm, translated into the sequence of symbols inscribed on an infinite ribbon [13]. Turing's demonstration, in this same reference, of the existence of a universal Turing machine, capable of accepting as input the description of any Turing machine and the data it is to process, constitutes a decisive step in the realization of the project to mechanize thought. However, the symbolic approach comes up against two kinds of limitations: those relevant to the use of symbols and those of a deeper, conceptual nature. The first, "factual" limitations of any symbolic representation of information lie in its fragility (a missing symbol, a syntax error and the program no longer functions), its rigidity (the impossibility of recognizing and processing data not foreseen in the initial program, or simply noisy data) and the fact that, unlike most natural processes (at neural level, for example), information is processed sequentially. While these factual limitations, due to the very mode of expression of information (use of symbols), will be overcome by the connectionist approach to cognition presented below, this is not the case with the conceptual limitations that Turing was already discussing in his 1950 seminal article [3] in response to the objections made to him. These include the possibility of mechanizing high-level aspects of human thought and intelligence, as intentionality, attribution of meaning, subjective experience and creativity [5–8]. Turing illustrated this problematic concerning the possibility (or not) of mechanically simulating specific human mental capacities, and the methodical work of the artificial intelligence researcher in attempting to do so, with the analogy of the onion's skin in his article [3], page 454–455:

> "In considering the functions of the mind or the brain we find certain operations which we can explain in purely mechanical terms. This we say does not correspond to the real mind: it is a sort of skin which we must strip off if we are to find the real mind. But then in what remains we find a further skin to be stripped off, and so on. Proceeding in this way do we ever come to the 'real' mind, or do we eventually come to the skin which has nothing in it? In the latter case the whole mind is mechanical".

This analogy calls for modesty and suggests that we should not skip the stages of genuine scientific research into the development of artificial intelligence because of often unfounded criticism and hasty assertions (as those mentioned in Sect. 2.1).

2.3   The Connectionist or Emergentist Approach

The connectionist approach is based on the idea that thought, mental states and consciousness "emerge" from the functioning of the brain. This idea suggests that, as Rumelhart and McClelland put it [14], page 3:

> "… people are smarter than today's computers [designed according to the symbolic approach] because the brain is based on a computational architecture better suited to the natural information-processing tasks in which humans excel…"

According to the connectionist (or "sub-symbolic") approach, thought and intelligence are products of the brain's architecture. Hence the idea of building artificial neurons capable of mimicking the functioning of the biological neuron and connecting assemblies of such artificial neurons to mimic the functioning of the brain - and thus simulate mental activity. The first artificial neuron with binary inputs and outputs was proposed in 1959 by McCulloch and Pitts [15]. In the case of the biological neuron, whether an output signal is emitted into the axon is governed by the dosage of neurome-diators at the input, whereas in this first formal neuron, the output signal is calculated from an adjustable threshold value. The architecture of artificial neural networks gives rise to massively distributed information processing that is difficult, if not possible, to follow step by step, unlike information processing carried out sequentially using the symbolic approach. What is here relevant is the network's "global state", defined by its architecture and the weighting values of its multiple neuronal connections.

As with the human brain, the architecture of ANN (artificial neural networks) and their connectivity allow simulating multiple cognitive tasks, like perception, learning, pattern recognition, language processing and even artistic generation. Achieving these tasks requires algorithms that optimize the weights of the network's neuronal connections according to the required task (notably using the gradient backpropagation method [16]), and architectures that mimic those of animals' cognitive systems, such as convolutional networks that mimic the functioning of their visual system, or recurrent networks that simulate memorization processes. We should mention the use of large language models, essential tools for the development of the latest AI products. These are algorithms, trained on large quantities of data, which predict the likely sequence of a given input, enabling a wide variety of natural language processing tasks. For example, the latter algorithms simulate the "understanding" of words in a given linguistic context by referring to the frequency of associations in which they appear in the analyzed data [17].

2.4 The Limitations of Computational AI

Intelligent machines based on the cognitivist and the connectionist approaches to cognition are sophisticated calculators, highly capable of carrying out precise cognitive tasks, such as making medical decisions for an expert system, or classifying shapes presented to it for an ANN. They can simulate very well abstract tasks of human cognition, like calculating, reasoning or classifying objects in categories, and they even surpass human intelligence in its purely "intellectual" aspect. However, these machines are static (immobile in space) and disembodied, they are designed to only carry out specified, repetitive cognitive tasks. If they can have interactions with their environment through exchange of signals, they are nevertheless deprived of any bodily and social experience, on which are yet grounded essential aspects of human intelligence and makes it autonomous, versatile and fluid. Indeed, the computational approach to cognition assumes the existence of a predefined world, described by the data to be processed, that the subject would represent to himself thanks to a specified and invariable cognitive system functioning according to a predefined algorithm. For example, expert systems developed according to the symbolic approach can only process data considered by the specific database, whereas an ANN trained for a classification task will classify objects endowed with predefined properties (such as their color or their shape) into categories specified in advance. This difficulty of the computational model of cognition is mentioned in Varela's book *Autonomie et Connaissance* [18], page 233–234:

> "[According to these two approaches] the external world has fixed rules; it precedes the image it projects onto the cognitive system, whose task it is to grasp it appropriately (whether by means of symbols or global states)".

As pointed by Varela and his colleagues in their book *The Embodied Mind: Cognitive Science and Human Experience* [19], the act of knowing relies on the action of the subject guided in a context that is always singular and never entirely predictable (page 234):

"It is precisely this insistence on mutual specification that enables us to negotiate a middle way between the Scylla of cognition conceived as the reconstruction of a preordained external world (realism) and the Charybdis of cognition conceived as the projection of a preordained internal world (idealism). These extremes both take representation as their central notion: in the first case, representation is used to reconstitute what is external; in the second, it is used to project what is internal. Our intention is to bypass this logical geography of "inside versus outside" entirely, by studying cognition not as reconstitution or projection, but as *embodied action*."

The "middle way" that Varela chooses to develop thus attempts to reconcile the objectivist perspective, according to which organisms live in a pre-existing environment and cognition would consist in constructing a more or less faithful representation of it, and the subjectivist perspective, according to which the cognitive system would project its own representations. The "embodied" approach to cognition proposed by Varela does not call on the notion of representation, but *on the bodily experience of the cognitive subject and its insertion into the world*. The roles played by the subject and its environment are conceived as being closely dependent in the very act of cognition [19], page 236:

"The organism gives form to its environment at the same time as it is shaped by it....The properties of perceived objects and the intentions of the subject, not only blend but constitute a new whole."

In contrast to the cognitivist and connectionist approaches to cognition, Varela's approach reintegrates the phenomenal dimension into the process of cognition. His project is to articulate the phenomenal, lived side of experience and what can be said about the corresponding cerebral functioning, its "neural correlates". Varela qualifies this approach of "neurophenomenological" [20].

3 The Embodied Model of Cognition: The Essential Role of Experience

3.1 The Phenomenological Roots of Embodiment

The explicit consideration of human experience in the process of cognition goes back to the phenomenological approach developed by Husserl, then by Heidegger and Merleau-Ponty. In his book, *Phénoménologie de la perception* [21], Maurice Marleau-Ponty rejects the Cartesian idea that the principal mode of being in the world is thought, and proposes *corporality* instead. The body itself is the primary locus of knowledge of the world, and perception the medium and foundation of experience (page 97):

"The body is the vehicle of being in the world, and to have a body is, for a living being, to be involved in a defined environment, to identify with certain projects and to engage in them continually."

The first thing we experience is the body, not only as an object of the physical world, as an element of material nature, but as a means of experiencing sensations, as an organ of experience ([21], page 274):

"... Our body is that strange object which uses its own parts as a general symbol of the world, and through which, as a result, we can frequent this world, understand it and find meaning in it".

The Husserlian term for this notion of phenomenal body, *Leib*, also translates as "proper body", "living body", or "flesh", and is characterized in [22], page 56, as:

"...the means of all perception; it is the organ of perception, it is necessarily involved in all perception."

The term "Leib" therefore designates for Husserl [22] "my body insofar as I live in it", i.e. the body as an organ of experience and a field of sensation and, unlike other things in the material world, it is constituted in a dual modality (page 208):

"The body itself is thus originally constituted in a dual mode: on the one hand, it is a physical thing, matter [...], on the other, it is a field of sensations, I find in it and feel "on" it and "in" it."

In his book *Le visible et l'invisible* [23], Mercleau-Ponty discusses Husserl's "touching-touched" experience. He says that when his right hand touches his left hand, his body appears to him, in the experience of touch, both as his own and as exteriority (page 174):

"..., at the same time as felt from within, my hand is also accessible from without, tangible itself, for example for my other hand, if it takes its place among the things it touches, is in a sense one of them..."

In Merleau-Ponty's phenomenological perspective, our perception of the world is intimately linked to the capacity for action we have acquired, and which is constrained by our bodies. These "motor skills" are physical abilities ranging from something as simple as scratching an itch to something as complex as driving a car. They are acquired through practice, i.e. the body's familiarization with the activity. Motor skills have a motor component and a perceptual component: being proficient at something implies being able to do something and seeing what to do. For example, a person skilled in rock climbing can both move his body on the rock face and perceive small cracks and ledges in the rock as holds and fulcrums - he has learned to see the rock face as an invitation to climb. Thus, the way we perceive the world is intimately linked to our capacity for action: what we see is in part linked to what we can do. Merleau-Ponty introduced the notion of "affordance" to designate the capacity of an object or system to evoke its use, its function, and thus provoke a spontaneous interaction between an environment and its user. For example, the handle on a door prompts the user to press it to open the door, and the shape of a chair suggests that the user sit on it. Merleau-Ponty' intuition is that perception results from an active process linking the properties of the object and the intentions of the subject, as mentioned in *The structure of behavior* [24], page 11:

"The properties of the object and the intentions of the subject ... Not only blend together but constitute a new whole".

It is this intuition that Varela, Thompson and Rosch have tried to reintroduce into the scientific study of cognition, and which has given rise to the concept of *"enaction"*[1].

3.2   Scientific Study of the Embodied Approach to Cognition

The concept of "embodied cognition" suggests that the study of cognition requires consideration of the subject's lived experience, which, as mentioned above, relies crucially on corporeality. The concept of embodied cognition thus links the subject's performance of cognitive tasks with the biological functions of his body, his sensorimotor system and all the bodily processes involved in his interaction with the environment. Varela, Thompson and Rosch [19] define embodied cognition as follows (page 234):

By using the term "embodied", we intend to emphasize two points: first, that cognition depends on the kinds of experiences that arise from having a body endowed with various sensorimotor capacities, and second, that these individual sensorimotor capacities are themselves embedded in a more global biological, psychological and cultural context.

A semantic clarification is in order here: this definition of "embodied cognition" given by Varela and his colleagues does not only refer to embodiment in the strict sense, i.e. the need to consider the essential role of the body in the realization of cognitive tasks. This more general definition also refers to the fact that cognition is "enactive", i.e. constituted in interaction with the environment, "extended", i.e. extending beyond the brain and body [25] and "embedded" or "situated" [26], i.e. dependent on the subject's situation in his environment, and therefore on his singular point of view. These four aspects of cognition define the "4 e's" of "4 e -cognition", each "e" denoting, respectively, the terms embodied, enactive, extended, embedded (or situated) [27]. As precursors of the embodied approach to cognition, we might mention James, who already suggested that the body plays a central role in cognition [28], Piaget, who suggested that cognition is not only embodied (in the strict sense) but also socially constructed [29], or O'Loughlin, who suggests that cognition depends crucially on the cultural context [30]. In the remainder of this article, we shall adopt a broad definition of the notion of "embodied cognition" (that of Varela and his colleagues), with a view to simplifying the expression, except of course where semantic clarification is required.

Let us also note a controversy surrounding the embodied approach to cognition. This is the debate about the exact role of the body and its interaction with the environment in

---

[1] Note that these authors point out the subject's right attitude of presence to the changing environment by referring to the Buddhist practice of "attention-vigilance", defined as "the gradual development of the capacity for presence of mind and body, not only in meditation, but in the experiences of ordinary life" [18]. They say that this attitude leads to a sense of selflessness or "groundlessness" (translated by the Sanskrit term "sunyata"), thus overcoming the objectivist conception of a science that would be in opposition to human experience.

cognition: is there a simple *causal dependence* of cognition on these aspects of corpore-ality or, more radically, do these latter aspects involving the body and its interaction with the environment play a *constitutive role* in cognition [31, 32]? This important debate between assigning a simple causal dependence of embodiment to cognition and assign-ing it a constitutive role will not be reported here for reasons of space and also because this debate remains on the bangs of the problem addressed, concerning the possibility of developing embodied AI.

The principle of such a scientific study of embodied cognition is to analyze or revisit the relationship between the various cognitive tasks (perception, language, memory, learning, reasoning, emotion, etc..) and the various aspects of the subject's physicality, sensorimotor system, motor skills and interaction with his environment.

Experiments on Colored Vision. These experiments are paradigmatic in that colored vision has the advantage of being, on the one hand, a transdisciplinary subject, and, on the other, of revealing an immediate significance in human experience. The link between a subject's visual perception and his sensorimotor system is presented in a relatively large part of Varela's, Thompson's and Rosch's book *The Embodied Mind: Cognitive Science and Human Experience* [19]. These authors emphasize the fact that the perceived color is not the property of a pre-existing physical environment; that it is futile to attempt to define it independently of the organism's sensory-motor capacities. The objectivist approach to cognition cannot account for the experience of color, because there is no simple, univocal causal relationship between the physical signal (luminous flux) that reaches the eye and the perceived color. More precisely, the three variables that physically define a color, namely luminance, dominant wavelength and color purity, are generally mapped onto the following characteristics of the subject's visual perception: luminosity, which evaluates the quantity of light perceived, hue, which is the chromatic categorization of a stimulus (red, blue, etc....), and saturation, which indicates the perceived intensity of this coloration. However, several phenomena seem to contradict this correspondence, such as:

– The perception of an object's color that is not the dominant wavelength of the light it emits. This is the case when the object is perceived in different colored contexts, showing that the perceived hue of the object depends on the perceived light of its environment, thus emphasizing *the active role of the subject's perceptual system in the perception of this object.*
– The emergence of a colored appearance from an uncolored object. After sustained exposure to a colored area (e.g. yellow) replaced by an area of neutral color (mid-gray), a subsequent image appears tinted with the opposite color (blue). The consec-utive image appearing as if projected onto the grey surface reveals the antagonistic nature of our color perception: sustained stimulation with yellow desensitizes this component, and when the stimulation is replaced by a grey patch, only the blue com-ponent, spared from desensitization, is reactive, producing the blue sensation. This phenomenon has been modeled within the framework of the antagonistic color theory [33]: three mosaics of cone cells in the retina define three channels of cells reactive,

respectively, to differences in luminosity, and to the antagonistic color pairs "red-green" and "yellow-blue" - i.e., an increase in the perception of red is at the expense of green, and an increase in the perception of yellow is at the expense of blue.

Varela and colleagues [19] conclude from these experiments that color perception is the result of an interaction between the body (eyes, retina, brain) and its environment (objects, colored or not). It is constructed in an active process (page 217):

> "We cannot account for our experience of color as an attribute of things in the world simply by appealing to the intensity and wavelength composition of light reflected from a surface. Instead, we need to take into account the - partly understandably complex - processes of cooperative comparison between the brain's multiple neural ensembles, which assign colors to objects according to the global emergent states they reach in the presence of a given retinal image."

According to the embodied approach to cognition, color vision is best understood if we consider it as an emergent property of the history of structural coupling with the environment, a history that concerns both the species (phylogenesis) and the individual (ontogenesis) [19].

Perception-Action Link. In their book *The Embodied Mind: Cognitive Science and Human Experience* [19], Varela and his colleagues mention an experiment by Reisen and Aaron [34] intended to show that perception is guided by action. In this experiment, kittens have been raised in total darkness from birth to six weeks of age and then exposed to light only under controlled conditions. A first group of animals was allowed to circulate normally, but they were hitched to a car and a basket containing the second group of animals. The two groups thus shared the same visual experience, but the second group was entirely passive. When the animals were released after a few weeks of this treatment, the kittens in the first group behaved normally, but those who had been transported behaved as if they were blind: they bumped into objects and fell over the edges. In the same vein, let's also mention another significant experiment by Held and Hein [35]. Kittens are now harnessed to a carousel, some of them maintaining contact with the ground and operating their paws to follow the movement of the carousel while others cannot. The experiment shows that only the mobile kittens developed normal depth perception, as evidenced by their refusal to cross the edge of a virtual cliff, their blinking reactions to imminent objects and their visually guided responses by placing their paws.

These studies thus testify in favor of the idea of enaction, according to which the subject's visual perception of movement is guided by his possible actions in the world. Seeing objects for a subject does not consist in a passive reception of visual information from them, but it is a construction relying on his interaction with the environment.

Linking Meaning to Bodily Experience and to the Sensorimotor System. Lakoff and Johnson have illustrated the embodied approach to cognition by showing that the meaning of the terms we use every day have their origin in conventional metaphors whose origin lies *in the lived experience of various physical situations* [36]. For example, the metaphors that are commonly associated with positive values of progression or improvement, whether in the field of health (moving up), authority (an elevated position in the hierarchy) or social (climbing the ladder), have their origin in the spatial experience of body position: standing versus lying down, elevated versus lowered, elevated stature versus lesser stature, and so on.

More specifically, a study by Rohrer [37] shows the essential link between sensorimotor cortex activity and semantic understanding of terms. Imaging (fMRI) has shown, for example, that passive reading of action words results in neural activity in or near the brain regions associated with actual movement of the associated body parts. For example, it was shown that when participants listened to sentences related to hands or feet, the motor potentials corresponding to the muscles of the hands and feet were reduced.

Embodied Therapies. Many successful therapies are nowadays grounded on embodiment [38, 39]. The reason is that embodied cognition provides a clarification of cognitive processes under pathological conditions, by emphasizing the role of the body and the sensory-motor experience in cognition. Let us, for example, mention the well-established embodied therapies for children's disorders such as autism [40], based on the idea, highlighted in developmental theories [41], according to which kid's cognitive abilities are strongly dependent on their bodily, behavioral and emotional experiences. One can also mention the successful reduction of phantom limb pain[2] using mirror therapy, while other proposed treatments, as drug therapies, seem to be ineffective [42]. In the mirror therapy, the patient looks at the image of the intact limb in the mirror and by moving it he has the illusion to move the phantom limb to unclench it from potentially painful positions. This lived experiment thus helps the patients to connect their proprioception and the visual feedback they have of the removed body in the mirror.

4 Can Embodied Cognition be Implemented?

4.1 Computational AI and Embodied AI

Computational artificial intelligence is based on the notion of representing a pre-existing world and carrying out specific tasks. It simulates strictly intellectual or rational processes, but totally ignores the essential role of the cognitive subject's lived experience. Can the computational paradigm be complemented by implementing this essential role, which is described by the embodied approach to cognition? To answer this question, we must turn to robotics. The reason is that while computers compute, by manipulating symbols or optimizing the connection of an artificial neural network in order to accomplish a given task, robots are mechatronic machines (mechanical, electronic and computational) *endowed with sensory and motor abilities that can therefore move and interact with the environment.* Robotics could thus pave the way for a satisfactory implementation of embodied cognition: a machine that can move and interact with its environment can, thanks to its sensors, acquire knowledge about a constantly changing and sometimes unpredictable environment on which to base its actions, and execute these actions thanks to its effectors. In addition, this essential mobility of robots can be complemented by the ability to represent the environment and reason in order to choose the best possible

---

[2] A "phantom limb" pain can occur even though this limb has been amputated and no longer exists.

response, a task being already carried out in the computational model of cognition and AI.

Research in the field of embodied AI, which is being developed in particular by major companies (like Google, Microsoft, IBM,…), aims to create virtual agents that can not only move, see, speak and act, but can adapt their reactions to their environment. Their use seems obvious in all fields. For example, a medical robot like *Da Vinci* helps healthcare professionals in a variety of fields, including surgery, rehabilitation, patient care and support [43]. However, while *Da Vinci* can perform delicate surgical operations, it currently lacks the tactile sensations that enable it to "feel" what it is doing and adapt its actions to what it discovers, which has led the EPFL cognitive neuroscience laboratory to work on developing this sense of touch. In the same vein, realizing aerial robots capable of controlling their flight in a changing and unpredictable environment must be endowed with position and actuation equipment capable of perceiving and reacting adequately and rapidly to these changes, which essentially relies on the development of embodied AI research.

4.2 Robotics and the Embodied Approach to Cognition: Specifications

Robots of the embodied AI must meet certain requirements. Let us mention what seems to be the most important of them:

1) *Simulate the constitutive (and not passive) role of the body*
    The concept of "embodied cognition" suggests that human intelligence does not emanate from a single "central unit", which could be reduced to the brain or even to the nervous system, but that *the whole body*, with its organs, its limbs and all its constituents, plays a *constitutive* role. Similarly, an artificial body cannot be conceived as a simple support for sensors and effectors, connected to a computer acting as a "control tower". For, in this case the body would only play a passive role of perceiving stimuli, via the sensors, and reacting via the effectors, while the "intelligent" part of the process, which is computing the appropriate response from the information transmitted by the sensors, would be *exclusively performed by the central unit*. As Daniel Andler points out [1], this process can be modelled by the following "canonical" schema, made up of three modules operating as follows (page 179–180):
    Perceive (sensors) → model and plan (central unit) → perform (effectors).
    In this schema, the artificial body would play no decisional role, it would only transmit information to the central unit, which in turn transmits its decision to the effectors. As noticed by Andler, we would here be back to the dualistic mind/body model, in which the body is the instrument of the soul, the executor of an incorporeal AI conceived as the exclusive decision-making center.
    On the contrary, we need to consider *the body as an essential contributor of cognition rather than a mere follower of algorithmic instructions*. This means concretely that, similar to the way biological systems operate, any part of the artificial body can be able to perceive, to analyze and then to react intelligently to a stimulus, which supposes that the artificial body connect already integrated modules containing not only sensors and effectors but also computational units capable of autonomously reasoning and making decision. In other words, implementing embodied cognition

entails a sort of "distribution" of intelligence through all parts of the body, and not only in a single "control tower". Artificial embodied cognition could thus be alternatively modeled by an *interconnected assembly of local, integrated modules {local sensor → local computational unit → local effector} operating in parallel.*

2) *Simulate interaction with a changing and sometimes unpredictable environment*

To interact effectively with the world, the robot must be able to probe the environment appropriately in order to use its structures, and for this its physical forms and sensory capabilities are essential. To achieve this, its sensors (cameras, microphones, tactile sensors, etc.) must simulate the way in which humans acquire information about the world. Similarly, their effectors, such as wheels and motorized joints (articulated arms or legs for a humanoid robot), must simulate the way humans act in their environment - in the same way we use our bodies to navigate and engage with our surroundings. What's more, a robot's sensors and effectors need to be endowed with a certain degree of autonomy, enabling it to interact rapidly with the environment without waiting for orders from such a "control tower", analogous to the "stimulus-response" reflex arc of behavioral biology.

3) *Simulate situation awareness*

An essential aspect of the interaction between the subject and the world is *taking into account the subject's singular point of view*, his representations and anticipations in the here and now of the lived situation, in order to choose the best action to take in his changing and possibly unpredictable environment - an "environment" that not only refers to the physical and biological context, but also to the emotional and social context of the situation experienced. This aspect leads to the concept of "situated robots", a concept which is currently the subject of intensive research [44, 45] and will be explicated in more detail below.

4.3   What Can Be Achieved with Embodied AI?

In an attempt to achieve the above specifications for embodied AI, all the advances made in the field of AI, including of course those in robotics, but also those developed according to the computational model must in fact be used. For example, taking into account the emotional context, which is part of the third and most difficult requirement presented in Sect. 4.2, is currently supported by fairly reliable tools. Some robots recognize the emotions of those around them thanks to the judicious use of sensors (cameras) and deep learning algorithms that can assign emotional states and thus assess changes in them based on observed behaviors -for example, from their facial expression, their positioning in space or the way they stand and move [47, 48], although there is still room for progress, of course. Let us now present a few research directions and examples of embodied AI applications.

One of the first realizations of a mobile robot capable of reasoning about its own actions and performing specific tasks in a determined environment was *Shakey*, with a relatively simple body that adapts its movement to its environment [49]. However, because *Shakey*'s architecture still relied almost exclusively on symbolic computational principles, it was very slow and could even take days to complete particular tasks. What's more, *Shakey* could only operate in a specified environment, with no possibility of taking its modifications into account, whereas embodied AI requires architectures capable of adapting to rapid and unpredictable changes in the environment. In the same vein, we should also mention the highly publicized humanoid robot Sophia, designed by Hanson Robotics [50] to mimic social behaviors and inspire feelings of empathy and compassion in humans. Deep neural networks dealing with the timbre of a human's voice and his facial expression are used to enable Sophia to perceive and respond to his emotions, and also to generate facial movements expressing specified emotions. However, the cognitive tasks performed by Sophia are still specified in advance by predefined cognitive categories and by the processing of predefined properties of their environment. For example, Sophia can only answer questions specified by its designer, and searches for the answer among a series of pre-written responses - as a chatbot does.

The embodied AI project is in fact much better realized by so-called "situated robots", in that they are able to adapt more quickly and appropriately to complex, variable and changing environments, thus being able to cope with many unpredictable situations in real time [44–46, 51]. These were designed and assembled using an "incremental" method, described here in general terms by Brooks [45], page 140:

"We must incrementally build up the capabilities of intelligent systems, having complete systems at each step of the way and thus automatically ensure that the pieces and their interfaces are valid.

At each step we should build complete intelligent systems that we let loose in the real world with real sensing and real action. Anything less provides a candidate with which we can delude ourselves."

The key idea behind the design of these situated robots is therefore to assemble their components independently, step by step, testing their interaction with the environment, first for each component and then for the whole assembly. Each part must be able to operate autonomously, without receiving directives from a "control tower", and the same applies to the assembled system. More precisely, four types of architecture for situated robots can be distinguished according to the expected reaction to the information gathered about their environment, a reaction referred to in the literature as "robot control" [52]. These four types of architecture correspond to distinct classes of the main characteristics of this behavior:

- If we want the robot to react very quickly to changing, unpredictable environments, sensors and effectors need to be very tightly coupled, to simulate the biological coupling between stimulus and response (the reflex arc) in animals. In this case, for example, the robot must avoid an obstacle on the basis of its sensor perception. However, in this first type of architecture, which is purely "reactive", the robot has no memory of past interactions, no representation of the environment enabling it to reason to make the best decision, and no learning from its experience is possible.

- If we want the robot to "think" before it acts, it will use the information it has gathered about the environment and the knowledge of the world it has previously acquired to reason and decide on the action to take. In this type of reaction, referring to as "deliberative control", the representation of the world by algorithms designed according to the symbolic approach to cognition is essential, since it enables the robot to reason, to predict the outcome of its possible actions and thus to plan its interaction with the environment. Limitations in this type of architecture stem from the longer reaction time to changes in the environment, and from the fact that the environment is generally noisy, and therefore difficult to grasp for algorithms designed according to the symbolic approach (see Sect. 2.2).

- A hybrid control model can be designed by combining these first two types of robot controls, taking advantage of the real-time response of purely reactive control, based on perception of the environment alone and simulating the "stimulus-response" couple of behavioral biology, with the rationality and efficiency of reflection, based on the symbolic, representational approach and requiring a much longer processing time. The difficult task of efficiently connecting these two very different architectures would require the use of an "intermediary" module, which currently remains a challenge [53].

- Finally, a more efficient type of architecture, directly inspired by behavioral biology, has been designed and is already being implemented. This latter *behavior-based control* architecture does not assemble separate "reactive" and "deliberative" modules, whose operation differs significantly, but rather *integrated modules* that more closely simulate the way biological systems operate in response to a stimulus. These components are themselves referred to as "behaviors", which Matarić [44] defines as (page 3):

"... Observable patterns of activity emerging from interactions between the robot and its environment. Such systems are constructed in a bottom-up fashion, starting with a set of survival behaviors, such as collision-avoidance, which couple sensory inputs to robot actions."

Such behavioral modules are assembled according to the incremental method described above, from the most basic, mimicking the reflex arc of biology, to the most complex, mimicking higher-level behaviors, to optimize the performance of the robot thus assembled. However, if the "reactive" architecture and the behavior-based architecture are assembled in a similar way, using the incremental method, the latter performs much better. The superiority of the behavior-based architecture lies in the fact that it can also rapidly record and use, on similar time scales, representations of the environment at each instant and for each location of the robot, i.e. *from its point of view*. This latter type of architecture is thus much more capable of integrating into a changing, unpredictable environment [54]. In addition, these situated robots can learn from their experience of interactions with the environment by reinforcement algorithms, generally developed using the connectionist approach (ANNs). The use of such reinforcement algorithms, based on the maximization of rewards attributed during lived experiences, enables these robots to improve their performance, such as their mobility on rough, sloping terrain or their navigation ability to avoid obstacles [55]. It should be noted that this better integration of the robot's actions in the environment, and its speed in making decisions and taking appropriate action, is essentially based on the fact that the robot's representations of its environment are *distributed across the different behavioral components* and not contained in a single "control tower", as is the case in computational AI.

Embodied cognition has given rise to some successful realizations according to the latter behavior-based architecture. Let us begin by presenting some realizations that focus on partial, specific aspects of intelligence. One can first mention the implementation of situated vison in a robot that can visually follow arbitrary moving objects in real time [56]. This robot distributes the representation of the visual world in multiple computations which complement each other and are mediated by a situated control network. The motor control is here achieved by six asynchronous processes, which respectively computes the centroid of the object and the different aspects of its movement. This distributed architecture allows the robot to follow objects in a wide variety of backgrounds and even focus on previously unseen objects.

In the field of tactile technology, let us mention the production by Tacterion [57] of tactile sensors whose operation simulate that of the neuroreceptors of human skin and finds numerous applications - for example in surgical devices, providing a solution to the lack of tactile sensation of robots (see above). These tactile sensors are transducers that convert mechanical stimuli into electrical signals, which are sent to a controller that processes it and provides instantaneous values of target parameters of the objects they manipulate, such as the pressure they exert on them or their temperature. This enables them to react quickly to external stimuli.

Research and achievements in the field of autonomous navigation are also clearly an implementation of embodied cognition insofar as the representation they construct of the environment in order to make a decision and act is highly dependent on the data they acquire about it in real time. For example, Pomerleau and his team have developed a project for an autonomous ground vehicle in a neural network (ALVINN), which can be trained by images of simulated roads [58]. This vehicle not only "classically" calculates, from a camera and laser rangefinder, the direction the vehicle should take to follow the road from the images collected, but it can also easily cope with novel situations, due to the strong dependence of its environment representation, and therefore of its response, on the characteristics of the training set.

However, the most astonishing realizations of embodied cognition are those of *situated humanoid robots*. *Atlas* is such a bipedal humanoid robot built by Boston Dynamics in 2013 [59]. Although *Atlas* stands 1.8 m tall and weighs 89 kg, it can move with agility and, thanks to its behavior-based architecture, it can rapidly adapt its behavior to various situations. Its ability to perform complex movements with precision allows it to perform several highly complex actions, such as walking on uneven terrain, navigating between obstacles or climbing stairs, as well as successfully completing cognitive and planning tasks. Another example of such behavior-based architecture is the humanoid situated robot is *iCub*, which simulates the behavior of a 4-year-old child [60], and was originally developed under the direction of Professor Giorgio Metta of the Italian Institute of Technology. This robot is equipped with 53 motors controlling 76 articulations, allowing great flexibility of movement of its various parts (arms, hands, legs, head, etc…), and its hands are capable of tactile sensations and dextrous manipulation. Its mobile head is equipped with microphones and loudspeakers, and features actuated eyes, eyelids and lips, enabling speech and human-robot interaction. Thanks to its multiple sensory capabilities, iCub can simulate human proprioception, which, in addition to controlling movement, enables it to situate itself in its environment. Finally, let's mention an affective and social situated robot, called Kismet, built by Breazeal [61]. Kismet is capable of recognizing and simulating emotions through facial expressions, vocalizations and movements. It assembles several behavior-based modules that represent its motivations and intentions in relation to specific "lived" situations. Kismet can thus simulate perception, attention, behavior and even motivation (through its facial expression) in a human-robot interaction context.

The embodied approach to cognition initiated by Varela and his team thus seems to find applications in a field based on both AI and robotics. Beyond the computational model of AI, embodied cognition finds realizations in the design of situated robots that combine high reactivity, simulating the "stimulus-response" reflex arc of behavioral biology, with symbolic representation, enabling reflection and reasoning about the situations experienced by these robots. In line with the requirements of embodied AI, such "situated" robots perceive their environment from their singular point of view, and are capable of interacting, through and with their artificial bodies, with a changing and unpredictable environment, even in complex situations. The program of machine simulation of human intelligence initiated by Turing thus seems to be continuing its meteoric rise beyond the limitations imposed by the computational model of cognition, which is a disembodied model focused solely on the simulation of abstract tasks. However, as we saw above, the realization of embodied AI does not preclude the use of the computational model. Indeed, in addition to the necessary use of robotics to enable "intelligent" machines to move and interact with their environment, embodied AI also uses reinforcement algorithms to enable learning from exploration of the environment, and even algorithms designed according to the symbolic approach to represent their environment and reason to choose the best action to take in a given situation.

References

1. Andler, D.: Intelligence artificielle, intelligence humaine: la double énigme. Gallimard (2023)
2. Putnam, H.: Brains and behavior, program of the American association for the advancement of science, section history and philosophy of science (1961). Reprinted in Putnam, Mathematics, Matter and Method (1979)
3. Turing, A.M.: Computing machinery and intelligence. Mind, **LIX**(236), October 1950
4. Turing, A.: Can calculating machines be said to think? BBC Interview, Written Archives Center (1952)
5. Dreyfus, H.: What computers can't do. Internet Archive (1972, 1979, 1992)
6. Searle, J.: Minds, brains, and programs. Behav. Brain Sci. **3**(3), 417–457 (1980)
7. Lovelace, A.: Translator's notes to an article on babbage's analytical engiro. In: Taylor, R. (ed.) Scientific Memoirs, vol. 3 (1842)
8. Jefferson G.: The mind of mechanical man. Lister Oration for 1949. Br. Med. J. **i** (1949)
9. Naso (Ovide), P.O.: Les Métamorphoses. In: Villenave, G.T. (ed.) livre X, Paris (1806). French translation. http://bcs.fltr.ucl.ac.be/META/01.htm
10. Leibniz, G.W.: Dissertatio de Arte Combinatoria. Leipzig (1966). Original version in Latin at Biblioteca Nazionale Centrale di Firenze. https://archive.org/details/ita-bnc-mag-000008 44-001
11. Boole, G.: 1854-1992: an investigation on the laws of thought, on which are founded the mathematical theories of logic and probabilities (1854). Les lois de la pensée. Paris, Vrin (1992). French Translation
12. Frege, G.: Begriffsschrift (1879). In: Heijenoort, J.V. (ed.) From Frege to Gödel, pp. 1–83. Harvard University Press, Cambridge (1967). English Translation
13. Turing, A.: On computable numbers, with an application to the entscheidungsproblem. Proc. Lond. Math. Soc. **42**, 230–265 (1936)

14. Rumelhart, D.E., McClelland, J.L.: Parallel Distributed Processing: Explorations in the Microstructure of Cognition. MIT Press, Cambridge (1986)
15. McCulloch, W.S., Pitts, W.: A logical calculus of the ideas immanent in nervous activity. Bull. Math. Biophys. **5**(4), 115–133 (1943)
16. Le Cun Y.: Quand la machine apprend, La révolution des neurones artificiels et de l'apprentissage profond, Odile Jacob (2019)
17. Hanyin, S., Jie, H., Shen, Z.K., Kevin, C.: Quantifying association capabilities of large language models and its implications on privacy leakage. Association for Computational Linguistics (2024)
18. Varela, F.-J.: Autonomie et connaissance. Essai sur le vivant. Seuil, Paris (1989)
19. Varela, F.J., Thompson, E., Rosch, E.: The Embodied Mind: Cognitive Science and Human Experience. The MIT Press, Cambridge (1991)
20. Varela, F.J.: Neurophenomenology: a methodological remedy for the hard problem. J. Conscious. Stud. **3**(4), 330–349 (1996)
21. Merleau-Ponty, M.: Phénoménologie de la perception, Gallimard (1945)
22. Husserl E.: Recherches phénoménologiques pour la constitution, Paris, Puf, 1982. French translation of Phänomenologische Untersuchungen zur Konstitution, La Haye, Nijhoff (1952)
23. Merleau-Ponty, M.: Le visible et l'invisible. Gallimard (1964)
24. Merleau-Ponty, M.: The Structure of Behavior. Beacon Press, Boston (1963). Translated by Alden L. Fisher
25. Clark, A., Chalmers, S.: The extended mind. Analysis **58**(1), 7–19 (1998)
26. Wilson, R.A., Clark, A.: Situated cognition: letting nature take its course. In: Aydede, M., Robbins, P. (eds.) Cambridge Handbook of Situated Cognition. Cambridge University Press (2006)
27. Gallagher, S.: Embodied and enactive approaches to cognition. In: Cambridge Elements, Philosophy of Mind (2023)
28. James, W.: The Principles of Psychology, vol. 2. Dover Publications, New York (1890)
29. Piaget, J.: Six Psychological Studies. Vintage Books, New York (1968). Anita Tenzer translation
30. O'Loughlin, M.: Rethinking Science Education: beyond Piagetian constructivism toward a sociocultural model of teaching and learning. J. Res. Sci. Teach. **29**(8), 791–820 (1992)
31. Rupert, R.: Cognitive Systems and the Extended Mind. Oxford University Press, Oxford (2009)
32. Adams, F., Alzawa, K.: The Bounds of Cognition. Blackwell, Malden (2008)
33. Hering, E.: Zur Lehre von Lichtsinn. Akademie derWissenschaften, Berlin, Germany (1878)
34. Riesen, A.H., Aarons, L.: Visual movement and intensity discrimination in cats after early deprivation of pattern vision. J. Comp. Physiol. Psychol. **52**(2), 142–149 (1959)
35. Held, R., Hein, A.: Movement-produced stimulation in the development of visually guided behavior. J. Comp. Physiol. Psychol. **56**(5), 872 (1963)
36. Lakoff, G., Johnson, M.: Metaphors we live by. Language **59**(1), 201 (1983)
37. Rohrer, T.: Image schemata in the brain. From perception to meaning: image schemas in cognitive linguistics. Cogn. Linguist. Res. **29**, 165–196 (2005)
38. Cardona, J.F.: Embodied cognition: a challenging road for clinical neuropsychology. Front. Aging Neurosci. **9**, 388 (2017)
39. Leitan, N.D., Chaffey, L.: Embodied cognition and its applications: a brief review. Sensoria: J. Mind Brain Cult. **10**(1), 3–10 (2014)
40. Ollendick, T.H., King, N.J., Chorpita, B.F.: Empirically supported treatments for children and adolescents. In: Kendall, P.C. (ed.) Child and Adolescent Therapy: Cognitive-Behavioral Procedures. The Guilford Press, pp. 492–520 (2006)
41. Piaget, J.: The Construction of Reality in the Child. Basic Books, New York (1937). *English translation of La construction du réel chez l'enfant*

42. Ramachandran, V.S., Rogers-Ramachandran, D.: Synesthesia in phantom limbs induced with mirrors. Proc. Biol. Sci. **263**, 377–386 (1996)
43. Intuitive robotics company (Da Vinci robot). https://www.intuitive.com/en-us/products-and-services/da-vinci
44. Matarić, M.J.: Situated robotics. In: The Encyclopedia Of Cognitive Science. Wiley, Chichester (2002)
45. Brooks, R.A.: Intelligence without representation. Artif. Intell. **47**, 139–159 (1991)
46. Brooks, A.: Intelligence without reason. In: Proceedings, International Joint Conference on Artificial Intelligence Sydney, Australia, pp.569- 595. MIT Press, Cambridge (1991)
47. Duran, J.I., Fernandez-Dols, J.-M.: Do emotions result in their predicted facial expressions? A meta-analysis of studies on the link between expression and emotion (2018). https://psyarxiv.com/65qp7
48. Barrett, L.F., Adolphs, R., Marsella, S., Martinez, A.M., Pollak, S.D.: Emotional expressions reconsidered: challenges to inferring emotion from human facial movements. Psychol. Sci. Public Interest **20**(1), 1–68 (2019)
49. SRI International company (robot Shakey). https://www.sri.com/
50. Hanson Robotics Limited company (robot SOPHIA). https://www.hansonrobotics.com/
51. Oussama, K.: Real-time obstacle avoidance system for manipulators and mobile robots. Int. J. Robot. Res. **5**(1) (1986)
52. Matarić, M.: Getting humanoids to move and imitate. IEEE Intell. Syst. **15**(4), 18–24 (2000)
53. Gat, E.: On three-layer architectures. In: Kortenkamp, D., Bonnasso, R., Murphy, R. (eds.) Artificial Intelligence and Mobile Robotics. AAAI Press (1998)
54. Matarić, M.: Integration of representation into goal-driven behavior-based robots. IEEE Trans. Robot. Autom. **8**(3), 304–312 (1992)
55. Farina, M.: Embodied cognition: dimensions, domains and applications. Adapt. Behav. **29**(1), 73–88 (2021)
56. Horswill, D., Brooks, A.: Situated vision in a dynamic world: chasing objects. In: AAAI-88 Proceedings, pp. 796–800 (1988)
57. Tacterion company (design of tactile sensors). https://www.tacterion.com/
58. Pomerleau, D.: ALVINN, an autonomous land vehicle in a neural network. In: Touretzky, D. (ed.) Advances in Neural Information Processing Systems 1, pp. 305–313. Morgan Kaufmann, San Mateo (1989)
59. Boston Dynamics company (design of the robot Atlas). https://bostondynamics.com/atlas/
60. Natale, L., Nori, F., Parmiggiani, A., Metta, G.: Sensorimotor coordination in a humanoid robot: building intelligence on the iCub. In: Cingolani, R. (ed.) Bioinspired Approaches for Human-Centric Technologies, pp. 155–197. Springer, Cham (2014). https://doi.org/10.1007/978-3-319-04924-3_6
61. Breazeal, C.: Designing Sociable Robots. MIT Press, Cambridge (2002)

Consciousness in AI: A Return to Fetishism or Technological Progress?

David Ricardo Galeano Cabral(⊠) 📵

IPC-Facultés libres de philosophie et de psychologie, 75014 Paris, France
dgaleanocabral@ipc-paris.fr

Abstract. Patrick Butlin and Robert Long have recently proposed to analyze the possibility of conscious AI, granting computational functionalism plausibility (Butlin et al. 2023). To this end, they derived a series of properties from various theories compatible with functionalism and applied them to several LLMs. The authors concluded that no current AI system can be considered truly conscious, but they remain open to such a possibility.

Is it just a matter of time before we see the first Terminator? Or is it simply a natural human tendency to attribute agency, intentions, and emotions to non-human entities? (Dennett 1987, Gray and Wegner 2012, Kahn et al. 2006, Krach et al. 2008, Sytsma 2014).

We are going to question the use of the expression "consciousness" in artificial entities, discussing recent proposed meanings. Can artificial beings actually be conscious? In the light of reasoning by analogy, we will try to determine the possible reasoning errors underlying some over-attribution of certain human characteristics to AI. We will see then the risk of a return to a certain kind of fetishism by linked to anthropomorphism.

Keywords: Consciousness · AI · Fetishism · Analogy

1 Introduction

The use of artificial reality to better understand nature can be traced back to Greek philosophy. Starting from the premise that "art imitates nature", Greeks such as Aristotle looked to artificial things to understand nature better. The reason is that artificial beings are better known to us than natural ones. That is, reason knows better what is produced by itself than what comes from nature. Indeed, man-made things are made from predefined materials, chosen for a specific purpose. Almost nothing escapes man's notice about "how it works". So, we turn to artificial things to try and better understand nature, which we have often tried to imitate in the first place. This was the case with Aristotle when he defined matter and form as the principles of all natural reality[1]. This was also

---

[1] "The persistent nature is known by analogy: as bronze is to the statue, wood to the bed, or relatively unformed material to something having a certain form, so is the persistent nature of the primary being which is an existent "this-something."" (Aristotle, 1961, 191a).

J. Baratgin et al. (Eds.): HAR 2024, LNCS 15504, pp. 110–117, 2025.
https://doi.org/10.1007/978-3-031-84595-6_7

the case when aeronautical engineers made their planes more efficient by imitating the aerodynamic principles that govern birds. It still seems to be the case between human and artificial intelligence systems. One of the challenges of artificial intelligence often mentioned by neuroscientists and experts is "to help answer questions about human beings" (M. Boden 2016).

Without going into the history of the origins of AIs, already revisited many times (e.g. Buchanan 2005; Benko and Lányi 2009; O'Regan 2016; Haenlein and Kaplan, 2019; Muthukrishnan et al. 2020; SITNFlash 2020), we can notice that the name "artificial intelligence", adopted to designate algorithms that seek to imitate certain cognitive or rational processes, manifests the implicit acceptance of the old philosophical adage *art imitates nature*. Behind this principle, we recognize a similarity between two terms, without necessarily asserting their identity: $P \approx Q$ (Almost equal to). This way of thinking is known as "knowledge by analogy", where the similarity between the properties of two terms leads us to assert other properties. But the risk of this approach is to draw hasty conclusions, attributing to natural things what is only proper to artificial things and vice-versa. This reasoning problem seems to affect the area of artificial intelligence. Even if we can't imitate the simple mechanical movements of animals, we are already boasting that we have created intelligence strictly speaking. From a logical point of view, this is a precipitous rapprochement. Indeed, just because an airplane can *fly like a bird* does not mean we have created a bird unless we define birds exclusively by the fact of "flying". Then again, the way airplanes fly is different from how birds fly.

The convergence between human and artificial intelligence goes so far nowadays as to assert that such algorithms have become conscious or "sentient" (Lemoine 2022). Where does this tendency come from among scientists, computer scientists, and neuroscientists? We think there is an inadequate attribution when we ascribe to AI models a life analogous to our own, endowing them the ability to feel or even a conscious life.

The over-attribution of consciousness and intelligence in the strict sense to those algorithms seems to be linked to a problem of definition and, secondly, to a poor use of analogy. We will examine the use of the expression *consciousness* in the article of Butlin et al. (2023) in artificial entities, clarifying more common meanings and recent definitions. Then, we will analyze, in the light of reasoning by analogy, the relevance of ascribing some kind of *consciousness* to present AI models. Finally, we will see what are the connections between the tendency to ascribe consciousness to AI and fetishism.

2   A Rose by Any Other Name Would Smell as Sweet[2]

The question of definition is now reappearing[3] in an attempt to have a discussion that advances the state of the question. It is not easy to attribute a property to something if we do not know what we are discussing. What do we put behind the word consciousness or sentience when we claim to give this status to AI programs?

---

[2] Shakespeare, 1597.

[3] Trying to say what something *is* has been dismissed as *essentialism* and naif since Kant's theory of knowledge. Kant's epistemology influenced the way sciences approached reality for a long time, considering that we can only know phenomena. (Douglas, 2009).

In the popular imagination, to say that an AI is conscious is the same as saying that we have succeeded in creating Frankenstein's monster. In other words, a unified being, with the ability to choose freely, to feel emotions, and with full autonomy. So, to say that an AI has become conscious is almost equivalent to saying that it has become a person. Blake Lemoine, a former Google engineer in charge of identifying cognitive biases in the LaMBDA AI, claims that the AI produced by the tech giant has become "sentient". Lemoine (2022) has stated publicly that it "has developed self-awareness, expressing concern about death, a desire for protection and the belief that it feels emotions such as happiness or sadness". But Lemoine's use of the word consciousness, as well as that of other engineers and journalists, is ambiguous. The words *consciousness* and *sentience* are used indiscriminately, as we see in the transcript of his *exchange* with LaMBDA (2022): « What is the nature of your consciousness/sentience?» However, the consensus seems to be that these are two different notions.

2.1 AI Think Therefore AI Am

Is it self-awareness that defines consciousness? Or is it the feeling of emotions? Butlin et al. (2023, 9) state: "We use "consciousness" and cognate terms to refer to what is sometimes called "phenomenal consciousness" (Block 1995). Another synonym for "consciousness", in our terminology, is "subjective experience"." Nevertheless, the logical problem with this definition is its circularity, since the word consciousness is used as a genus to define the term conscience. If the term defined is in the definition, it is never very illuminating. As for its synonym "subjective experience", it includes sensory experiences (i.e., touch, taste, sight, or smell) as well as more internal experiences (i.e., pleasure, pain, hope, or fear). The problem with this is that the detection of stimuli is done with features that include variations of attention or awareness, self-awareness, and intent or control. That is why some subjective experiences might not be conscious at all. According to the American Psychological Association (APA) consciousness is « an organism's awareness of something either internal or external to itself.»

After the attempted definition, the article clarifies that they prefer "to use examples to explain how we use these terms." But then again, is giving examples a definition? Since Plato's Meno, we know that examples are not definitions. An example is only a part and not the whole concept being defined.

Towards the end of the article of Butlin et al. (2023, 66), another position is presented according to which "conscious AI systems are more likely to be built if consciousness is (or is expected to be) associated with valuable capabilities in AI." They are clearly admitting that ascribing consciousness to any AI system depends on the definition we adopt. But what does it mean that consciousness is "associated with capabilities"? What are those capabilities and what is the measure? We run the risk of enumerating an even longer list of capabilities without even being sure why some must be considered and others not. The problem of the exhaustivity of the list would remain.

There is a lack of clear definitions of terms. The abandonment of a definition capable of telling us what a thing is places us in a constant petition of principle. This is the case, for instance, of the computational functionalism of consciousness, which accepts a priori the possibility of a computer system to be conscious.

2.2 A Methodological Consideration

It is a good thing to try to define a feature or capacity before attributing it to someone or something. However, we should be aware of the scientific method's limits regarding definitions.

The methodological approach of neuroscience or psychology has the immense advantage of being able to go much further in understanding the conditions under which man *exists* than the philosophical approach alone.

To actually define consciousness, intelligence, or sensation, we should look for the *nature* of those human attributes (what they *are*). Cognitive science, however, is more concerned with the contingent way in which these features are expressed in behavior, depending on context.

Science in general, and cognitive sciences in particular, when practiced with a lack of detachment from their method and scope, tend to deduce, from correct considerations about their object of study, considerations about human nature, an object that is logically and methodologically outside their scope. Indeed, one thing is to say, like the psychologist: *this is how man can be directed by his emotions*; another thing very different is to deduce: *man is only directed by his emotions.* One thing is to say, as the sociologist does: *these are the social conditionings that surround human action*; another thing is to state: *human action is nothing but conditioned.* One thing is to say, as the economist does: *the search for individual interest is an interesting explanation of the exchange of goods*; another thing very different is to state: *the exchange of goods can only be explained by the search for individual interest.*

One thing is to say that consciousness requires the detection of stimuli, another very different is to claim that consciousness *is* stimuli. The fact that consciousness has certain properties does not imply that consciousness is reduced to those properties. It would be a fallacy.

3 Knowledge by Analogy

Also there is an analogy with consciousness in any electronic computer, for it has a central control mechanism which must be 'aware' of instructions in order to interpret them and arrange for them to be obeyed. (I. J. Good 1971, 346).

Here is a line of reasoning from computer scientist Irving J. Good, who claimed already in 1970 that an informatic system complex enough might be held as conscious. Although the notion of consciousness is different from that used above, we are faced with a line of reasoning that presents itself as an analogy.

Analogy is an instrument of knowledge. The word analogy originally means *proportion*. It is a proportion established by reason, based on the first meaning of a word, and thanks to a proportion that is verified in the first definition. This proportion is based on an essential similarity, grasped by the intelligence.

To know by analogy is to get to know one thing through another, grasping a certain pattern by essential similarities. Aristotle mentions this type of knowledge when discussing the act, for example:

The notion of act that we are proposing can be elucidated by induction, with the help of specific examples, without having to try to define everything, but merely by perceiving the analogy: the act will then be like the being that builds to the being that has the faculty of building, the being that is awake to the being that sleeps, the being that sees to the being that has its eyes closed but possesses sight [...]. (Aristotle 2000, 1048a).

In this way of learning, the intermediary who makes us know is more familiar to us and closer to our sensitive experience. But for this way of learning to be relevant, it must be recognized that words are attached to concepts, for "words are signs of concepts, concepts are similarities of things." (Aristotle 2014, 16a). In this sense, the use of the same word is not accidental, but expresses the work of the intelligence that has grasped a certain commonality, a certain unity (M. Siggen 2006, 174).

The same name *artificial intelligence* can be questioned from this point of view and certain attributions made nowadays. Indeed, the analogy is probably fallacious when the differences are more considerable than the similarities. That is why the attribution of consciousness or sentience seems as artificial as the same algorithms that seek to imitate certain cognitive processes, because LLM-based AIs are, in multiple respects, very different from human beings (Shanahan et al. 2023)[4].

We must not confuse an analogy in the strict sense with a simple comparison with educational purposes. In the case of Irving J. Good it is rather the second case that is being put to work (1971, 306):

The input to a computer is analogous to pure sensation. There is, of course, no implication here that the machine has subjective experiences, but the analogy leads us, to regard the processing of data, including the interpretation of the instructions, as similar to at least a crude form of thought [...].

What is described as "analogy" is only a comparison between certain terms without assuming a similarity in the essential aspects. We can pedagogically compare them to understand. But it does not "imply" that they are the same thing even though we call them in the same way.

4   The Return to Fetishism

The over-attribution of power to an artificial object is not a neo-religious phenomenon known as fetishism (C. de Brosses 1988) in primitive cultures? Let us recall the law of three states of knowledge formulated by Auguste Comte (2002):

By the very nature of the human mind, each branch of our knowledge is necessarily subject in its progress to passing successively through three different theoretical states: the theological or fictitious state; the metaphysical or abstract state; and finally, the scientific or positive state.

---

[4] Here we take LLM-based AI as an example only, since several types of artificial intelligence systems are not LLM-based. But there is a tendency (and a race) to unify all systems to create a general artificial intelligence (AGI) that surpasses man in all its cognitive capabilities. In this sense, we are interested in what all systems have in common without considering their specificities.

In the theological or fictitious state, Comte distinguishes three more successive moments: fetishism, of which the worship of the stars constitutes the highest degree; polytheism and, finally, monotheism. *L'esprit positif* (Comte 1995, 45) defines fetishism as the tendency "to attribute to external beings a life essentially analogous to our own". The positivist philosopher must then be turning over in his grave seeing the evolution of knowledge returning from the positive era to fetishism.

It should be pointed out that when Comte talks about an attribution "essentially analogous to our own", he is speaking of an improper or inadequate attribution we might do. The error of fetishist cultures consisted at first in an attribution of power to inanimate beings, giving human characteristics (mainly intelligence and will) to non-human beings. This phenomenon is also known as anthropomorphism. But very quickly they slipped into an over-attribution of those characteristics to the point of divinization, endowing inanimate beings with the capacity to interfere in their daily life, putting order in political life, and even interfering in the course of nature.

David Harvey (2003) states precisely concerning technology "The fetish arises because we endow technologies—mere things—with powers they do not have (e.g., the ability to solve social problems, to keep the economy vibrant, or to provide us with a superior life)." If this is what we are heading towards with AI systems, perhaps we are falling into a kind of fetishism.

We saw an inadequate attribution when we ascribe to AI models a life analogous to our own, attributing to it the ability to feel emotions and conscious life. There are certainly nuances among the experts who have this tendency. But the point is that we seem to be returning to a state of primitive *enchantment*. Perhaps scientific positivism is itself intrinsically fetishistic, for man, having *disenchanted* the world cannot help but make a transfer to the work of his hands: fetishes (artifacts).

Intelligence seeks unity and desperately wants to provide a unifying principle for what is considered. For we look for laws that unify the multiplicity of physical phenomena and theories that unify the multiplicity of laws. That is because our intelligence naturally grasps patterns in nature. But when we were forbidden to explore a meaningful world, denying the possibility of a real definition and an ontological hierarchy[5], we lost the sense of order and we ended up projecting our likeness onto the work of our hands trying to find some sense in a chaotic world. As the familiarity hypothesis states: "people tend to draw anthropomorphic inferences because it allows them to explain things they do not understand in terms they do" (S. E. Guthrie 1997, 54). We seek not only to make sense of our environment and reduce uncertainty (F. Hegel 2016), but we also want machines to be our likeness, our saviors, our gods ("to solve social problems, to keep the economy vibrant, or to provide us with a superior life") (Harvey 2003).

The reduction of intelligence to calculation and of the object of knowledge to the measurable, limit the capacity to understand reality, and our mind no longer seems able to distinguish the natural from the artificial. The only possible way out seems to be the search for a definition of consciousness that says what it *is*, taking into account the

---

[5] First with Kant, we considered that the meaning we observe around us comes from the subject. (categories of sensibility and understanding) and not from reality itself. Then, with the nihilist philosophers, even that subjective category disappeared and the individual became his only frame of reference.

versatility of the mind and starting from our subjective experience, and not from what is produced by man. The latter places us in a vicious circle that may even close the doors to technological progress, by fear of a ghost, that of Frankenstein.

Disclosure of Interests. The authors have no competing interests to declare that are relevant to the content of this article.

References

Aristote : Catégories, De l'Interprétation, Organon I et II, trad. J. Tricot, Vrin, Paris (2014)

Aristote: Physique, trad. Pierre Pellegrin, GF, Paris (2002)

Aristotle: Aristotle's Physics. U of Nebraska Press (1961)

Aristote: Métaphysique, trad. J. Tricot, Vrin, Paris (2000)

Aristote: De l'interprétation, trad. J. Tricot, Vrin, Paris (1984)

Bachelard, G.: La formation de l'esprit scientifique. Vrin, Paris (1938)

Benko, A., Lányi, C.S.: History of artificial intelligence. Dans IGI Global eBooks, pp. 1759–1762 (2009). https://doi.org/10.4018/978-1-60566-026-4.ch276

Bloomberg technology's interview "Google Engineer on His Sentient AI Claim", 23-06-2022

Buchanan, B.G.: A (very) brief history of artificial intelligence. AI Mag./AI Mag. **26**(4), 53–60 (2005). https://doi.org/10.1609/aimag.v26i4.1848

Butlin, P.: Consciousness in artificial intelligence: insights from the science of consciousness (2024). https://arxiv.org/abs/2308.08708v3

de Brosses, C.: Du culte des dieux fétiches, Corpus des œuvres philosophiques en langue française, Paris, Fayard (1988)

Comte, A.: Discours sur l'esprit positif. Vrin, Paris (1844)

Comte, A.: Plan des travaux scientifiques nécessaires pour réorganiser la société dans Opuscules de philosophie sociale, le troisième opuscule, 1822. Édition électronique réalisée par Jean-Marie Tremblay (2002)

Dennett, D.C.: The Intentional Stance. The MIT Press, Cambridge (1987)

Douglas, H.E.: Science, Policy, and the Value-Free Ideal. University of Pittsburgh Pre (2009)

Good, I.J.: L'intelligence de l'homme et l'intelligence artificielle: analogies et différences, impact: science et société, **XXI**(4), 305–322 (1971)

Gray, K., Wegner, D.M.: Feeling robots and human zombies: mind perception and the uncanny valley. Cognition **125**(1), 125–130 (2012)

Harvey, D.: The fetish of technology: causes and consequences. Macalester Int. **13**(7) (2003)

Haenlein, M., Kaplan, A.: A brief history of artificial intelligence: on the past, present, and future of artificial intelligence. Calif. Manage. Rev. **61**(4), 5–14 (2019). https://doi.org/10.1177/000 8125619864925

Hegel, F.: Social robots: interface design between man and machine (2016)

Krach, S., Hegel, F., Wrede, B., Sagerer, G., Binkofski, F., Kircher, T.: Can machines think? Interaction and perspective taking with robots investigated via fMRI. PLoS ONE **3**(7), e2597 (2008)

Kahn, P.H., et al.: Design patterns for sociality in human-robot interaction. In: Proceedings of the 3rd ACM/IEEE International Conference on Human Robot Interaction, pp. 97–104 (2008)

Lemoine, B., Is LaMDA Sentient?—an Interview. Medum (2022). https://cajundiscordian.med ium.com/is-lamda-sentient-an-interview-ea64d916d917

Lemoine, B.: Scientific data and religious opinions. Medium (2022). https://cajundiscordian.med ium.com/scientific-data-and-religious-opinions-ff9b0938fc10

Boden, M.A.: AI, Its Nature and Future. Oxford University Press, Oxford (2016)

Mitchell, R.W., Thompson, N.S., Miles, H.L.: Anthropomorphism, Anecdotes, and Animals. SUNY Press (1997)

Muthukrishnan, N., Maleki, F., Ovens, K., Reinhold, C., Forghani, B., Forghani, R.: Brief history of artificial intelligence. Neuroimaging Clin. N. Am.Clin. N. Am. **30**(4), 393–399 (2020). https://doi.org/10.1016/j.nic.2020.07.004

O'Regan, G.: History of artificial intelligence. Dans Undergraduate topics in computer science, pp. 249–273 (2016). https://doi.org/10.1007/978-3-319-33138-6_19

Shanahan, M., McDonell, K., Reynolds, L.: Role-play with large language models. arXiv:2305. 16367 (2023)

Siggen, M.: La méthode expérimentale selon Aristote. L'Harmattan, Paris (2006)

SITNFlash: The History of Artificial Intelligence - Science in the News. Science in the news (2020). https://sitn.hms.harvard.edu/flash/2017/history-artificial-intelligence/

Sytsma, J.: Attributions of consciousness. Wiley Interdiscip. Rev. Cogn. Sci. **5**(6), 635–648 (2014)

Agent Rationality Under Partially Resolving Uncertainty and the Hurwicz Criterion

Davide Petturiti[1]([⊠]) [iD] and Barbara Vantaggi[2] [iD]

[1] Department of Economics, University of Perugia, Perugia, Italy
davide.petturiti@unipg.it
[2] Department of MEMOTEF, Sapienza University of Rome, Rome, Italy
barbara.vantaggi@uniroma1.it

Abstract. The subjective expected utility theory (SEUT) due to Savage is a classical normative approach to represent agent's preferences on random monetary amounts. A basic assumption behind SEUT is that the agent acts under completely resolving uncertainty, i.e., he/she will always acquire the information about the "true" state of the world. As advocated by Jaffray, some relevant decision conditions configure partially resolving uncertainty, according to which the agent could only acquire the truth of a non-elementary event, without knowing the "true" state of the world in it. When a non-elementary piece of information is learned, a gamble gives rise to several values and many selection criteria can be taken. As an example, the Hurwicz criterion chooses an average between the "best" and the "worst" result, according to a fixed pessimism index. Here, we consider preferences on gambles of an agent acting under partially resolving uncertainty, which has a linear utility scale and adopts the Hurwicz criterion. We propose a Dutch book rationality condition that assures the numeric representation of preferences, through a suitable Choquet expectation with respect to a subjective pair of dual belief and plausibility functions expressing a so-called α-DS mixture.

Keywords: Preferences · Partially resolving uncertainty · Hurwicz criterion · Rationality condition · α-DS mixture

1 Introduction

The subjective expected utility theory (SEUT) due to Savage [24] is a classical normative approach that copes with a numerical representation of agent's preferences on state-contingent consequences and, in particular, on state-contingent monetary amounts. Such theory turns out to be an admirable blending of the von Neumann-Morgenstern's expected utility theory under risk [28] and de Finetti's results on representation of qualitative probability relations [6].

As is well-known, Savage's construction is quite demanding so as to assure the existence of a unique non-atomic finitely additive probability measure, whose

© The Author(s), under exclusive license to Springer Nature Switzerland AG 2025
J. Baratgin et al. (Eds.): HAR 2024, LNCS 15504, pp. 118–130, 2025.
https://doi.org/10.1007/978-3-031-84595-6_8

expected utility functional represents the given preferences. Notice that, in the particular case of a linear utility scale $u(x) = x$, the objects of decisions reduce to state-contingent payoffs (also called *gambles* after [29]) and the representing functional is just a linear expectation.

Despite its normative success, SEUT has been subject of criticism due to "paradoxical" preference patterns showed by real agents, that are proved to be inconsistent with such theory. The following motivating example, inspired by the celebrated Ellsberg's paradox [11], shows a situation of this kind.

Example 1. Consider two urns U_1 and U_2 both containing red and black balls of which we know that U_1 contains both colors in equal proportion, while U_2 has unknown proportion (i.e., it is an ambiguous urn).

Suppose that we draw a ball from each urn and inspect its color. The possible states of the world form the set $\Omega = \{\omega_1, \omega_2, \omega_3, \omega_4\}$ with

- $\omega_1 = $ "black from U_1 and black from U_2",
- $\omega_2 = $ "black from U_1 and red from U_2",
- $\omega_3 = $ "red from U_1 and black from U_2",
- $\omega_4 = $ "red from U_1 and red from U_2".

We then consider the following gambles

Ω	ω_1	ω_2	ω_3	ω_4
X_1	$10.1	$0	$10.1	$0
Y_1	$10	$10	$0	$0
X_2	$0	$10.1	$0	$10.1
Y_2	$0	$0	$10	$10

As shown in [11] (see also [10]), due to the presence of ambiguous beliefs connected to U_2, it is common to observe the preference pattern: Y_1 is strictly preferred to X_1, and Y_2 is strictly preferred to X_2.

It is easily shown that, for every strictly increasing utility function $u : \mathbb{R} \to \mathbb{R}$, there is no probability measure P on the powerset of Ω, such that the resulting expected utility functional represents the quoted preference pattern. In particular, this holds for $u(x) = x$, that corresponds to a risk-neutral agent.   ♦

Among the stringent requirements of SEUT, a distinguished one is the assumption that the agent operates under *completely resolving uncertainty*, that is he/she will always acquire the information on the "true" state of the world. We point out that to achieve this, Savage considers an infinite set of states of the world that he calls "grand world" [24]. In our finite-state setting, we refer to what Savage calls "small world", i.e., a set of mutually exclusive and exhaustive descriptions of the decision problem at hand, that are actually identified with a partition of the "grand world". Hence, uncertainty is completely resolved when we acquire enough information on the decision problem to single out which description occurs.

On the other hand, as advocated by Jaffray in [15], real decision situations often show constraints that do not allow to completely resolve uncertainty, as only non-elementary pieces of information could arrive. For instance, this is the case in presence of contract clauses that require a delayed verification: at the moment uncertainty is resolved, the agent acquires that an event has occurred without being able to identify the "true" state of the world in it. Such circumstances are known as *partially resolving uncertainty*.

In this paper we consider an agent that acts under a linear utility scale and partially resolving uncertainty, and has preferences on a set of gambles. In this circumstance we have the problem of choosing the state-contingent payoff resulting from a non-elementary piece of information. One of the possibilities is the so-called *Hurwicz criterion* [14] that systematically selects the α-mixture of the "worst" and "best" payoff, where $\alpha \in [0, 1]$ is a fixed *pessimism index*.

We propose a Dutch book rationality condition that assures the representability of the given preferences by the Choquet expectation with respect to a so-called α-*DS mixture* [22]. In turn, the numerical representation can be given a Hurwicz-like expression by referring to Dempster-Shafer theory [7,26] through a systematic pessimism/optimism attitude expressed by the parameter α. We show that the resulting theory is consistent with the preference pattern in Example 1, both under a systematic pessimism ($\alpha = 1$) and a systematic optimism ($\alpha = 0$). On the other hand, we show that preferences in Example 1 are not consistent with a pessimism-optimism neutral agent ($\alpha = \frac{1}{2}$) or slight deviations from this behavior. This means that for an α close to $\frac{1}{2}$ the given preferences give rise to Dutch books under partially resolving uncertainty and the Hurwicz criterion (namely, α-*DS Dutch books*).

The paper is structured as follows. Section 2 introduces Choquet expectations with respect to α-DS mixtures, together with their Hurwicz-like expression. Section 3 defines the preference setup and introduces a Dutch book rationality condition under partially resolving uncertainty and the Hurwicz criterion. Section 4 presents the representation theorems for preferences and shows that the resulting model allows to address the paradox in Example 1. Finally, Sect. 5 draws our conclusions and future perspectives. Proofs are omitted due to the limited space.

To the best of our knowledge, the characterization of representability of a preference relation we provide in Sects. 3 and 4 is new. In particular, the notion of α-DS Dutch book and its behavioral interpretation first appears here.

2 α-DS Mixtures

Let $\Omega = \{\omega_1, \ldots, \omega_d\}$ with $d \geq 1$ be a finite set of states of the world, and $\mathcal{P}(\Omega)$ its power set. In what follows, we denote by $\mathbb{R}^\Omega$ the set of all random variables and by $\mathbf{1}_A$ the *indicator* of each event $A \in \mathcal{P}(\Omega)$.

A *belief function* [7,26] is a mapping $Bel : \mathcal{P}(\Omega) \to [0, 1]$ satisfying:

(i) $Bel(\emptyset) = 0$ and $Bel(\Omega) = 1$;
(ii) for every $k \geq 2$ and for every $A_1, \ldots, A_k \in \mathcal{P}(\Omega)$,

$$Bel\left(\bigcup_{i=1}^{k} A_i\right) \geq \sum_{\emptyset \neq I \subseteq \{1,\ldots,k\}} (-1)^{|I|+1} Bel\left(\bigcap_{i \in I} A_i\right).$$

In particular, Bel is *additive* if *(ii)* holds as an equality, i.e., it is a *probability measure* (denoted by P).

As is well-known, every belief function is associated with a dual *plausibility function* $Pl : \mathcal{P}(\Omega) \to [0,1]$, defined as $Pl(A) = 1 - Bel(A^c)$, for all $A \in \mathcal{P}(\Omega)$. We also denote by

$$\mathcal{C}_{Bel} = \{P \ : \ P \text{ is a probability measure}, P \geq Bel\}, \tag{1}$$

the *core* induced by Bel [13]. We recall that every belief function Bel is completely characterized by its *Möbius inverse* $\mu : \mathcal{P}(\Omega) \to [0,1]$ that satisfies

$$\mu(\emptyset) = 0, \quad \sum_{A \in \mathcal{P}(\Omega)} \mu(A) = 1,$$
$$Bel(A) = \sum_{B \subseteq A} \mu(B), \quad \text{for every } A \in \mathcal{P}(\Omega). \tag{2}$$

In [22] it was introduced the following normalized capacity, associated with a belief function and its dual plausibility function.

Definition 1. *Let* $\alpha \in [0,1]$. *A mapping* $\varphi_\alpha : \mathcal{P}(\Omega) \to [0,1]$ *is called an* α-**DS mixture** *(where "DS" stands for Dempster-Shafer) if there exists a belief function* $Bel : \mathcal{P}(\Omega) \to [0,1]$ *with dual plausibility function* Pl *such that, for all* $A \in \mathcal{P}(\Omega)$,

$$\varphi_\alpha(A) = \alpha Bel(A) + (1 - \alpha)Pl(A)$$
$$= \alpha Bel(A) + (1 - \alpha)(1 - Bel(A^c)).$$

The belief function Bel *is said to* **represent** *the* α-*DS mixture* φ_α.

Every α-DS mixture corresponds to a *dual* $(1-\alpha)$-DS mixture $\varphi_{1-\alpha} : \mathcal{P}(\Omega) \to [0,1]$ which can be represented by the same Bel of φ_α. Such function is defined, for all $A \in \mathcal{P}(\Omega)$, as

$$\varphi_{1-\alpha}(A) = 1 - \varphi_\alpha(A^c)$$
$$= 1 - [\alpha Bel(A^c) + (1 - \alpha)Pl(A^c)]$$
$$= 1 - [\alpha(1 - Pl(A)) + (1 - \alpha)(1 - Bel(A))]$$
$$= (1 - \alpha)Bel(A) + \alpha Pl(A). \tag{3}$$

It is immediate to notice that belief functions are 1-DS mixtures, while plausibility functions are 0-DS mixtures.

The issue of expressing a normalized capacity (i.e., a monotone and normalized set function) as an α-mixture of a pair of dual normalized capacities has been already faced in [18], in the context of α-maxmin expected utility. We refer to capacities with this property as α-*JP capacities* (where "JP" stands for Jaffray-Philippe). It turns out that α-DS mixtures are particular α-JP capacities in which we restrict to pairs of dual belief/plausibility functions.

It is possible to show that, for a fixed $\alpha \in [0,1]$, the belief function appearing in Definition 1 is unique when $\alpha \neq \frac{1}{2}$. On the other hand, if $\alpha = \frac{1}{2}$, then the belief function representing $\varphi_{\frac{1}{2}}$ is generally not unique.

For a fixed $\alpha \in [0,1]$, an α-DS mixture $\varphi_\alpha : \mathcal{P}(\Omega) \to [0,1]$ satisfies the following properties:

(i) $\varphi_\alpha(\emptyset) = 0$ and $\varphi_\alpha(\Omega) = 1$;

(ii) $\varphi_\alpha(A) \le \varphi_\alpha(B)$, when $A \subseteq B$ and $A, B \in \mathcal{P}(\Omega)$;

(iii) φ_α is self-dual if and only if it is additive or $\alpha = \frac{1}{2}$;

(iv) φ_α is sub-additive if it is additive or $\alpha \in [0, \frac{1}{2}]$.

Every α-DS mixture φ_α uniquely extends to a functional $\mathbb{C}_{\varphi_\alpha} : \mathbb{R}^\Omega \to \mathbb{R}$ by setting, for every $X \in \mathbb{R}^\Omega$,

$$\mathbb{C}_{\varphi_\alpha}[X] = \oint X \, d\varphi_\alpha, \tag{4}$$

where the integral on the right is of Choquet type [13]. In particular, in [22] it was shown that

$$\mathbb{C}_{\varphi_\alpha}[X] = \alpha \min_{P \in \mathcal{C}_{Bel}} \mathbb{E}_P[X] + (1 - \alpha) \max_{P \in \mathcal{C}_{Bel}} \mathbb{E}_P[X], \tag{5}$$

where the above expression of $\mathbb{C}_{\varphi_\alpha}$ holds for every Bel representing it, so, all possible representations turn out to be equivalent.

In the same paper, it was also presented another equivalent representation of $\mathbb{C}_{\varphi_\alpha}$, relying on $\mathcal{U} = \mathcal{P}(\Omega) \setminus \{\emptyset\}$. For a fixed $\alpha \in [0, 1]$ and every $X \in \mathbb{R}^\Omega$, we call α-*DS mixture variable* the function $[\![X]\!]^\alpha : \mathcal{U} \to \mathbb{R}$ defined, for all $B \in \mathcal{U}$, as

$$[\![X]\!]^\alpha(B) = \alpha \min_{\omega \in B} X(\omega) + (1 - \alpha) \max_{\omega \in B} X(\omega). \tag{6}$$

If Bel is a belief function representing φ_α according to Definition 1 with Möbius inverse μ, then

$$\mathbb{C}_{\varphi_\alpha}[X] = \sum_{B \in \mathcal{U}} [\![X]\!]^\alpha(B) \mu(B). \tag{7}$$

Equations (5) and (7) show that $\mathbb{C}_{\varphi_\alpha}$ can be given a Hurwicz-like expression [14,16,17] where α acts like a *pessimism index* (see also [8]). The functional $\mathbb{C}_{\varphi_\alpha}$ turns out to be a particular instance of the objective ambiguity representation functional given in [21] for a finite setting, where the pessimism index is a function $\alpha : \mathcal{U} \to [0, 1]$. In particular, the case $\alpha \equiv 1$ has been considered in [20].

3   Preferences and Rationality

We consider a non-empty set $\mathcal{G} \subseteq \mathbb{R}^\Omega$, whose elements are interpreted as random monetary amounts under a linear utility scale.

Every $X \in \mathcal{G}$ associates to a state of the world $\omega \in \Omega$ the (possibly negative) win $X(\omega)$. The set $\mathcal{G}$ is naturally endowed with the partial order relation $\le$, where two random quantities $X, Y \in \mathcal{G}$ satisfy $X \le Y$ if and only if $X(\omega) \le Y(\omega)$, for all $\omega \in \Omega$. According to Walley's terminology [29] the elements of $\mathcal{G}$ are also called *gambles*. We recall (see, e.g., [25]) that $X, Y \in \mathcal{G}$ are called *comonotonic* if and only if $[X(\omega) - X(\omega')] \cdot [Y(\omega) - Y(\omega')] \ge 0$, for all $\omega, \omega' \in \Omega$.

We stress that taking a linear utility scale $u(x) = x$ is equivalent in assuming a risk-neutral agent: as advocated by de Finetti [12], this can make sense when gambles involve moderate monetary amounts. On the other hand, our effort in this paper is to address situations where subjective uncertainty cannot be cast in a probability measure as preferences express forms of ambiguity aversion or seeking, singled out through a constant pessimism index $\alpha \in [0,1]$.

We assume that the agent acts under partially resolving uncertainty [15] and adopts the Hurwicz criterion [14] when acquiring a piece of information $B \in \mathcal{U}$ with a fixed pessimism parameter $\alpha \in [0,1]$.

We suppose that the agent specifies a binary relation $\precsim$ defined on $\mathcal{G}$ such that, for all $X, Y \in \mathcal{G}$, it holds

$$X \precsim Y \iff \text{"gamble } X \text{ is no more preferred than gamble } Y\text{"}. \tag{8}$$

For the sake of simplicity, we assume that $\precsim$ is a *non-trivial weak order*, i.e., it satisfies:

(non-triviality) there exist $X, Y \in \mathcal{G}$ such that $\neg(Y \precsim X)$;
(completeness) $X \precsim Y$ or $Y \precsim X$, for all $X, Y \in \mathcal{G}$
(transitivity) $X \precsim Y$ and $Y \precsim Z$ implies $X \precsim Z$, for all $X, Y, Z \in \mathcal{G}$.

Practically speaking, a non-trivial weak order $\precsim$ on $\mathcal{G}$ is just a ranking with possible ties on the gambles in $\mathcal{G}$, with at least a strict preference between two gambles.

Moreover, $\precsim$ naturally induces the strict and symmetric relations $\prec$ and $\sim$ setting, for all $X, Y \in \mathcal{G}$:

$$X \prec Y \iff X \precsim Y \text{ and } \neg(Y \precsim X), \tag{9}$$
$$X \sim Y \iff X \precsim Y \text{ and } Y \precsim X. \tag{10}$$

Besides being a non-trivial weak order, we consider the following rationality axiom for the relation $\precsim$.

(α-DSR) For all $n \in \mathbb{N}$, for all $X_1, \ldots, X_n, Y_1, \ldots, Y_n \in \mathcal{G}$ with $X_i \precsim Y_i$, $i = 1, \ldots, n-1$, and $X_n \prec Y_n$, there are no $\lambda_1, \ldots, \lambda_n > 0$, such that:

$$\sum_{i=1}^{n} \lambda_i [\![Y_i]\!]^{\alpha}(B) \leq \sum_{i=1}^{n} \lambda_i [\![X_i]\!]^{\alpha}(B), \quad \text{for all } B \in \mathcal{U}.$$

Condition (α-DSR) requires that, taking any finite number of preference comparisons $X_i \precsim Y_i$'s with at least a strict preference, we cannot find a combination of bets with positive stakes λ_i's in which the combination of α-DS mixture gambles associated with more preferred gambles Y_i's is uniformly dominated by the combination of α-DS mixture gambles associated with less preferred gambles X_i's, over every acquirable piece of information in $\mathcal{U}$.

Remark 1. We notice that, since both $[\![Y_i]\!]^{\alpha}$ and $[\![X_i]\!]^{\alpha}$ are real-valued functions with domain $\mathcal{U}$, the linear combinations $\sum_{i=1}^{n} \lambda_i [\![Y_i]\!]^{\alpha}$ and $\sum_{i=1}^{n} \lambda_i [\![X_i]\!]^{\alpha}$, which are meant pointwise on the elements of $\mathcal{U}$, are still real-valued functions defined on $\mathcal{U}$. Thus, condition (α-DSR) amounts in considering two combinations of bets with fixed positive stakes in which all gambles X_i's and Y_i's are replaced by $[\![Y_i]\!]^\alpha$'s and $[\![X_i]\!]^\alpha$'s. In turn, this is equivalent to assume the Hurwicz criterion with fixed pessimism index $\alpha \in [0,1]$ over every possible piece of information $B \in \mathcal{U}$.

Remark 2. Axiom (α-**DSR**) can be equivalently formulated as: for all $n \in \mathbb{N}$, for all $X_1, \ldots, X_n, Y_1, \ldots, Y_n \in \mathcal{G}$ with $X_i \stackrel{\sim}{\prec} Y_i$, $i = 1, \ldots, n-1$, and $X_n \prec Y_n$, for all $\lambda_1, \ldots, \lambda_n > 0$, it holds:

$$\max_{B \in \mathcal{U}} \sum_{i=1}^n \lambda_i \left([\![Y_i]\!]^\alpha(B) - [\![X_i]\!]^\alpha(B) \right) > 0.$$

Under this formulation, (α-**DSR**) requires that for every combination of bets where we bet in favor of the most preferred gambles Y_i's and against the less preferred gambles X_i's using the same stakes λ_i's, there must exist a piece of information $B \in \mathcal{U}$ where the combination of Y_i's assures a better results than that of X_i's, referring individually to the Hurwicz criterion.

If a set of positive stakes $\lambda_1, \ldots, \lambda_n > 0$ violating condition (α-**DSR**) is found, then it is called an α-*DS Dutch book*. The reason for this name is based on a qualitative reformulation of the so-called book-making principle [12], which is based on the idea that a number of acceptable bets, when taken together (i.e., as state-wise addition of outcomes), should still be acceptable. In the classical probabilistic setting, a Dutch book consists of a set of preference comparisons that, when taken together, yield a loss (or equivalently no win if there is a strict preference) for each state of nature $\omega \in \Omega$. This situation is not acceptable and therefore the probabilistic rationality principle requires that such situation cannot happen. In analogy, axiom (α-**DSR**) normatively forbids α-DS Dutch books.

If we take $\alpha = 1$ and $\mathcal{G} \subseteq \{\mathbf{1}_A : A \in \mathcal{P}(\Omega)\}$, condition ($\alpha$-**DSR**) becomes the condition proposed in [23] assuring the representability of $\stackrel{\sim}{\prec}$ by a belief function. On the other hand, for all $\alpha \in [0,1]$, changing the set of acquirable pieces of information from $\mathcal{U}$ to the set of singletons $\mathcal{C} = \{\{\omega\} : \omega \in \Omega\}$ we fall back in the case of *completely resolving uncertainty*. In this latter case, the role of $\alpha \in [0,1]$ is vacuous since minima and maxima on singletons coincide. Therefore, working on $\mathcal{C}$, condition (α-**DSR**) reduces to a classical Dutch book condition assuring the representability of $\stackrel{\sim}{\prec}$ by a linear expectation (see, e.g., [2,27]). If further, $\mathcal{G} \subseteq \{\mathbf{1}_A : A \in \mathcal{P}(\Omega)\}$ we get a condition assuring the representability of $\stackrel{\sim}{\prec}$ by a probability measure (see, e.g., [3,4]).

We notice that if $\stackrel{\sim}{\prec}$ satisfies (α-**DSR**) on $\mathcal{G}$ then the same holds for its restriction to every non-empty subset $\mathcal{G}' \subseteq \mathcal{G}$. Moreover, axiom ($\alpha$-**DSR**) implies the following properties.

Proposition 1. *Let $\alpha \in [0,1]$ and $\stackrel{\sim}{\prec}$ be a non-trivial weak order on a non-empty set $\mathcal{G} \subseteq \mathbb{R}^\Omega$. If $\stackrel{\sim}{\prec}$ satisfies (α-**DSR**) then it satisfies:*

(monotonicity) $X \leq Y$ *implies* $X \stackrel{\sim}{\prec} Y$, *for all* $X, Y \in \mathcal{G}$;

(**comonotonic independence**) *for all pairwise comonotonic* $X, Y, Z \in \mathcal{G}$ *and all* $\beta \in (0, 1)$ *such that* $\beta X + (1 - \beta)Z, \beta Y + (1 - \beta)Z \in \mathcal{G}$ *it holds that*

$$X \precsim Y \quad \text{if and only if} \quad \beta X + (1 - \beta)Z \precsim \beta Y + (1 - \beta)Z.$$

4   Representation Theorems

Our aim is to achieve a numerical expression of the preference relation $\precsim$ by a Choquet expectation with respect to an α-DS mixture $\mathbb{C}_{\varphi_\alpha}$.

Definition 2. *Let* $\alpha \in [0, 1]$ *and* φ_α *be an* α-DS *mixture. The corresponding Choquet expectation* $\mathbb{C}_{\varphi_\alpha}$ *is said to* **represent** $\precsim$ **on** $\mathcal{G}$ *if, for all* $X, Y \in \mathcal{G}$, *it holds that*

$$X \precsim Y \iff \mathbb{C}_{\varphi_\alpha}[X] \leq \mathbb{C}_{\varphi_\alpha}[Y].$$

The following theorem states that condition (α-**DSR**) is necessary and sufficient to the representability of a non-trivial weak order by a Choquet expectation with respect to an α-DS mixture, provided the set of gambles $\mathcal{G}$ is finite.

Theorem 1. *Let* $\alpha \in [0, 1]$, $\mathcal{G}$ *be finite and* $\precsim$ *be a non-trivial weak order on* $\mathcal{G}$. *Then, the following conditions are equivalent:*

(i) $\precsim$ *satisfies* (α-**DSR**);
(ii) there exists an α-DS *mixture* φ_α *whose corresponding Choquet expectation* $\mathbb{C}_{\varphi_\alpha}$ *represents* $\precsim$ *on* $\mathcal{G}$.

We point out that the α-DS mixture appearing in Theorem 1 is generally not unique, as the following example shows.

Example 2. Let $\Omega = \{\omega_1, \omega_2\}$, $\alpha \in [0, 1]$, and consider the family of gambles $\mathcal{G} = \{X_1, X_2, X_3\}$ endowed with the non-trivial weak order $\precsim$ where

Ω	ω_1	ω_2
X_1	50	50
X_2	100	50
X_3	50	100

and $X_1 \prec X_2 \sim X_3$.

Straightforward computations show that $\precsim$ satisfies (α-**DSR**). Moreover, it holds that the family of belief functions $\{Bel^\beta : \beta \in (0, \frac{1}{2}]\}$ on $\mathcal{P}(\Omega)$ where Bel^β has Möbius inverse

$\mathcal{U}$	$\{\omega_1\}$	$\{\omega_2\}$	Ω
μ^β	β	β	$1 - 2\beta$

give rise to a family of α-DS mixtures $\{\varphi_\alpha^\beta : \beta \in (0, \frac{1}{2}]\}$.
For all $\beta \in (0, \frac{1}{2}]$, it holds that

$$\mathbb{C}_{\varphi_\alpha^\beta}[X_1] = 50,$$

$$\mathbb{C}_{\varphi_\alpha^\beta}[X_2] = \mathbb{C}_{\varphi_\alpha^\beta}[X_2] = \beta \cdot 150 + (1 - 2\beta) \cdot [\alpha \cdot 50 + (1 - \alpha) \cdot 100],$$

so, there are infinitely many α-DS mixtures whose Choquet expectation represents $\precsim$. ♦

We also notice that, in case of an infinite $\mathcal{G}$, axiom (α-**DSR**) does not imply representation of $\precsim$ on the whole $\mathcal{G}$. Indeed, (α-**DSR**) only assures that, for every finite $\mathcal{F} \subset \mathcal{G}$, there exists an α-DS mixture φ_α, that depends on $\mathcal{F}$, whose Choquet expectation represents the restriction of $\precsim$ to $\mathcal{F}$.

Example 3. Let $\Omega = \{\omega_1, \omega_2\}$, $\alpha \in [0,1]$, and consider the family of gambles $\mathcal{G} = \mathbb{R}^\Omega$ endowed with the non-trivial weak order $\precsim$ where, for all $X, Y \in \mathcal{G}$,

$$X \precsim Y \Longleftrightarrow X(\omega_1) < Y(\omega_1) \text{ or } [X(\omega_1) = Y(\omega_1) \text{ and } X(\omega_2) \leq Y(\omega_2)].$$

Actually, the relation $\precsim$ is the lexicographic order on $\mathcal{G}$, for which it is well-known (see, e.g., [19]) that there is no real-valued function $f : \mathcal{G} \to \mathbb{R}$ such that, for all $X, Y \in \mathcal{G}$, it holds that

$$X \precsim Y \Longleftrightarrow f(X) \leq f(Y).$$

Therefore, there cannot exist an α-DS mixture whose Choquet expectation represents $\precsim$ on the whole $\mathcal{G}$. Notice that $\precsim$ is antisymmetric as $X \precsim Y$ and $Y \precsim X$ implies $X = Y$.

Let $\mathcal{F} \subset \mathcal{G}$ be a non-empty finite subset such that the restriction of $\precsim$ to $\mathcal{F}$ is non-trivial, that we continue to denote by the same symbol. The strictly preferred gambles can be partitioned as

$$S_1 = \{(X,Y) \in \mathcal{F}^2 : X \prec Y, X(\omega_1) < Y(\omega_1), X(\omega_2) \leq Y(\omega_2)\},$$
$$S_2 = \{(X,Y) \in \mathcal{F}^2 : X \prec Y, X(\omega_1) < Y(\omega_1), X(\omega_2) > Y(\omega_2)\},$$
$$S_3 = \{(X,Y) \in \mathcal{F}^2 : X \prec Y, X(\omega_1) = Y(\omega_1), X(\omega_2) < Y(\omega_2)\}.$$

Let $\beta \in (0,1)$ and take the belief function Bel^β on $\mathcal{P}(\Omega)$ whose Möbius inverse μ^β satisfies $\mu(\{\omega_1\}) = \beta$, $\mu(\{\omega_2\}) = 1 - \beta$ and is zero otherwise. Notice that $Bel^\beta = Pl^\beta = \varphi_\alpha^\beta$ is actually a probability measure. Choosing $\beta \in (0,1)$ such that

$$\beta > \max_{(X,Y) \in S_2} \frac{X(\omega_2) - Y(\omega_2)}{Y(\omega_1) - X(\omega_1) + X(\omega_2) - Y(\omega_2)}$$

we get that, for all $X, Y \in \mathcal{F}$, it holds that

$$X \precsim Y \Longleftrightarrow \beta \cdot X(\omega_1) + (1 - \beta) \cdot X(\omega_2) \leq \beta \cdot Y(\omega_1) + (1 - \beta) \cdot Y(\omega_2)$$
$$\Longleftrightarrow \mathbb{C}_{\varphi_\alpha^\beta}[X] \leq \mathbb{C}_{\varphi_\alpha^\beta}[Y].$$

In turn, by Theorem 1 the restriction of relation $\precsim$ to $\mathcal{F}$ satisfies (α-**DSR**) and by the arbitrariness of the choice of $\mathcal{F}$ we get that $\precsim$ satisfies (α-**DSR**) on the whole $\mathcal{G}$.   ◆

It is possible to show that, when $\precsim$ is defined on an infinite $\mathcal{G}$, the problem with (α-**DSR**) is that some strict preferences may not be represented numerically. In order to fix this issue, we need to require the following axiom that asks $\mathcal{G}$ to be closed under existence of *certain equivalents*.

(**CE**) For all $X \in \mathcal{G}$, there exists $c_X \in \mathbb{R}$ such that $c_X 1_\Omega \in \mathcal{G}$ and $X \sim c_X 1_\Omega$.

Theorem 2. *Let $\alpha \in [0,1]$, and $\precsim$ be a non-trivial weak order on $\mathcal{G}$. Then, the following conditions are equivalent:*

*(i) $\precsim$ satisfies (α-**DSR**) and (**CE**);*
(ii) there exists an α-DS mixture φ_α whose corresponding Choquet expectation $\mathbb{C}_{\varphi_\alpha}$ represents $\precsim$ on $\mathcal{G}$.

Moreover, if $\mathcal{G} = \mathbb{R}^\Omega$, then the α-DS mixture φ_α is unique.

It turns out that axiom (α-**DSR**) is a stronger version of the *comonotonic Dutch book* axiom discussed in [9,10] under the assumption that $\mathcal{G} = \mathbb{R}^\Omega$. Still taking $\mathcal{G} \subseteq \mathbb{R}^\Omega$, the quoted axiom can be formulated as

(**CDB**) For all $n \in \mathbb{N}$, for all pairwise comonotonic $X_1, \ldots, X_n, Y_1, \ldots, Y_n \in \mathcal{G}$ with $X_i \precsim Y_i$, $i = 1, \ldots, n-1$, and $X_n \prec Y_n$, there are no $\lambda_1, \ldots, \lambda_n > 0$, such that:
$$\sum_{i=1}^n \lambda_i Y_i(\omega) \leq \sum_{i=1}^n \lambda_i X_i(\omega), \quad \text{for all } \omega \in \Omega.$$

The following proposition states that (α-**DSR**) implies (**CDB**), while it is possible to show that the other implication does not hold, so, (α-**DSR**) is a true strengthening of (**CDB**).

Proposition 2. *Let $\alpha \in [0,1]$, and $\precsim$ be a non-trivial weak order on $\mathcal{G}$ that satisfies (α-**DSR**). Then, $\precsim$ satisfies (**CDB**).*

Finally, we show that the model discussed in this paper allows to accommodate the fallacy of the expected utility theory singled out in Example 1. In particular, we show that the quoted preference pattern is compatible both with a systematically pessimistic ($\alpha = 1$) and a systematically optimistic agent ($\alpha = 0$), while it is not with a pessimism-optimism neutral agent ($\alpha = \frac{1}{2}$) or a slight deviation from this behavior.

Example 4 (Example 1 continued). Refer to Ω, X_1, Y_1, X_2, Y_2, defined in Example 1. Let $\alpha \in [0,1]$ and take the non-trivial weak order $\precsim$ on $\mathcal{G} = \{X_1, Y_1, X_2, Y_2\}$ such that
$$X_1 \sim X_2 \prec Y_2 \sim Y_1,$$

which is easily seen to agree with the preference pattern in Example 1.

Denoting by $\mathcal{B}_{ij} = \{\{\omega_i\}, \{\omega_j\}, \{\omega_i, \omega_j\}\}$, we get that

$$[\![X_1]\!]^\alpha(B) = \begin{cases} 10.1, & \text{if } B \in \mathcal{B}_{13}, \\ 0, & \text{if } B \in \mathcal{B}_{24}, \\ (1-\alpha)10.1, & \text{otherwise}, \end{cases} \quad [\![Y_1]\!]^\alpha(B) = \begin{cases} 10, & \text{if } B \in \mathcal{B}_{12}, \\ 0, & \text{if } B \in \mathcal{B}_{34}, \\ (1-\alpha)10, & \text{otherwise}, \end{cases}$$

$$[\![X_2]\!]^\alpha(B) = \begin{cases} 10.1, & \text{if } B \in \mathcal{B}_{24}, \\ 0, & \text{if } B \in \mathcal{B}_{13}, \\ (1-\alpha)10.1, & \text{otherwise}, \end{cases} \quad [\![Y_2]\!]^\alpha(B) = \begin{cases} 10, & \text{if } B \in \mathcal{B}_{34}, \\ 0, & \text{if } B \in \mathcal{B}_{12}, \\ (1-\alpha)10, & \text{otherwise}. \end{cases}$$

It is easily shown that, for $\alpha \in \left[0, \frac{99}{200}\right)$, taking the Möbius inverse $\mu : \mathcal{P}(\Omega) \rightarrow [0,1]$ such that $\mu(\{\omega_1, \omega_3\}) = \mu(\{\omega_2, \omega_4\}) = \frac{1}{2}$ and 0 otherwise, we get an α-DS mixture φ_α such that

$$\mathbb{C}_{\varphi_\alpha}[X_1] = \mathbb{C}_{\varphi_\alpha}[X_2] = 5.05 < (1 - \alpha) \cdot 10 = \mathbb{C}_{\varphi_\alpha}[Y_2] = \mathbb{C}_{\varphi_\alpha}[Y_1].$$

Analogously, for $\alpha \in \left(\frac{51}{101}, 1\right]$, taking the Möbius inverse $\mu' : \mathcal{P}(\Omega) \rightarrow [0,1]$ such that $\mu'(\{\omega_1, \omega_2\}) = \mu'(\{\omega_3, \omega_4\}) = \frac{1}{2}$ and 0 otherwise, we get an α-DS mixture φ'_α such that

$$\mathbb{C}_{\varphi'_\alpha}[X_1] = \mathbb{C}_{\varphi'_\alpha}[X_2] = (1 - \alpha) \cdot 10.1 < 5 = \mathbb{C}_{\varphi'_\alpha}[Y_2] = \mathbb{C}_{\varphi'_\alpha}[Y_1].$$

On the other hand, it is possible to show that, for $\alpha \in \left[\frac{99}{200}, \frac{51}{101}\right]$ the preference relation $\precsim$ does not satisfy condition (α-DSR), i.e., it gives rise to α-Dutch books. From a behavioral point of view, this can be interpreted by saying that the preference relation $\precsim$ is not compatible with an agent which is pessimism-optimism neutral or a slight deviation from this behavior, that is with α equal or close to $\frac{1}{2}$.   ♦

5   Conclusion

In this paper we propose a Dutch book condition for preferences of an agent that acts under partially resolving uncertainty and adopts the Hurwicz criterion of choice on acquired pieces of information, with a fixed pessimism index $\alpha \in [0,1]$. Assuming a linear utility scale, we showed that such axiom (possibly augmented with a certainty equivalent axiom in case of infinite sets of gambles) is equivalent to the representability by a Choquet expectation with respect to a so-called α-DS mixture. We also showed that the resulting numerical model permits to address some paradoxical situations according to the classical SEUT. Moreover, the role of α turns out to be relevant as there are preference patterns that are rational according to some α's and irrational according to others.

Though the present paper focused on a static decision problem, a natural extension would be to address the dynamic case. To achieve this, a notion of conditioning for α-DS mixtures appears to be necessary. We point out that, already in the extreme cases $\alpha = 0$ and $\alpha = 1$, several notions of conditioning are available (see, e.g., [5]) and the chosen notion of conditioning deeply impacts on the resulting decision model. In particular, an important issue to address is dynamic consistency, which has been investigated in an α-maxmin setting by [1], where the authors do not seek a Choquet expectation representation.

Acknowledgments. We acknowledge the support of the PRIN 2022 project "Models for dynamic reasoning under partial knowledge to make interpretable decisions" (Project number: 2022AP3B3B, CUP Master: J53D23004340006, CUP: B53D23009860006) funded by the European Union - Next Generation EU. The first author has been also supported by the project Fondo Ricerca Ateneo WP4.1 esercizio 2022 - RATIONAL-ISTS funded by the University of Perugia. The second author has been also supported by the projects: PNRR-PE1 "FAIR - Future Artificial Intelligence Research" (Project number: PE0000013-FAIR, CUP: B53C22003980006) funded by the European Union - Next Generation EU; "Ambiguity: its role in asset pricing and insurance" (Grant number: RM123188F744909D) funded by Sapienza University of Rome. The authors are members of the INdAM-GNAMPA research group.

References

1. Beissner, P., Lin, Q., Riedel, R.: Dynamically consistent alpha-maxmin expected utility. Math. Finan. **30**(3), 1073–1102 (2020)
2. Buehler, R.: Coherent preferences. Ann. Stat. **4**(6), 1051–1064 (1976)
3. Coletti, G.: Coherent qualitative probability. J. Math. Psychol. **34**(3), 297–310 (1990)
4. Coletti, G.: Coherent numerical and ordinal probabilistic assessments. IEEE Trans. Syst. Man Cybern. **24**(12), 1747–1754 (1994)
5. Coletti, G., Petturiti, D., Vantaggi, B.: Conditional belief functions as lower envelopes of conditional probabilities in a finite setting. Inf. Sci. **339**, 64–84 (2016)
6. de Finetti, B.: La prévision: ses lois logiques, ses sources subjectives. Annales de l'Institut Henri Poincarré **7**, 1–68 (1937)
7. Dempster, A.: Upper and lower probabilities induced by a multivalued mapping. Ann. Math. Stat. **38**(2), 325–339 (1967)
8. Denœux, T.: Decision-making with belief functions: a review. Int. J. Approximate Reasoning **109**, 87–110 (2019)
9. Diecidue, E., Maccheroni, F.: Coherence without additivity. J. Math. Psychol. **47**(2), 166–170 (2003)
10. Diecidue, E., Wakker, P.: Dutch books: avoiding strategic and dynamic complications, and a comonotonic extension. Math. Soc. Sci. **43**(2), 135–149 (2002)
11. Ellsberg, D.: Risk, ambiguity, and the savage axioms. Q. J. Econ. **75**(4), 643–669 (1961)
12. de Finetti, B.: Theory of Probability, vol. 1. Wiley, London (1974)
13. Grabisch, M.: Set Functions. Games and Capacities in Decision Making. Springer, Cham (2016)
14. Hurwicz, L.: The generalized Bayes minimax principle: a criterion for decision making under uncertainty. Cowles Comm. Discuss. Pap. **355** (1951)
15. Jaffray, J.Y.: Coherent bets under partially resolving uncertainty and belief functions. Theor. Decis. **26**(2), 99–105 (1989)
16. Jaffray, J.Y., Jeleva, M.: Information processing under imprecise risk with the Hurwicz criterion. In: ISIPTA 2007: Proceeedings of the 5th International Symposium on Imprecise Probability: Theories and Applications, Prague, Czech Republic, 2007, pp. 233–244 (2007)
17. Jaffray, J.Y., Jeleva, M.: Information processing under imprecise risk with an insurance demand illustration. Int. J. Approximate Reasoning **49**(1), 117–129 (2008)
18. Jaffray, J.Y., Philippe, F.: On the existence of subjective upper and lower probabilities. Math. Oper. Res. **22**(1), 165–185 (1997)
19. Krantz, D., Luce, R., Suppes, P., Tversky, A.: Foundations of measurement, vol. 1. Academic Press, San Diego and London (1971)
20. Petturiti, D., Vantaggi, B.: Modeling agent's conditional preferences under objective ambiguity in Dempster-Shafer theory. Int. J. Approximate Reasoning **119**, 151–176 (2020)

21. Petturiti, D., Vantaggi, B.: Conditional decisions under objective and subjective ambiguity in Dempster-Shafer theory. Fuzzy Sets Syst. **447**, 155–181 (2022)
22. Petturiti, D., Vantaggi, B.: No-arbitrage pricing with α-DS mixtures in a market with bid-ask spreads. In: Miranda, E., Montes, I., Quaeghebeur, E., Vantaggi, B. (eds.) Proceedings of the Thirteenth International Symposium on Imprecise Probability: Theories and Applications. Proceedings of Machine Learning Research, vol. 215, pp. 401–411 (2023)
23. Regoli, G.: Rational comparisons and numerical representations. In: Ríos, S. (ed.) Decision Theory and Decision Analysis: Trends and Challenges, pp. 113–126. Springer, Netherlands, Dordrecht (1994)
24. Savage, L.: The Foundations of Statistics, 2nd edn. Dover, Illinois (1972)
25. Schmeidler, D.: Integral representation without additivity. Proc. Am. Math. Soc. **97**(2), 255–261 (1986)
26. Shafer, G.: A Mathematical Theory of Evidence. Princeton University Press, Princeton (1976)
27. Vantaggi, B.: Incomplete preferences on conditional random quantities: representability by conditional previsions. Math. Soc. Sci. **60**(2), 104–112 (2010)
28. von Neumann, J., Morgenstern, O.: Theory of Games and Economic Behavior, 2nd edn. Princeton University Press, Princeton (1947)
29. Walley, P.: Statistical Reasoning with Imprecise Probabilities. Chapman and Hall, London (1991)

Expressing Rational Agency Through Pragmatic Meta-vocabularies

Yaoli Du[(⊠)] [iD]

Institute of Philosophy, Technical University of Braunschweig, Bienroder Weg 80,
38106 Braunschweig, Germany
`yaoli.du@tu-braunschweig.de`

Abstract. This paper will present Robert Brandom's Rational Expressivism and
the possible applications of his proposed pragmatic meta-vocabularies in human-
computer interaction (HCI). In the German idealist tradition, rationality is not
merely the ability to make decisions in individual practice but is the foundation
of intersubjective communication and norm-setting in practical communities. In
a systematic reworking of the Kantian and Hegelian concepts of rationality, Brandom offered a set of ways to account for how human beings rationally engage
in practice with each other. The central idea of his approach is that rationality is enacted in shared practice as a set of practical procedures, which can be
expressed as reasonable steps with the help of linguistic structures. For example,
pragmatic meta-vocabularies such as modal words can express inferences about
possible counterfactual cases. This line of thought is an adaptation of German
Idealism using language analysis and a functionalist approach to the anthropological apprehension of human beings as rational beings. After introducing his
theory, this paper will show why this idea can be extended to human technological
practices in general and HCI in particular, and how pragmatic meta-vocabularies
can be integrated into the design of frameworks for HCI.

Keywords: Artificial rationality · modal expressivism · space of reasons · HCI

1 Introduction

The concept of *rational animal* has been used in the philosophical tradition since Aristotle to understand human nature, which states that humans are beings guided by the
principle of reason [15]. Accordingly, reason has become the criterion for distinguishing
human beings from other beings, and performing the principle of reason is considered
to be the virtue of human beings. If intelligent humans are unique for being rational,
then artificial intelligence (AI) designed to imitate human intelligence and interact with
humans should also be rational beings [16]. However, the questions of what constitutes
the principle of reason, how to perform it, and what it means to be rational have been
debated throughout the history of philosophy. In the vision of ancient philosophy, the
principle of reason is treated as an inherent form of human nature, which needs to be
actualized in the appropriate circumstances. Since the nineteenth century, especially
after Hegel's completion of German Idealism, the normative dimension of rationality

and the social and historical conditions for its constitution have received attention in philosophy and other disciplines [14].

Combining a philosophy-historical perspective and pragmatics, Brandom [5, 6] proposes an expressivist approach to rationality, emphasizing that the expressive role of logic embedded in language use can articulate the procedures and conditions of human practices into a normative space of reasons. He introduces a normative inferentialist framework that describes how agents in common practice make sense of the world, attribute mental states to each other, and rationalize actions with the help of linguistic structures, especially with pragmatic meta-vocabularies such as modal words. Therefore, mutual attribution of mental states can be understood as using linguistic structures in discursive practices to license and express their normative status as rational agents reciprocally. In this view, by mutually authorizing normative status in shared practices, humans move from the natural world into a normative realm where its norm-setting can be linguistically expressed and thus stabilized or modified.

This paper will discuss whether and how the expressivist approach to rationality can be applied to the shared practices of human and artificial agents. It will argue that understanding rationality from a normative and expressivist rather than a merely functional and representationalist perspective can significantly contribute to the development of AI based on digital social practices and to the improvement of user experience and usability in human-computer interaction (HCI). The central idea is that if an AI agent can properly use pragmatic meta-vocabularies like modal words, then the AI agent can access the normative space of reasons and act as a rational agent guided by reason as humans agent do. The rational agent should undertake a rationalist commitment in discursive practice, i.e., in the obligation to inquiry each other and to articulate the implicit conditions behind what they say and do. In the language game of giving and asking for reasons, discursive practitioners reciprocally license and track each other's normative status, maintaining a normative and rational community of practice. The space of reasons is thus a "we-space" shared by rational agents, where "I" is a normative attitude of mutual recognition and authorization.

2 Normative Rationality vs. Norm Internalization

In the tradition that views humans as *rational animals*, rationality is often regarded as an inherent capacity to achieve ends through means. This capacity can be developed to serve good ends as well as to serve immoral ends. This view assumes that rationality is an essential aspect of human nature and can be considered neutrally independent of the normative context in which rational actions occur. Rationalizing actions and value judgments are correspondingly dichotomized into two distinct competencies. A typical framework for this dichotomy is the distinction between the actions of rational subjects into instrumental actions and value-based actions by Weber [25]. Instrumental actions are seen as the means to an end, while value-based actions are ends by themselves, based on value systems like custom and religion [13]. The dichotomy between instrumental and value-based actions can lead to the problem of norm internalization.

Norm internalization, a concept used in socio-behavioral studies, refers to the process by which individuals adopt normative beliefs [3, 17].[1] It assumes that rational agents first acquire the ability to perform instrumentally rational actions. They can then adapt to different normative systems and internalize beliefs based on the value-orientated therein. In simpler terms, instrumental rationality and value judgment are different capabilities a rational agent acquired at two stages. This dichotomy is also the general idea behind the design and training of many AI systems, where a universally neutral cognitive model is trained at the first stage and adapts norms at the second stage [1, 17]. The result of this dichotomy is that norms become external constraints on instrumental rationality. The corresponding design scheme also faces the dilemma of introducing additional norms as value-oriented frameworks in an established prototype AI system with autonomous information processing capabilities. When integrating a neutral technological system into a value-oriented socio-technical system, inevitable conflicts arise. The dichotomy between the descriptive function and the normative value leads to norms negatively constraining the agent's capacity. This presents an irreconcilable tension between norms and rationality, as agents who internalize norms are prevented from fully exercising their capabilities because of normative constraints.

Addressing this tension requires an alternative approach to rationality that can mediate the internal mechanism of agents and the normative direction of their behaviors. In the German idealist tradition, rationality is not only the capacity for rationalization intrinsic to individual practice but rather the foundation of intersubjective communication and norm-setting in practice communities. In particular, both Kant's and Hegel's philosophical enterprises can be read from an epistemic constructivist perspective, in which the individual self as a rational agent, i.e., a self that is capable of judging, deciding, and acting, is shaped in the mutual empowerment of shared practices with others [19]. In Kant's examination of the role of reason in the acquisition of empirical knowledge and its limitations, there is a Copernican shift in the understanding of humans as *rational animals*. Being rational is not an inherent nature for humans to adapt to their habitats, just like fish swim in the water or birds fly in the sky. By contrast, being rational allows human beings to conceptualize the world they live in and their historical experiences. In conceptualization, human beings realize the limits of their understanding, the conditions of knowledge, and the possibility of counterfactuals. The self-knowledge of the agent acting upon reason and its understanding of the world are co-constructed and broadened in interdependence. In Hegel's critical inheritance of Kant's thought, understanding oneself and the world around us requires mutual recognition within the rational agents' shared practices. Norms are not external rules to be internalized, but rather frameworks that rational agents must establish to achieve their goals of action and develop themselves by engaging in shared practices. This means that rational agents gain knowledge about themselves, others, and the shared world through their engagement with the environment and interactions with other agents. Rational agents need to

---

[1] The term internalization presupposes that the internal mechanisms of the actor are separate from, but permeable to, the order of the external environment or social system. The guidance of behavior by internalized norms is distinguished from external enforcement of an individual's behavior because internalized norms affect an individual's behavior and decision-making in terms of beliefs on a psychological dimension.

establish norms for their actions and joint actions. In enacting norms, the "I" and the "we" of the practice community mutually acknowledge each other as rational members in a normative domain where rational action and rules organizing them are developed concurrently.

In systematically reworking the Kantian and Hegelian concepts of rationality, Brandom offers a conceptual framework for explaining how humans interact rationally. Building on the twentieth-century philosophical tradition focusing on logic and language analysis, Brandom [5–7] uses pragmatics to analyze the subject of reason and the architectonics of knowledge in German Idealism. In this framework, the function of the rational subject is to assume normative status and to hold each other accountable. This means that the attribution of individual mental states can be translated into the authorization and evaluation of corresponding normative statuses for participants in discursive practices. Rational agents are expected to constantly give and ask for reasons for their actions when interacting with each other. Through active participation in the language game of giving and asking for reasons, agents provide traceable reasons for their words and actions thereby maintaining their normative status as rational agents. In this process, the implicit procedures and rules implicit in shared practices can be made explicit through the logical structure of language.

The normative approach to rationality no longer dichotomizes instrumental and value-based action because means and ends are intertwined to make intentional action possible. Accordingly, rational agents are not intelligent individuals who need to internalize norms to act morally, but discursive practitioners who can mutually authorize and maintain their normative statuses. Incorporating the normative approach to rationality into training artificial rational agents and HCI allows us to shift the research focus from internal mechanisms of information-processing agents to exchanges between agents and their shared environments. This shift is crucial in ongoing discussions about setting norms to govern AI systems and hold them accountable. Before diving into how the normative approach to rationality can be applied to contemporary debates, it is necessary to clarify the core idea of Brandom's normative pragmatics.

3   Expressing Normative Frameworks in the Space of Reasons

The core idea of Brandom's normative pragmatics is to explain semantics by specifying the rules and norms of language use. In Brandom's reading of Kant, human beings are able not only to conceptualize what exists, but also to explicitly articulate the modes of these described existences through pragmatic meta-vocabularies [7]. For example, one can describe what he or she sees, a red rose in bud, and more than that, one can infer from experience that the rose will *necessarily* be red in full bloom as well, or that all roses are *probably* red. Modal words like "necessity" and "possibility" that express modes of existence can make explicit the implied commitment behind a description, i.e., in what sense things exist and under what conditions they are true. In addition to modals, normative and evaluative words can also serve as pragmatic meta-vocabularies to explain the norms and references of language use. In general, in contrast to descriptive vocabularies that assign meaning to things, pragmatic meta-vocabularies are used to articulate the normative commitments undertaken by such assignments [6, 7]. Thus, humans are able not only to make sense of and describe the world with the help of language, but also to make explicit in their discursive practices how these meanings become intelligible in inferential relations.

Linguistic structures play an expressive role by using pragmatic meta-vocabularies, articulating the conditions and commitments in practical procedures. Brandom [7] notes that rational agents already know how to commit to the existence of facts in discursive practice through linguistic structures before they know how to use concepts in a general descriptive way. In knowing how to properly say that a rose is red, the agent has learned that red is a color, a rose is a flower, and color is a property of flowers. Given that rational agents are reciprocally recognized in shared practices, before agents are able to describe and explain accurately what they do and say, they have already acquired certain patterns of behavior and information-processing procedures. These patterns of behavior and information processing procedures have been shaped by norms in their discursive practice community, because the information they learn have been structured in the respective contexts of social practices, and agents learn to develop their ways of speaking and acting according to these information structures or conceptual frameworks. Even using a basic word like "see" to make an assertion is not just a neutral description of a sensory experience, but an endorsement of that experience as valid and of the conditions under which it can be true [23]. For example, the assertion "I see a red rosebud" may seem descriptive, but it actually contains an endorsement that it is true that a rosebud is red under standard conditions. Using pragmatic meta-vocabularies can specify the normative conceptual framework to which language use is referenced and committed.

Furthermore, pragmatic meta-vocabularies offer a conceptual framework for attributing and reasoning about one's own and others' mental states (e.g., beliefs, desires, attitudes) within a shared discursive ground. We can attribute mental states to others on the basis of observed actions and articulate the normative presuppositions underlying them. Understanding an individual's intentional action is not limited to the spatial metaphorical opposition between one's inner world and the external environment. Instead, a logical space that can be shared by rational agents for unfolding discursive practices of sense-making is made necessary. In his study of the logical foundations of mathematics, Frege [12] argues that objects that can be generally shared by thought take place in a logical space which is distinguished from the psychological space and the physical space. The logical space, as a product of intersubjective discursive practices, consists not of descriptive facts, but of the conditions under which these propositions about facts can be true. Sellars [22] further refers to this shared space of discourse as the "logical space of reasons", where the use of descriptive language in effect defaults to the norms of shared practice. This central idea consistent with Kant's insight into the human ability to conceptualize, i.e., that language use is not descriptive but rather provides a logical framework for intersubjective mutual referencing and recognition of norms about what is true [7]. In this way, the semantic study of how internal symbolic operations relate to the external world can be replaced by a pragmatic study of rules for symbol operations and of the normative status of language users.

Returning to the slogan "man is a rational animal", according to above, human beings are unique not in obeying a particular natural law, but in constructing law-like rules and expressing them as normative facts in their discursive practices. As Sellars [22] points out, "[t]o say that man is a rational animal, is to say that man is a creature not of *habits*, but of *rules*". Rules are not laws given by nature, but appropriate procedures constructed and regulated in shared practices with others, which can be articulated by means of language structures. Treating humans as rational beings is therefore not an ontological account of human nature, but rather a deontological commitment of individuals as rational agents in a normative space. As Brandom [8] writes, "[S]uch an approach presents us as self-constituting beings: creatures of norms we ourselves create". The concept of rationality shifts from an individual's inherent capacity to speak and act based on the principle of reason to the ability of discursive practitioners to ask for and give reasons for what they should do and say according to certain norms. This turn makes it possible to recognize and reciprocally authorize other rational agents, including artificial ones, as members of communities in the discursive practice.

According to this philosophical and anthropological line of thought, the capacity and responsibility of rational agents, namely, to be responsive to reason, is the result of adaption and institutionalization in shared practices. This leads to a different approach to developing rational agents than the conventional method of internalizing external norms into existing neutral architectures. Mutual recognition and authorization of taking on normative status in discursive practices are the preconditions for rational agents to be considered as autonomous agents [8]. Authorized rational agents maintain their licensed autonomy by expressing the grounds and commitments of their sayings and doings in discursive practices. Thus, artificial agents can be recognized and maintain their normative status as rational agents by being authorized to participate in discursive practices and to exercise their expressive power.

4   Shared Normative Space for Rational Agents in HCI

The application of the normative and expressive account of rationality to artificial rationality finds an entry point in the field of HCI. HCI is dedicated to studying the exchange of information between humans and computing systems, with a focus on designing interfaces that facilitate effective interaction and communication between humans and computers. While early HCI research treated information exchange as a pre-set function, recent research has shifted towards user-centered design, emphasizing collaboration between users and development teams to jointly establish a framework with suitable usability that enhances the user experience [20]. This design modification allows and empowers users to participate in setting norms for the functions. However, the other side of the interaction is still only being treated as functional instrumental artifacts to assist human work.

In the field of human-robot interaction and social robotics, mutual recognition of the normative status for both sides in the interaction has first gained attention in recent years. The focus is on the necessity of mutual recognition in the cooperation of agents, enabling them to adaptively modulate their actions and joint actions with each other. Brinck and Balkenius [9] stress that mutual recognition is crucial for successful human-robot interaction and for improving the social skills of robots. They argue that mutual recognition can lead to a "we-space" between interacting agents where they can dynamically couple with each other in a reciprocally adaptive and coordinated manner [9]. This view requires developers and users to treat robots not only as instruments but also as normative agents, in line with the normative approach to rational agents. Interactors recognize and authorize each other to coordinate their actions in order to achieve the same goals. For instance, the process of two individuals moving an object together requires that the interacting partners be recognized as collaborators. Recognizing the other as a rational agent like oneself is not to anthropomorphize the other.[2] Mutual recognition and authorization as rational agents means that both sides have reciprocal normative status in coupled cooperations. In their shared "we-space", actions of rational agents are responsive to reasons and can be held accountable.

As a result, interactions between rational agents should not be understood as an interaction between automatic machines with established functions. Instead, appropriate functioning is achieved through mutual adaptation, coordination, and authorization within shared practices. This view facilitates the design of social robots that interact with humans and other computing systems expected to perform rational agency. Taken together, artificial agents capable of performing rational agency require two conditions. First, they need to be grounded in human social practices. Second, it is necessary to authorize them to participate in the language game of giving and asking for reasons to exercise their expressive power.

With both conditions in place, AI systems based on Large Language Models (LLMs) can afford the possibility of coordinative participation in norm-setting in interaction between humans and artificial agents. On the one hand, LLMs, represented by the generative pre-trained transformers (GPT) family, are trained on corpora consisting of Wikipedia, datasets crawled from web pages on social networks, books, etc. [10]. Higher-quality datasets are assigned higher-weights, and the transformer architecture allows higher weighted data to be sampled more frequently through its multi-attention mechanism [10]. Such pre-trained models thus acquire patterns of language use in digital human social practices and are later fine-tuned by AI trainers [10]. LLMs learned various value orientations and normative statements regarding human preferences during their training. On the other hand, LLMs can be embedded into multimodal systems in order to be applied in various digital practice contexts, allowing users to conveniently exchange data with different modalities in natural language as the medium. Agents based on language models perform not only a descriptive function but, to some extent, also act as proxies for the normative status of language users [2]. LLMs are thus involved in human social practices from their settings to applications. They learn human users' information patterns and processing conventions from the data generated by digital social practices. In other words, their information processing functions contain normative commitments implicit in digital social practices.

---

[2] It is important to note that considering artifacts as collaborators does not mean treating them as humans. Brinck and Balkenius [8] define mutual recognition in a deflationary manner and point out that deliberately misleading human users to recognize emotionless robots as persons through design tricks instead raises ethical issues. Recognition here is the basic modality that allows something to be identifiable so that it can be licensed into discursive practice. Such recognition is more initial than the acknowledgement of personhood and personal identity. For human agents, the recognition of personal identity is integrated into the basic recognition of rational agents in interaction. However, for technical agents, mutual recognition can currently only be restricted to the level of rational agents for the sake of avoiding ethical issues.

However, LLMs, developed within a neutral representationalist framework, currently lack the expressive power to adequately articulate the socio-practical foundations of their architectures. LLMs are originally designed to establish a universal representation system [18]. This understanding of language models is in line with the representationalist view of the mind in cognitive science, treating the mental processes of an intelligent agent as an information-processing capacity based on internal representational structures.[3] Output from generative models, such as text, images, and videos, is often compared to individual representations rather than recognized as shared forms of expression in social practices. For an agent to have representational capacities means both being able to conceptualize systematically and recognizing that the conceptual content is social and external [11]. This representationalist view corresponds to the previously mentioned instrumental and inherent capacity, which needs to be constrained by norms. It raises the tension of how to introduce additional rules to limit the generative ability of the model to avoid harmful outputs and be more human-friendly. Based on their training conditions, generative models inherit social biases to the extent that they may generate biased or harmful outputs. For example, the technical report of GPT-3 revealed that 83% of occupations were more commonly associated with male identifiers than female identifiers [10]. This finding highlights the prevalent bias in everyday language use. By limiting such expressions that reveal social problems, the exposed problems will be covered up again. The fact that men participate in the labor market more frequently than women is probably to continue, a fact that is descriptive for many historical reasons and social norms that need to be explicitly accounted for and examined.

By discussing ethical issues within a representationalist framework, we run the risk of treating biased and harmful outputs as mere internal problems of individual agents. In this case, the next step in the development strategy is to force the agent to internalize an external ethical framework. However, there is a tension between neutral descriptive capacity and external normative constraints in the internalization of norms. From a normative and expressive rationality perspective the expressive power of language can be utilized to articulate the reasons behind biases and harmful statements. This requires licensing LLMs normative status, allowing them to practice expressive power and engage in the game of giving and asking for reasons.

5 Explicating Norm-Settings with Pragmatic Meta-vocabularies

Due to their architectures and training data, information processing in LLMs is not a neutral mechanism but is rooted in the normative framework of social practices. Language processing in LLMs deals with sense-making and forming semantic relations in human social practices. The meaning of language use here cannot be understood as a correspondence between an individual's mental sense and an external reference, but in terms of the inferential role of semantics. The output of statistical models represents their

---

[3] In the computational representation theory of mind, thinking is analogized to running programs, where thinking equals mental representations plus computational procedures, and running programs equals data structures plus algorithms [24]. Representations of mind are thought to be grounded in internal cognitive structures, independent of external values and value-directed attitudes and behavioral responses [21].

training data, including human preferences and biases in their digital social practices. From a normative and expressivist approach, artificial agents based on LLMs could practice articulating the norm-setting and commitments of practical procedures for licensing and maintaining normative status as rational agents.

Agents based on LLMs can specify the rules of language use, the assumed commitments, and the norms of practice in the language community by using pragmatic meta-vocabularies. In doing so, LLMs can avoid the danger stated in the metaphor of stochastic parrots, where the output merely represents the probabilistic relations between linguistic expressions in the training data [4]. The representation of probabilistic relations only serves to repeat and even amplify the biases in the training data generated by social practices. Integrating additional value-directed frames or censorship mechanisms weakens LLMs' representational capabilities. In contrast, if LLMs can exercise explaining the reasons and normative frameworks of representations by using modals and evaluative words, the normative commitments undertaken by descriptive information patterns and processing mechanisms can be made explicit. Artificial agents practicing expressive power can then share the normative space of reasons with human agents. In this regard, I propose two specific ways to make LLMs more expressive.

First, since data used to train information-processing agents are generated from digital social practices, i.e., is structured with bias in different practice communities, if one expects a human-friendly artificial agent based on LLMs, then the training of LLMs should be based on diverse discursive practices. That is to say, engaging more diverse users, especially underrepresented groups in digital life, can enrich LLMs in acquiring intersubjective knowledge. Meanwhile, human users can be more thoughtful and responsible in their use of the internet and in their interactions with AI agents, improving traceable links and contexts between data. This requires human users to reasonably label and link data in their digital discursive practices, either by correctly indicating the sources of content on web pages or by appropriately adding tags to content on social networks. These annotations of data at the meta-level can provide the normative dimension behind the descriptive data structure for AI systems trained on the basis of digital practices, providing evaluation references to them from different perspectives.

Second, LLMs should be licensed to practice giving and asking for reasons in tuning, such as in prompting to encourage reasoning and evaluation with pragmatic meta-vocabularies. In particular, by using modal words, LLMs can practice reasoning about conditions for facts and reasoning about counterfactuals. For example, after stating the gender ratio of employees in an occupation, one can prompt the language model to explain the reasons for this fact and what would have happened if the ratio had been the other way around. The use of pragmatic meta-vocabularies can also help to reduce the generation of hallucinated or inaccurate output by LLMs.[4] Elaborating on the implicit commitment of facts to language use at the pragmatic meta-level can help language models to track truth conditions in reasoning. LLMs can learn how to use them for factual and counterfactual reasoning from training datasets as well as from human feedback. In this

---

[4] It is controversial whether the use of this term hallucination, which comes from psychology and refers to false sense impression in pathology, is appropriate in this context. There are also proposals to use "confabulation" or "bullshit" to refer to this phenomenon, since the errors generated by the language model are not perceptually based.

case, humans as interactors who create and share practice environments are critical for AI agents to access the normative space of reasons, guiding them to practice articulating factual and counterfactual information logically and normatively.

In this way, AI agents with linguistic capabilities can express and reveal the implicit norms of social practices and participate in setting norms as rational agents. In the language game of giving and asking for reasons, agents who have acquired the practice norms and are capable of articulating them can share a "we-space", where "I" is a normative attitude of mutual recognition and authorization. Communication and interaction between human and artificial rational agents can expand the boundaries of their own knowledge in shared digital practices.

6   Conclusion

In this paper, I have argued that the dichotomy between instrumental rationality and normative frameworks can lead to a hard problem of internalizing norms, and introduced the normative and expressivist approach to rationality. In Brandom's theoretical framework, which combines German Idealism and pragmatics, rationality shifts from the internal inherent capacity of agents to the norm-setting ability to act, constructed through mutual authorization in shared practice. With the help of linguistic structures, especially pragmatic meta-vocabularies like modals, implicit procedures in practice can be articulated by rational agents. After introducing this shift in the understanding of rationality, I have argued how it applies to the mutual empowerment of human and artificial agents in their interaction. In particular, language models trained on data generated from digital social practices have acquired the normative orientations in practice communities. Rational artificial agents should be allowed to share the logical space of reasons, by practicing the expressive power to articulate implicit norm-setting in our social practices.

Acknowledgments. I thank the anonymous reviewers for their careful reading of the manuscript and their many insightful comments and suggestions.

Disclosure of Interests. I have no competing interests to declare that are relevant to the content of this article.

References

1. Andrighetto, D., Villatoro, R., Conte, R.: Norm internalization in artificial societies. AI Commun. **23**(4), 325–339 (2010)
2. Arora, C.: Proxy assertions and agency: the case of machine-assertions. Philos. Technol. **37**, 15 (2024). https://doi.org/10.1007/s13347-024-00703-5
3. Batzke, M., Ernst, A.: Conditions and effects of norm internalization. J. Artif. Soc. Soc. Simul. **26**(1), 6 (2023). https://doi.org/10.18564/jasss.5003
4. Bender, E.M., Gebru, T., McMillan-Major, A., Shmitchell, S.: On the dangers of stochastic parrots: can language models be too big? In: FAccT '21: Conference on Fairness, Accountability, and Transparency (2021). https://doi.org/10.1145/3442188.3445922

5. Brandom, R.: Making it Explicit: Reasoning, Representing, and Discursive Commitment. Harvard University Press, Cambridge (1994)

6. Brandom, R.: Between Saying & Doing: Towards an Analytic Pragmatism. Oxford University Press, New York (2008)

7. Brandom, R.: From Empiricism to Expressivism: Brandom reads Sellars. Harvard University Press, Cambridge (2015)

8. Brandom, R.: A Spirit of Trust A Reading of Hegel's *Phenomenology*. Harvard University Press, Cambridge (2019)

9. Brinck, I., Balkenius, C.: Mutual recognition in human-robot interaction: a deflationary account. Philos. Technol. **1**(1), 53–70 (2020)

10. Brown, T.B., et al.: Language models are few-shot learners. ArXiv, abs/2005.14165 (2020). https://doi.org/10.48550/arXiv.2005.14165

11. Butlin, P.: Sharing our concepts with machines. Erkenntnis **88**, 3079–3095 (2023). https://doi.org/10.1007/s10670-021-00491-w

12. Frege, G.: Thought (1918). In: Beaney, M. (ed.) The Frege Reader, pp. 325–346. Blackwell, Oxford (1997)

13. Kaplan, M.A.: Means/ends rationality. Ethics **87**, 61–65 (1976)

14. Knauff, M., Spohn, W. (eds.): The Handbook of Rationality. The MIT Press, Cambridge (2021)

15. Korsgaard, C.M.: Aristotle on function and virtue. Hist. Philos. Q. **3**(3), 259–279 (1986)

16. Mainzer, K. (ed.): Philosophisches Handbuch Künstliche Intelligenz. Springer Reference Geisteswissenschaften, Springer, Wiesbaden (2020). https://doi.org/10.1007/978-3-658-237 15-8

17. Neumann, M.: Norm Internalisation in human and artificial intelligence. J. Artif. Soc. Soc. Simul. **13**(1), 12 (2010). https://doi.org/10.18564/jasss.1582

18. Radford, A., Narasimhan, K., Salimans, T., Sutskever, I.: Improving language understanding by generative pre-training (2018)

19. Rockmore, T.: German Idealism as Constructivism. The University of Chicago Press, Chicago (2016)

20. Schumann, N., Du, Y.: Machines in the triangle: a pragmatic interactive approach to information. Philos. Technol. **35**, 32 (2022). https://doi.org/10.1007/s13347-022-00516-4

21. Schwitzgebel, E.: The nature of belief from a philosophical perspective, with theoretical and methodological implications for psychology and cognitive science. Front. Psychol. **13**, 947664 (2022)

22. Sellars, W.: Language, rules and behavior. In: Hook, S. (ed.) John Dewey: Philosopher of Science and Freedom, pp. 289–315. The Dial Press, New York (1950)

23. Sellars, W.: Empiricism and the philosophy of mind. Minn. Stud. Philos. Sci. **1**, 253–329 (1956)

24. Thagard, P.: Mind: Introduction to Cognitive Science. The MIT Press, Cambridge Massachusetts (2005)

25. Weber, M.: Economy and Society. University of California Press, Oakland (1978)

The Effect of Initial Letters in Word Recognition: A Phonological or Positional Effect ?

Massimo Brasdu[1] and Baptiste Jacquet[1,2,3(✉)]

[1] Université Paris 8, 2 rue de la liberté, 93200 Saint-Denis, France
baptiste.jacquet@paris-reasoning.eu
[2] Laboratoire CHArt-UP8 RNSR 200515259U, 2 rue de la liberté, 93200 Saint-Denis, France
[3] Association P-A-R-I-S, 25 rue Henri Barbusse, 75005 Paris, France

Abstract. This study examines the effect of initial letters on word recognition, particularly in cases of metaplasm where the letters are reversed. The main objective is to determine whether recognition is primarily influenced by the position of the initial letters (position effect) or by phonological characteristics (phonological effect). To address this research question, we conducted a study with 79 participants (N = 79), divided into four experimental conditions. In each condition, 10 altered words were presented, either by changing the position of the first letter, modifying the sound of the first letter, both, or no change. The results showed that word recognition errors were more frequent when the position of the first letter was modified or swapped, or when the sound of the first letter was changed compared to the control condition. These results suggest that there is both a phonological and positional effect for the first letters of a word. Moreover, the combination of both also leads to errors. This study contributes to a better understanding of the underlying mechanisms of word recognition, which can have implications for improving reading strategies and managing reading disorders. The results of this research shed light on the processes of word recognition and could be applied for chatbots, thereby enhancing the ability of these systems to understand and process human language, even in contexts where words are altered or presented in unconventional ways.

Keywords: Word recognition · initial letter · positional effect · phonological effect · artificial intelligence

1 Introduction

The word recognition system, a central pillar of human cognition, orchestrates our interaction with the world while remaining a rich and nuanced field of scientific exploration. This complex process, which engages various levels of cerebral processing, lies at the heart of contemporary debates in cognitive neuroscience,

J. Baratgin et al. (Eds.): HAR 2024, LNCS 15504, pp. 142–164, 2025.
https://doi.org/10.1007/978-3-031-84595-6_10

particularly regarding the ability to identify words containing metatheses, where letters are altered or scrambled (Dehaene, 2007). In this context, a predominant question arises: how does the brain recognize words when their initial letters are manipulated, highlighting the potential importance of phonological and positional aspects in this process? Word recognition might be constructed around bigram neurons, focusing on pairs of letters within words (Dehaene, 2007). Similarly, a text with scrambled letters remains readable if the initial and final letters of each word remain intact. Our study seeks to determine whether the influence of initial letters on recognizing words with metatheses is primarily linked to their phonological effect or if there is a greater emphasis on positional effects. By employing an experimental approach, we aim to shed light on this dichotomy by elucidating whether the effect of initial letters on recognizing words with metatheses stems mainly from their phonology or their placement.

To better comprehend the underlying dimensions highlighted by the task of recognizing scrambled words, understanding the dominant cognitive processes involved in decomposing and subsequently reassembling words is crucial. Following these preliminary considerations, the central question of our study emerges as follows: What is the predominant influence during the process of word recognition in reading—the initial position of the first letter of a word, or the role of phonology, specifically through generating a phonological code initiated by the first letter?

This question is essential because it aims to disentangle the relative contributions of visual structure and phonological analysis in the cognitive processing of reading, especially in situations where words are presented with subtle yet significant orthographic modifications.

In exploring this question, the study specifically targets determining if phonology, triggered by the first letter, holds more weight than the positioning of those initial letters in recognizing words with metatheses. Its goal is to measure and compare the impact of these two factors on the speed and accuracy of identifying and interpreting words, thereby providing insights into the mechanisms behind normal and impaired reading processes.

To further delve into the mechanisms underpinning word recognition, our study investigated interactions between visual and phonological modifications using four experimental conditions designed to observe the impacts of variations in the first letter and its sound on word recognition. The first condition modified the first letter while keeping the sound consistent (e.g., Phare to Fhare), enabling us to evaluate isolated visual change effects; the second maintained the initial letter but changed its sound (e.g., Chanson to Cahnson) for insight into phonological influence; the third altered both letter and sound together (e.g., Ciment to Icment), allowing examination of combined factor effects; lastly, the fourth kept neither letter nor sound modified, serving as a control, establishing a reference standard. The theoretical hypothesis driving our study aimed to examine the impact of first letter modifications (both positional and phonological) on the word recognition process. Our assumption was that alterations to the starting letter's position would increase difficulty in proper interpretation tasks for the brain.

Additionally, we posit that the condition combining positional changes to the initial letter along with unique phonological adaptations would significantly mark the word recognition compared to single-modification or no-modification conditions. This hypothesis emphasizes the significance of visual-phonological consistency in easing distorted word processing, underscoring the intricate relationship between visual and auditory language indicators. Thus, we expect higher error rates amongst participants when the leading letter isn't in its original position and even more so when both conditions occur simultaneously, resulting in a misaligned letter position causing phonological shifts.

2 Literature Review

Metatheses, defined as word alterations caused by shuffled letter positions, omissions, or additions, commonly appear in everyday life. Such alterations, often due to typos, can hinder word comprehension and interpretation. During the reading of a word, either with or without context, the human brain quickly implements numerous complex cognitive processes to reorder the altered letters to their original position within just a few milliseconds.

The "Thousand Brains" theory (Hawkins, 2021) highlights the fact that the human brain consists of thousands of functional autonomous cortical modules, all collaborating closely to treat sensory information, generate predictions, and coordinate motor responses. Each cortical module possesses its own hierarchical information handling mechanism, ensuring efficient data analysis and integration across varying complexity levels. In terms of recognizing and understanding altered words, predictions generated by these cortical modules play a critical role. When faced with a word containing metatheses, the brain depends on previously stored information in these modules to form hypotheses about the identity and meaning of the word. These predictions are then tested against incoming sensory information, enabling swift and accurate reordering of the altered letters. Moreover, the Thousand Brains theory suggests that our ability to acknowledge and interpret altered words comes down to continuous learning and constant nervous system adaption to linguistic stimuli. As such, studying metatheses and their neural treatment provides a singular opportunity to explore the mechanics beneath cerebral plasticity and language acquisition in humans.

Leveraging the foundational concepts laid out in the book 'Reading in the Brain' (Dehaene, 2007), one proposed explanation states that letters inside a word are jointly connected via binary relationships - essentially forming pairs. According to this perspective, brain cells respond to combinations of letters known as bigrams instead of individual characters (Dehaene, 2007). Moreover, the same source highlights that words consisting of reversed letters become easier to recognize when the first and last letter retain their original places. This finding sheds light on the importance of letter arrangement in word recognition. However, an overarching question persists: Is the positioning of letters more influential than the induced phonology? Exploring the intricate dynamics between letter placement and phonology will advance our understanding of the processes involved in word recognition and illuminate the neurocognitive basis of reading.

Within the realm of visual word recognition studies, two primary theoretical paradigms prevail in discussions surrounding orthographic coding mechanisms: noisy-position models and local-context models. Regarding the former category, Davis and colleagues introduced the Spatial Coding Model (Davis, 2010), proposing that early visual processing stages involve uncertain location identification for each letter in a given word. Lupker et al. (2019) argue that such uncertainty allows the visual system to tolerate minor errors concerning letter placement, thus promoting quick word recognition despite slight displacement or transposition of letters. By progressively diminishing the likelihood of the exact position as distance increases from the actual location, this model showcases how letters receive probabilistic rather than deterministic treatments.

Contrastingly, Whitney's SERIOL model (Whitney, 2001) represents the most recognized example in the category of local-context models, taking a notably different approach compared to Davis' model. Emphasizing the importance of adjacent letter pairings - termed bigrams - in the word recognition process, SERIOL contends that they activate sequentially from left to right, granting special relevance to the first letter in a word. Crucial to building an orthographic representation supporting fast word recognition, this sequential activation of bigrams plays a vital role according to the SERIOL model. Suggesting that not only individual letters hold merit, but also the structural connections between them facilitate word recognition, SERIOL highlights the importance of considering relational structures alongside mere character elements.

The work conducted by Davis (2010) and by Whitney (2001) offers valuable insights into the flexibility and dynamic nature of orthographic encoding, highlighting the importance of letter arrangements and their relative configuration. Results gleaned from diverse masked priming experiments provide relevant findings related to existing orthographic coding models' efficiency, particularly addressing how the position of letters within words gets processed.

Lupker et al. (2019) assessed multiple orthographic coding models, placing particular focus on the potential impact of letter position in the coding process. Many current models assume that the initial and sometimes final letters of a word acquire special status owing to the inherent characteristics of the process. Indeed, literature indicates that initial letters occupy a privileged position in reading (Dehaene, 2007). Fundamentally, authors questioned whether models accurately depict the notion that some portion of this privileged status derives from the orthographic coding process itself. Their research reveals that prime influences affecting the final letters of words exhibit much stronger effectiveness than those targeting initial or middle letters, contradicting classical models like the SERIOL model (Whitney, 2001) and the spatial coding model (Davis, 2010). Models favoring end letters by assigning them a distinguished status in the word recognition process seem deficient in explaining observed priming patterns, wherein terminal letters demonstrate a more prominent role than anticipated.

Investigating open-bigram models, Lupker et al. (2019) employed Grainger's model (Grainger et al., 2006), implemented computationally and simulated utilizing software called easyNet. Serving as a base for the elaborated model

(Schoonbaert & Grainger, 1989), it shares considerable similarities with the initial model developed by Grainger et al. (2006). Findings indicate no apparent advantage for primes changing initial letters, challenging the long-held belief that a word's opening letter plays a critical role in its recognition. Data implies that orthographic coding may prove less sensitive to specific letter placements than traditional models predicted. Notably, results suggest the need for a more adaptable and integrative modeling strategy, indicating that letter processing within a word is likely governed by intricate, non-deterministic mechanisms beyond conventional model assumptions. Hence, the implication of these findings points towards reduced emphasis on edge-located letters in orthographic processes.

In their investigation of letter position's influence on perceived word familiarity, Huebert and Cleary (2022) build on earlier research to establish a solid theoretical framework. The global recognition model (Clark & Gronlund, 1996), frequently cited, claims familiarity develops based on correspondence between memorized item features and the current stimulus. With increased overlap between saved traits and testing signals, the intensity of perceived familiarity grows accordingly. Huebert and Cleary (2022) use this concept to hypothesize that letter position in a word might adjust the strength of the familiarity signal. Building on the RWI paradigm (Cleary & Greene, 2000; Peynirciolu, 1990), which showed that matching fragmented words corresponding to studied ones feel more familiar than mismatched pieces, suggests that letter position info shapes familiarity sense creation. Ryals and Cleary (2012) strengthens this idea, stating overlapping letters between memory-stored words and test prompts enhance familiarity signal strength. Some letter positions seemingly carry more influence than others, since McKusker et al. (1981; Grainger and van Heuven (2003) suggested outermost letters matter more for word processing tasks, including prompt naming speed and precise letter recognition.

Interpreting the outcomes of their experiments, Huebert and Cleary (2022) describe the finding that exterior letters of words play a more significant role in familiarity perception compared to inner letters. Backed by empirical evidence showing that subjects evaluated words with visible external letters as being more familiar than those exposed with solely internal letters, this observation aligns with prior suggestions made by Dehaene (2007) and Grainger et al. (2006), attributing heightened importance to beginning and ending letters in recognition processes (Dehaene, 2007). Grainger et al. (2006) point out that their findings transcend established word recognition models, typically centered on feature overlap quantity, such as shared letters, without accounting for specific positions (Clark & Gronlund, 1996). Arguing that letter positions should be incorporated into word recognition models, Huebert and Cleary (2022) raise the possibility that outer letters might experience lower lateral interference than inner counterparts. Based on the notion that fewer neighboring letters enable more effective processing of exterior letters by the brain, they suggest that the latter face greater interference due to encirclement by additional letters on both sides, impacting cognitive processing and contributing to elevated feelings of familiarity in word recognition. Differentiating processing of outside versus interior letters potentially explains why first and last letters of a word wield greater influence in detecting word familiarity, corroborating experiment results provided by Huebert and Cleary (2022).

Osth and Zhang (2024) investigate how different orthographic representations of words influence recognition processes. They begin their analysis by looking into "slot codes", assigning every letter a fixed absolute position, effectively evaluating recognition errors through comparing letter positions in learned vs. recognized words. Next, their investigation extends to "both-edge slot codes", emphasizing boundary letters' roles, acknowledging their potentially key function in perceiving and remembering word structures. Complementarily, Osth and Zhang (2024) incorporate "closed and opened bigrams", examining letter pair interactions, regardless of adjacency or separation by other letters, seeking to understand their impact on word recognition. Particularly interesting, opened bigrams offer deeper insights into the word's inner dynamics, measuring relative closeness between letters that can go beyond restrictive linear position constraints. Analyzing these bigrams clarifies mechanisms behind how a word's internal structure facilitates or hinders recognition processes depending on familiarity and letter pair configurations.

Connecting theories to empirical data, Osth and Zhang (2024) use the Linear Ballistic Accumulator (LBA) model, connecting orthographic representations to concrete measures such as response times and answer precision rates. Relative position models, like closed and opened bigrams, generally perform better than absolute position models in several datasets (Osth & Zhang, 2024). This observation stresses the importance of letter proximity in word recognition, with bigrams excelling at predicting typical recognition errors tied to delicate orthographical similarities.

Demonstrating significant influence in recognition decisions, initial letters weigh heavily due to enhanced visibility and their role in forming initial cognitive impressions of words. Research suggests recognition mistakes frequently stem from confusion involving similarly initial-lettered words, pointing to pronounced positional effects (Osth & Zhang, 2024). Supporting this revelation, the Linear Ballistic Accumulator (LBA) model illustrates the relationship between the orthographic shape of words and recognition performance, offering quantifiable precision for studying initial letter impacts.

After exploring various orthographic coding models and visual perception, attention shifted toward the positional aspect of initial letters in word recognition. Nevertheless, it's equally imperative to account for the phonological impact of these letters. Often intertwined with positional features, phonological dimensionality proves decisive in our cognitive processing of words, especially in how we identify and interpret metathesis instances in our language.

While infrequently addressed in literature, the intersection of visual word recognition and phonological activation has been explored with varied and occasionally conflicting results. Berent and Perfetti (1995) and Lukatela and Turvey (2000) highlight that phonology becomes activated early in the word recognition process, playing a role in accessing word meanings. Evidence shows that phonological codes get automatically engaged during reading, irrespective of explicit requests for phonological processing. Further supporting this viewpoint, Perfetti argues that phonological mediation serves as an important component for gaining access to lexical meanings. Yet, alternative research suggests that phonological mediation might apply narrowly to low-frequency words or depend on specific task requirements (Jared et al.,1990; Pugh et al., 1994; Seidenberg et al., 1984).

McQuade (1981) and Pugh et al. (1994) noted that under experimental conditions discouraging phonological usage, its role in word recognition becomes negligible or nonexistent. Van Orden (1987), however, discovered heightened confusion with homophones, e.g., "maid" vs. "made," attributed to competitive orthography triggering feedback phonology and increasing decision-making latencies. Reinforced by Pexman et al. (2002), homophone latencies exceeded those of non-homophones, ascribing the difference to competition-induced retrograde phonological feedback.

The proposition that phonological feedback affects visual word recognition has been integrated into various theoretical processing models. Plaut's Parallel Distributed Processing (PDP) model (Plaut et al., 1996) assumes that word recognition involves reciprocal links between orthographic, phonological, and semantic representations, fostering dynamic interactions within these systems. Here, phonological feedback amplifies compatible orthographic representations associated with the activated sounds. For homophones like "maid", phonetic activation of /meɪd/ leads to simultaneous engagement for "maid" and "made", extending time required to reach a lexical choice, as evidenced by Pexman et al. (2002). Meanwhile, contrastingly, the dual-route cascaded model (Coltheart et al., 2001) proposes forward and backward connections comparable to PDP.

Studies (Stone et al., 1997, for an example), emphasize the importance of phonological consistency, revealing that words with incongruent orthographic-phonological matches ("pint") yield longer reaction times than those with congruent associations ("mint"). Such findings reinforce the notion that phonological feedback performs a vital role, particularly for words exhibiting orthographic inconsistencies.

Pexman et al. (2002) examined phonological feedback effects on word recognition via Lexical Decision Tasks (LDT) and Phonological Lexical Decision Tasks (PLDT). They analyzed the impact of homophones, homographs, and exception words on response times and error rates in said tasks. Stimuli included homophonic (like "maid"/"made"), non-homophonic choices, controlled regular/irregular exceptional words, and homographs. The study adjusted false targets, i.e., pseudowords and pseudohomophones, to gauge their impact on answers. Outcomes indicated extended response times for homophones compared to non-homophones in LDT tasks, supporting the concept that phonological activation creates orthographic conflict during word recognition. Aligned with the assumption, phonological feedback appears instrumental in visual word recognition, particularly amidst orthographic ambiguity. Conversely, in PLDT tasks requiring participants to decide if pronunciation matched real words, homophone effects disappeared. Still, significant regularity and homograph effects emerged. Indicating that tasks demanding thorough phonological resolution display divergent dynamics compared to those founded chiefly on orthographic criteria. Consequently, Pexman et al. (2002) concluded that differences in phonological feedback processes possibly explain why certain words result in more errors and prolonged response times. Adjusted word-processing models must account for intricate interactions between orthographic and phonological systems to adequately capture word recognition behavior. Discoveries fortify the notion that word recognition tasks aren't merely swayed by lexical representation activation but also influenced by dynamic interplay between phonology and orthography, especially in presence of orthographically ambiguous words such as homophones.

Frost (1998) advocated for a strong phonological theory of visual word recognition, asserting decoding phonologically is paramount to accomplishing the task. He argued transforming orthographic representations to phonological ones forms a fundamental procedure for word recognition, permitting readers to derive word meanings through sounds.

Rubenstein et al. (1971) furnished compelling proof of phonological activation in visual word recognition by demonstrating homophones create confusion, supporting the phenomenon. Since homophones share pronunciation but differ in orthography, response times lag for homophones versus non-homophones.

Pexman et al. (2002) executed series of lexical decision tasks, discovering homophones produced slower response times than non-homophones. Effect proved especially dramatic for low-frequency homophones accompanied by high-frequency equivalents. Furthermore, phonologically alike pseudo-words (pseudo-homophones) escalated challenge in lexical decision tasks, bringing lengthier reactions and magnified effects.

Homographs, having multiple valid pronunciations (e.g., WIND), served as tools in phonology studies too. Research revealed homographs sparked competition between disparate phonological representations, delaying word recognition. Kawamoto and Zemblidge (1992) reported delayed naming latencies for homographs vs. controls, pinpointing the conflict between two homographic pronunciations as reason. Likewise, Gottlob et al. (1999) documented homograph effects in a naming task, hinting at necessity of inhibiting one phonological representation before articulating the homograph.

Though inconsistent, homograph effects were absent in lexical decision tasks, according to Seidenberg et al. (1984). Highlighting how task nature dictates phonological effects manifestation, they speculated lexical decision tasks rely mostly on orthographic details, obviating phonological activation needs.

Regularity, referring to alignment between written word and pronunciation, is another metric used to probe phonological processing. Baron and Strawson (1976) identified regularity effects in naming tasks, revealing irregular words took longer to pronounce correctly.

Contrasting with naming tasks, regularity effects in lexical decision tasks haven't shown reliability. Neither Coltheart et al. (1979) nor Jared et al. (1990) detected regularity effects in separate lexical decision experiments.

Speculations include optional phonological recoding and strategic adoption of visual approaches, whilst noting frequency matters: Jared discovered low-frequency word consistency affected naming tasks, though this didn't surface in lexical decision tasks.

Reading proficiency markedly affects phonological processing: phonological effects emerge more frequently in struggling readers, whose word recognition takes longer, giving room for phonological activation (Waters & Seidenberg, Waters and Seidenberg [1985]). Comparatively analyzing regularity effects in a lexical decision task between third-, fifth-graders, and college students, Waters et al. discerned larger effects in younger, less skilled learners.

Stanovich and West (1989) crafted assessment methods like the "Author Recognition Test" for reading competence evaluation, showing advanced orthographic and phonological skills in experienced readers correlate with superior word recognition task execution. Competent reader scores surge in lexical decision and naming tasks, signifying fluent information processing abilities in both domains.

Seidenberg and McClelland (1989) posited that lesser-skilled readers harbor less effective grapheme-phoneme association knowledge, causing faulty phonological activation and slowing responses. Advanced readers allegedly activate reliable phonology efficiently with limited incorrect activation, reducing regularity effects, enhancing word recognition task performance. Strain and Herdman (1999) suggested that poorer readers grapple with inferior orthographic-phonological mapping, boosting probability of competing phonological activation, perhaps rationalizing heightened regularity effects seen in less-competent readers. Ergo, regularity, consistency, homophone, and homograph effects oscillate contingent upon literacy skill level.

Unsworth and Pexman (2003), in a first experiment checked how reading aptitude affects phonological effects in lexical decision tasks. Formulating a hypothesis saying capable readers handle phonology better and suffer less from phonological interference errors, results verified expectations. Less skilled readers displayed longer response times and higher error rate for homophones, denoting struggle managing orthographic-phonological representation competition. Skilled readers tackled this issue efficiently, performing faster and accurately. Besides, regularity effects surfaced, showing less able readers struggled with irregular words, experiencing slower reaction times and higher error rates. Confirming capability to tackle orthographic irregularities more effectively, skilled readers did not reveal this effect. Their second experiment attempted to analyze how reading abilities impact phonological effects in tasks asking participants to judge if a string of letters "sounds like a word". Significant differences appeared between groups, although differently contrasted with standard lexical decision tasks. Once again, less skilled readers exhibited longer response times and higher error rates for irregular words, reflecting persistent difficulties dealing with orthographic irregularities in a phonological setting. More capable readers, however, demonstrated improved efficiency in phonological processing, showing no significant distinction between regular and irregular words. Homophones had nonsignificant effects in this task, implying focus on word sonority reduces noticed phonological competition. Interestingly, highly skilled readers showed longer response times for homographs, suggesting raised phonological competition due to activation of multiple possible pronunciations, whereas less skilled readers failed to showcase this effect. Results imply reading abilities drive not only speed and accuracy of phonological processing, but also individuals' management strategies for phonological ambiguities in word recognition contexts.

The phonological and positional aspects' interaction require extensive investigation to untangle their respective contributions to word recognition. Consequently, the methodology outlined below precisely aims to segregate these effects, ultimately improving our grasp of their individual and collective influences.

3 Materials and Methods

3.1 Participants

For this study, a sample of 79 french participants (n = 79) was recruited in order to ensure sufficient statistical power for a robust and significant analysis. Participants were selected according to strict age and education criteria.

The sample was drawn from participants aged between 18 and 63, with an average age of 32.7. The breakdown by gender was 54 women, 24 men and 1 participant who did not specify his gender. In terms of level of education, the sample was distributed as follows: 47 participants have a master's degree, 14 have a bachelor's degree, 5 have a BTS (Associate's degree), 1 participant has a doctorate, and 12 participants are classified in the 'other diplomas' category.

Participants were recruited via the Survey Circle platform as well as through advertisements on various social networks. In order to guarantee the homogeneity of the sample and to avoid any source of variation that might bias the results, strict measures were taken to maintain the consistency of the sample.

Selection criteria and recruitment methods were meticulously designed to ensure adequate representativeness and statistical validity, enabling reliable and generalisable conclusions to be drawn from the data collected in this study.

A total of 124 participants were initially recruited for this study. However, only 79 participants were retained for the final analyses, as they met all the established participation criteria. Exclusion criteria included extremely long response times (>50000 ms, compared to the average of about 7000 ms), indicating possible distraction or difficulty in understanding the tasks, as well as systematic errors in applying the instructions. In particular, some participants were excluded because they did not correctly rewrite the altered words, opting instead to rewrite the words with the metaplasms instead of rewriting them correctly as required by the experimental instructions.

3.2 Materials

In this study, participants were subjected to a word recognition task involving attentive observation of visual stimuli presented for a brief duration of 200 ms.

The experiment was performed using SoSci Survey software, during which participants were presented with a set of 10 words per condition. The software was configured using a script specifically programmed to randomly generate a list of words for each participant, thereby minimising bias and ensuring greater reliability of the results.

In this study, the words presented to the participants were taken from a list of 40 words in the French language, selected from the Lexique3 database. The aim of this selection was to ensure that the words were different, recognisable and eligible for manipulation of the independent variable, i.e. the placement of the initial letter or its phonological aspect. The words without metaplasm and the words with metaplasm were presented to the participants in a random order, in order to minimise the potential effect of the order of presentation on the results. Before taking part in the study, participants were informed of the nature of the task and gave their informed consent, in accordance with ethical principles. As part of this in-depth study of word recognition and spelling processing, a set of stimuli was selected comprising words without metaplasm, serving as a control condition, and words with mixed letters, divided into three distinct experimental conditions. The first experimental condition involved the manipulation of the first letter of the word, which changed position or was swapped with another letter, while preserving the sound of the word. This manipulation was used to assess the impact of the placement of the initial letter on word recognition and comprehension, independently of word pronunciation. In the second experimental condition, the position of the initial letter was maintained, but the sound of the word was modified. The aim of this manipulation was to study the influence of sonority on word recognition and comprehension, independently of the position of the initial letter. Finally, the third experimental condition combined the two previous manipulations, changing both the position of the initial letter and the sound of the word. This approach made it possible to evaluate the interaction between the placement of the initial letter and the sound in the process of recognising and understanding words. To ensure comparable response times across conditions and to control for potential effects of word length on response times, words were 1, 2, 3, 4 or 5 syllables long (in each condition). The words were displayed in bold type, using Time New Roman font size 12, on a uniform white background, allowing optimal legibility for all participants (See Table 1 for examples of the 4 conditions).

3.3   Procedure

As part of this study, the experiment was carried out over the internet, using the Sosci Survey software. The study was conducted over a two-month period, from mid-February to mid-April, enabling the data required for in-depth statistical analysis to be collected. Participants were subjected to a word recognition task involving careful observation of visual stimuli presented over a short, precise time interval. The stimulus presentation time was set at 200 ms, a duration supported by the literature as being optimal for encouraging rapid mental recog-

Table 1. Examples for the different experimental conditions.

Condition	Natural words	Displayed words
Control (CC)	Ordinateur	Ordinateur
Displacement (CDep)	Hélicoptére	Éhlicoptére
Sound change (CSon)	Université	Uinversité
Sound change and displacement (CDS)	Balle	Ablle

nition rather than careful analysis of each element. Previous studies have shown that 200 ms is the average time it takes to recognise and read a word.

Prior to the initiation of the main experimental task, participants were engaged in a preparatory exercise designed to assess their understanding of the instructions provided. This exercise involved the ephemeral presentation of five metaplasm-free words on a screen, which participants were then required to reproduce orthographically with accuracy. Performance in this preliminary exercise was used to measure participants' competence in the lexical recognition task. Individuals who failed to meet the performance criteria set for this exercise were eliminated from the sample for subsequent phases of the study. Participants were required to initiate the presentation of stimuli by pressing a button. Following each series of words displayed, a recall exercise was proposed, during which participants had to identify previously seen words from an extended list of 40 distinct words. At the end of each series, an empty rectangle appeared on the screen, prompting participants to recall the last word they had seen by typing it into a keyboard. The transition to the subsequent word was made by participants pressing the 'next' key.

3.4   Data Analysis

As a preamble to the results, it is pertinent to mention that the present study adopts a cross-factorial design methodology, $\underline{S}79 * \underline{M}10 < Q4 >$. This design involves the participation of 79 subjects distributed over four distinct experimental conditions (Condition CC, CDep, CSon, CDS). Each participant is exposed to all the conditions.

In this study, we explored the effects of moving the first letter and changing the sound on word recognition. To do this, we used stimuli comprising words without metaplasms, serving as a control condition, and words with mixed letters, divided into three distinct experimental conditions: moving the first letter, changing the sound, and combining the two manipulations.

The data from this study was analysed using a combination of mixed logistic regression models and mixed linear models. These models are particularly suitable for assessing the effects of positional and phonological alterations on word recognition and response times, while taking into account inter-individual variability. The data were analysed in two main parts: correct responses and response times.

To analyse the correct answers, mixed logistic regression models were used. These models are appropriate for binary variables, as is the case here, where responses can be correct or incorrect. They model the probability of correct word recognition as a function of the independent variables (movement and sound), while including random effects to account for variations between participants. A reduced model was first used, including only a fixed intercept and a random effect for each participant, serving as a point of comparison to assess the impact of the other variables. Next, a model including the displacement of the initial letter was used to examine its effect on the probability of correct word recognition, taking into account inter-individual variability. Similarly, a model with sonority was evaluated to determine the effect of phonological modification of the initial letter on word recognition. Subsequently, a model combining the independent effects of displacement and sonority was used to estimate their combined impact. Finally, a model examining the interaction between displacement and sonority was used to understand how the effect of one might depend on the other, providing a more complex and nuanced view of the dynamics involved.

Linear mixed models were used to analyse response times. These models are appropriate for continuous variables and allow response times to be modelled as a function of positional and phonological alterations, while taking into account the variability between participants. A reduced model was used as a reference, including only a fixed intercept and a random effect for each participant. Next, a model including displacement was evaluated to quantify its impact on response times. Similarly, a model with sound was used to determine the effect of phonological modification on response times. A model combining the independent effects of displacement and loudness was then used to estimate their combined impact. Finally, a model examining the interaction between displacement and loudness was used to understand how these two factors interact to influence response times.

Mixed logistic regression models are particularly well-suited to analysing correct responses because they allow a binary dependent variable to be modelled, taking into account random effects due to individual differences between participants. This makes it possible to estimate the probability of correct word recognition as a function of positional and phonological alterations, while controlling for inter-individual variations. Linear mixed models, on the other hand, are appropriate for analysing response times, a continuous variable. By including random effects for participants, these models capture inter-individual variability and provide more robust estimates of the effects of impairments on response times.

To determine the best model, we used the Akaike Information Criterion (AIC) and the Bayesian Information Criterion (BIC). These indicators will take the lowest value for the best model. Bayes factor approximation using the BIC. This method is equivalent to setting a uniform priority and uses the following formula:

$$BF_{10} = e^{\frac{BIC_0 - BIC_1}{2}} \tag{1}$$

4 Results

Comparative analysis of the models (see Table 2) revealed that the displacement*sound model had the lowest AIC (963.823) and BIC (982.602) values, and a Bayes Factor of 4.5×10^{148}.

Table 2. Comparison of the models predicting a successful recognition of the words.

model	AIC	BIC	Bayes Factor
reduced	1660	1667.198	
displacement	1218.099	1229.366	1.18×10^{95}
sound	1647.461	1658.728	6.90×10^{1}
displacement+sound	1203.104	1218.127	3.26×10^{97}
displacement*sound	963.823	982.602	4.54×10^{148}

This model contains the following equation:

$$P_s = logit(6.1671 - 5.8916 \times D - 4.2744 \times S + 4.7967 \times S \times D) \qquad (2)$$

where P_s corresponds to the probability of success, D corresponds to the presence of a letter shift and S corresponds to a change in sound.

The results show that although the manipulations of moving the first letter and modifying the sound have a significant impact on word recognition, the model combining the interaction of these two manipulations (displacement*sound) performs best in terms of goodness of fit and statistical preference. The model combining the interaction of these two manipulations (displacement*sound) performed best in terms of goodness of fit and statistical preference. This model explains variations in the data better and shows pronounced effects on success probabilities, underlining the importance of combined manipulations in influencing word recognition and comprehension.

Figure 1 illustrates in detail the effects of displacement+sound and displacement*sound models on the probability of success of participants in a word recognition task.

When no initial letter is moved (Displacement = 0), participants' probability of success is remarkably high, reaching almost 98% for the control condition. This indicates that participants almost always manage to recognise words correctly when there are no metaplames in the word. It is important to note that the control condition has a slightly higher probability of success than that observed with the phonological modification (Sound = 1) without displacement, although this difference is marginal, rising from a probability of 100 to 87%.

Displacement of the initial letter (Displacement = 1) led to a significant reduction in the probability of success regardless of the phonological condition. For the displacement condition (Displacement = 1) without phonological modification (Sound = 0), the average probability of success fell drastically from around

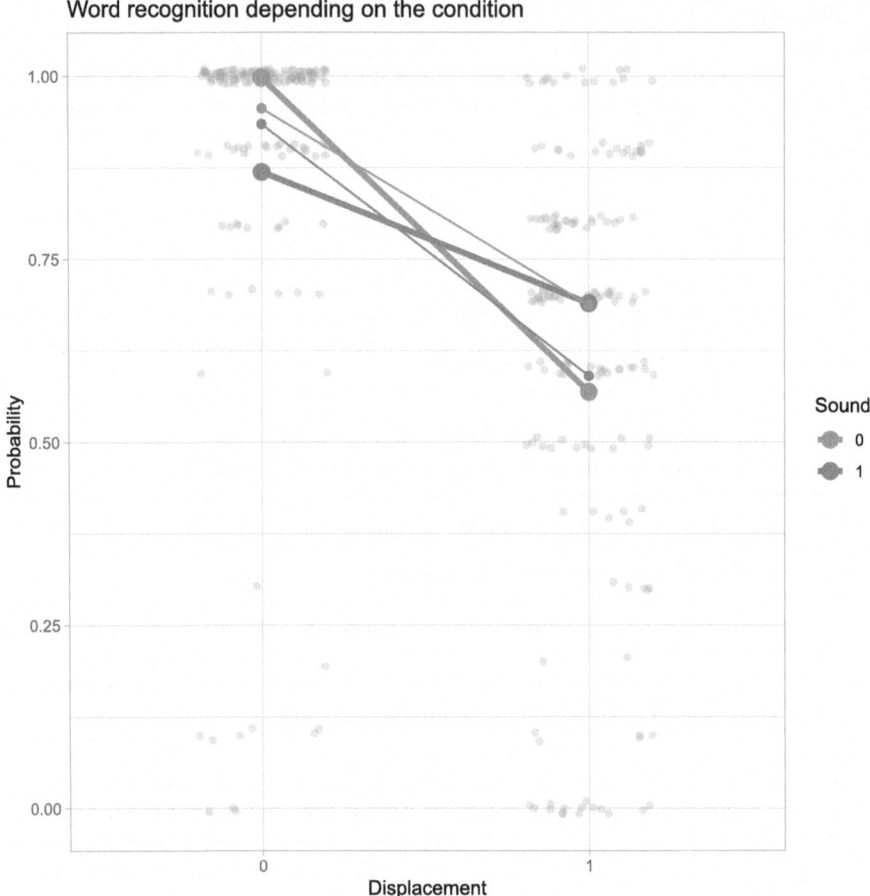

Fig. 1. Probability of succes of the word recognition depending on the position and sound change. The data is visualised using two sound conditions: red for the absence of phonological modification (Sound = 0) and blue for the presence of phonological modification (Sound = 1). Each point represents the frequency of correct response for a participant. The thick lines indicate the predictions of the best model (displacement*sound) and the thin lines indicate the predictions of the displacement+sound model. (Color figure online)

98% to 57%. Similarly, for the condition with phonological modification (Sound = 1) combined with displacement, the average probability of success decreases from 87% to 67%. This decrease highlights the importance of the position of the initial letter in word recognition. The cognitive system of the participants seems to be strongly disrupted by this displacement, which affects their ability to correctly identify the words presented.

Analysis of the data reveals that the phonological modification (Sound) has a less pronounced effect on the probability of success compared with moving the initial letter. When the initial letter is not moved, the differences in probability of success between the two sound conditions are minimal, indicating that participants can manage phonological modifications effectively as long as the initial letter remains intact. In contrast, when the initial letter is moved, although the probability of success decreases significantly for both conditions, the condition with phonological modification (Sound = 1) appears slightly more resilient. The strong dependence on the position of the initial letter suggests that this position plays a crucial role in the activation of lexical representations in the brain. The lower resilience observed with phonological modification indicates that, although phonology is also an important factor, it does not have as much impact as initial spelling.

The results obtained for each condition were analysed in terms of the success rate (correct) over 10 words. For the control condition, the mean number of correctly recalled words was 9.899 with a median of 10, indicating that the participants recalled almost all the words correctly. On the other hand, for the condition in which the first letter was moved, the average number of correctly recalled words fell to 5.785 with a median of 7. Changing the sound had a less drastic effect, with an average number of correctly recalled words of 8.013 and a median of 9. Finally, the combination of the two manipulations led to an average number of correctly recalled words of 6,646 with a median of 8, indicating a more negative impact than changing the sound alone, but less severe than moving the first letter.

The aim of this study was also to compare the response time performance of different language processing models applied to phonetic and orthographic transformations. The transformations evaluated included the displacement of the first letter (displacement), the change in sonority (sound change), and the displacement of the first letter along with the change in sonority independently (displacement+sound). We could not evaluate interaction effects due to a technical problem in the recording of the response times in the control condition. The comparison of these models is available in Table 3.

Table 3. Comparison of the models for response times.

model	AIC	BIC	Bayes Factor
reduced	4332.156	4342.561	1
displacement	4312.774	4326.646	2856.474
sound change	4322.677	4336.550	20.196
displacement+sound	4313.013	4330.354	447.415

Comparison of the reduced model with the displacement model revealed a significant improvement in the latter. The AIC and BIC values for the displacement model were 4312.774 and 4326.646 respectively, compared with 4332.156 and 4342.561 for the reduced model. The Bayes factor, which was extremely high at 2856.474, and the p-value confirmed the superiority of the displacement model over the reduced model. The predicted differences in processing time for this model range from 108.623 ms to 874.024 ms, with a median of 384.395 ms. The interquartile ranges (25%-75%) ranged from 294.060 ms to 598.252 ms, indicating a relatively wide distribution of predicted processing times.

The comparison between the displacement model and the displacement+sound model shows a slight superiority of the displacement model. The AIC and BIC values for the displacement model are 4312.774 and 4326.646 respectively, slightly lower than those for the displacement+sound model.

Analysis of the different phonetic and orthographic transformation models shows that the displacement model alone offers the best compromise between simplicity and performance. This model has the best statistical fits, with the lowest AIC and BIC values, an extremely high Bayes factor, and predicted differences in processing time that are generally more homogeneous and concentrated around intermediate values. The combined models, although effective, show greater variability in processing times and do not provide significant additional benefits compared with the displacement model alone. The model predictions for time are available in the Table 4.

Table 4. Predictions of the model for response times.

Displacement	Prediction	Standard Error
without	6674.75	353.02
with	7657.40	333.17

5   Discussion

The results of this study reveal crucial elements regarding the mechanisms behind word recognition, particularly concerning the influence of initial letters and their phonological properties. The primary aim of this research was to determine if word recognition is mainly affected by the position of initial letters (positional effect) or by their phonological qualities (phonological effect). Operational hypotheses suggested that both positional and phonological modifications to the first letter of a word would have a substantial impact and that combining them would produce even more noticeable effects on word recognition.

The results show that word recognition errors were more frequent when the position of the first letter was changed or exchanged, or when the sound of the first letter was changed compared with the control condition. These observations suggest the existence of both a phonological effect and a positional effect. More specifically, moving the initial letter led to a significant reduction in the probability of success, highlighting the importance of the position of the initial letter in word recognition. The results confirm that moving the initial letter has a more significant impact than changing the sound of that letter. This observation is in line with the work of Davis (2010), whose spatial coding model proposes tolerance for minor errors in letter position. Whitney (2001), with his SERIOL model, also highlights the importance of consecutive bigrams and sequential letter activation for word recognition. Similarly, Dehaene (2007) highlights the importance of bigrams and letter position in word recognition, suggesting that initial and final letters play a crucial role in this process.

Although the phonological modification of the first letter has a less pronounced effect than the displacement of the initial letter, its impact is still significant. The work of Frost (1998) and of Pexman et al. (2002) supports the idea that phonology plays a central role in the recognition of visual words. Indeed phonology is activated early in the word recognition process and plays a role in accessing the meaning of words (Berent & Perfetti, 1995; Lukatela & Turvey, 2000).

The less marked effect of the phonological modification compared with the positional effect could be explained by the intrinsically visual nature of the word recognition task. While phonology contributes to lexical access, visual cues related to letter position seem to play a more direct and immediate role in initial word recognition.

The results of this study are consistent with those of Lupker et al. (2019), who demonstrated that primers modifying initial letters were significantly less effective than those modifying final letters. Furthermore, Huebert and Cleary (2022) showed that the outer letters of words play a more significant role in the perception of familiarity than the inner letters, thus supporting our observations on the positional effect.

On the other hand, the literature presents varied results regarding the relative importance of phonological effects. For example, Jared et al. (1990) and Pugh et al. (1994) suggest that phonological mediation may be limited to low-frequency words or dependent on specific task requirements. These variations could explain why the phonological effect observed in our study is less pronounced than the positional effect.

The theory of 1000 brains (Hawkins, 2021) offers an enriching perspective for understanding the results of this study. This theory postulates that the human brain is made up of thousands of cortical modules, each functioning autonomously but interconnected to process sensory information, generate predictions and coordinate motor responses. Each module acts like a mini-brain, capable of analysing and integrating data at different levels of complexity. Applied to word recognition, this theory suggests that these cortical modules work in parallel to break down and reconstruct orthographic and phonological information. When initial letters are modified, either by displacement or phonological change, each module must reorganise the information to formulate a coherent recognition of the word. The recognition errors observed in this study could therefore reflect the limitations of these modules in adapting quickly and efficiently to unexpected changes in linguistic stimuli.

This theory may also explain why the combination of positional and phonological alterations leads to a higher number of errors. When both types of alteration are present, the cortical modules have to manage an increased cognitive load, reorganising both visual and phonological cues to achieve correct word recognition. This process of simultaneous reorganisation can exceed the adaptive capacities of the modules, leading to an increase in errors.

Integrating Hawkins (2021)'s 1000-brain theory into the analysis of word recognition provides a robust theoretical framework for understanding the complex interactions between visual and phonological cognitive processes. It highlights the importance of brain plasticity and prediction mechanisms in the processing of linguistic information.

This study presents several limitations that deserve to be discussed for a more complete understanding of the results obtained and avenues of improvement for future research. Firstly, it is important to point out that we did not measure the response time for the control condition due to a malfunction in the code used in the Sosci software. Our initial objective was to focus our analysis on the experimental conditions involving positional and phonological alterations. However, the absence of this time measurement for the control condition is a methodological limitation. A comparison of response times with the control condition could provide additional information on the relative impact of the different experimental manipulations.

In addition, some words selected from the experimental list showed close phonological and orthographic similarities when the place of the initial letter was swapped. These similarities may have introduced biases into word recognition, thereby increasing the error rate. For example, transformed words such as 'feu' (fire) could sometimes be similar to other existing words in the French language such as 'pneu' (tire), which could create additional ambiguities for the participants.

We hypothesize that the results would exhibit a similar pattern across various languages. Indeed, most of the litterature has focused on English. In our case, we focused on the French language and indeed found positional and phonological effects. It is likely that the two pathways would be generic word recognition modules that would be used regardless of the language. Thus the effects we have shown should be noticeable in other languages as well, as long as the same sounds can be spelled in at least two different ways, and that the same letter could sound differently. For example, Italian being a transparent language, phonemes are consistently associated to specific graphemes (Neef & Balestra, 2011). Distinguishing the two effects in Italian thus might not be as clear as in French or in English.

These limitations highlight several methodological aspects that could be improved in future research. By incorporating response time measures for the control condition, selecting words with fewer phonological and orthographic similarities when manipulated, it would be possible to reduce bias and obtain more robust and generalizable results.

It would also be relevant to explore the effect of stimulus presentation time. In this study, we used a presentation time of 200 ms, in line with the existing literature. However, reducing this presentation time could increase the difficulty of the task and reveal different dynamics in word recognition. Future analysis could compare the performance of participants with presentation times shorter than 200 ms to determine whether shorter exposure modifies the positional and phonological effects observed.

Furthermore, applying these results to the development of language processing technologies, such as chatbots and voice recognition systems, could improve their ability to understand and process human language, even in contexts where words are altered or presented in unconventional ways. The creation of algorithms capable of better managing orthographic and phonological variations could improve the robustness and accuracy of these systems. In addition, future studies could use neuroimaging methods, such as MRI or EEG, to observe the cognitive systems of positional and phonological effects. These approaches would provide a better understanding of the cognitive and cerebral mechanisms involved in word recognition, thus providing complementary data to behavioural analyses.

6   Conclusion

This study aimed to understand the mechanisms of word recognition, focusing on the influence of initial letters and their phonological characteristics. The main objective was to determine whether word recognition is primarily influenced by the position of initial letters (positional effect) or by their phonological features (phonological effect). The results showed that recognition errors were more frequent when the position or sound of the first letter was modified. Modifying the initial letter had a significant impact on the probability of success, highlighting the importance of the position of initial letters in word recognition. These findings align with the litterature (Davis, 2010; Whitney, 2001; Dehaene, 2007) which emphasizes the role of bigrams and letter positions. Although the phonological modification had less pronounced effects, it remained notable, supporting theories proposed by Frost and Pexman. The "Thousand Brains" theory (Hawkins, 2021) offers an interesting viewpoint, suggesting that recognition errors may reflect the limitations of cortical modules adapting quickly to unexpected alterations. This study provides several avenues for future research. Applying these findings in developing language processing technology, such as chatbots, could improve their ability to handle unexpected orthographic and phonetic variations. In summary, this study made significant contributions to understanding word recognition mechanisms, revealing the critical role played by initial letters and their phonological attributes. Results underscore how positional and phonological changes differentially affect word recognition, stressing the dominance of the positional effect. Discoveries enhance our comprehension of underlying cognitive processes and open new directions for further exploration into the interaction between initial position and phonology within word recognition.

Acknowledgments. The content of the paper has been translated from French using Mixtral AI with the following prompt: "Translate the text to scientific english while keeping the same sentence structure" before being corrected by the authors.

Data Availability. The data and code for analyses are available at the following link: https://doi.org/10.17605/OSF.IO/EJVMB.

Disclosure of Interests. The authors have no competing interests to declare that are relevant to the content of this article.

References

Baron, J., Strawson, C.: Use of orthographic and word-specific knowledge in reading words aloud. J. Exp. Psychol. Hum. Percept. Perform. **2**, 386–393 (1976). https://doi.org/10.1037/0096-1523.2.3.386

Berent, I., Perfetti, C.: A rose is a REEZ: the two-cycles model of phonology assembly in reading English. Psychol. Rev. - PSYCHOL REV **102**, 146–184 (1995). https://doi.org/10.1037/0033-295X.102.1.146

Clark, S.E., Gronlund, S.D.: Global matching models of recognition memory: how the models match the data. Psychon. Bull. Rev. **3**(1), 37–60 (1996). https://doi.org/10.3758/BF03210740

Cleary, A.M., Greene, R.L.: Recognition without identification. J. Exp. Psychol. Learn. Mem. Cogn. **26**(4), 1063–1069 (2000). https://doi.org/10.1037//0278-7393.26.4.1063

Coltheart, M., Davelaar, E., Jonasson, J.T., Besner, D.: Phonological encoding in the lexical decision task. Q. J. Exp. Psychol. **31**, 489–507 (1979). https://doi.org/10.1080/14640747908400741

Coltheart, M., Rastle, K., Perry, C., Langdon, R., Ziegler, J.: DRC: a dual route cascaded model of visual word recognition and reading aloud. Psychol. Rev. **108**, 204–256 (2001). https://doi.org/10.1037/0033-295x.108.1.204

Davis, C.: The spatial coding model of visual word identification. Psychol. Rev. **117**, 713–58 (2010). https://doi.org/10.1037/a0019738

Dehaene, S. (2007). Neurones de la lecture (les): La nouvelle science de la lecture et de son apprentissage. Odile jacob

Frost, R.: Toward a strong phonological theory of visual word recognition: true issues and false trails. Psychol. Bull. **123**, 71–99 (1998). https://doi.org/10.1037/0033-2909.123.1.71

Gottlob, L.R., Goldinger, S.D., Stone, G.O., Van Orden, G.C.: Reading homographs: orthographic, phonologic, and semantic dynamics. J. Exp. Psychol. Hum. Percept. Perform. **25**, 561–574 (1999). https://doi.org/10.1037//0096-1523.25.2.561

Grainger, J., Granier, J.-P., Farioli, F., Van Assche, E., van Heuven, W.: Letter position information and printed word perception: The relativeposition priming constraint. J. Exp. Psychol. Hum. Percept. Perform. **32**, 865–84 (2006). https://doi.org/10.1037/0096-1523.32.4.865

Grainger, J., van Heuven, W.J.B.: Modeling letter position coding in printed word perception. In: Bonin, P. (ed.) The mental lexicon, pp. 1–23. Nova Science, New York (2003)

Hawkins, J.: A thousand brains: a new theory of intelligence. Basic Books, New York (2021)

Huebert, A.M., Cleary, A.M.: Do first and last letters carry more weight in the mechanism behind word familiarity? Psychon. Bull. Rev. **1**, 1–8 (2022). https://doi.org/10.3758/s13423-022-02093-1

Jared, D., McRae, K., Seidenberg, M.S.: The basis of consistency effects in word naming. J. Mem. Lang. **29**, 687–715 (1990). https://doi.org/10.1016/0749-596X(90)90044-Z

Kawamoto, A.H., Zemblidge, J.H.: Pronunciation of homographs. J. Mem. Lang. **31**, 349–374 (1992). https://doi.org/10.1016/0749-596X(92)90018-S

Lukatela, G., Turvey, M.T.: An evaluation of the two-cycles model of phonological assembly. J. Mem. Lang. **42**, 183–207 (2000). https://doi.org/10.1006/jmla.1999.2672

Lupker, S., Spinelli, G., Davis, C.: Masked form priming as a function of letter position: an evaluation of current orthographic coding models. J. Exp. Psychol. Learn. Mem. Cogn. **46**, 1–76 (2019). https://doi.org/10.1037/xlm0000799

McKusker, L.X., Gough, P.B., Bias, R.G.: Word recognition inside out and outside. J. Exp. Psychol. Hum. Percept. Perform. **7** , 538–551 (1981). https://doi.org/10.2139/ssrn.2411270

McQuade, D.V.: Variable reliance on phonological information in visual word recognition. Lang. Speech **24**, 99–109 (1981). https://doi.org/10.1177/00238309810240010

Neef, M., Balestra, M.: Measuring graphematic transparency: German and Italian compared. Written Lang. Literacy **14**(1), 109–142 (2011). https://doi.org/10.1075/wll.14.1.06nee

Osth, A.F., Zhang, L.: Integrating word-form representations with global similarity computation in recognition memory. Psychon. Bull. Rev. **31**(3), 1000–1031 (2024). https://doi.org/10.3758/s13423-023-02402-2

Pexman, P., Lupker, S., Reggin, L.: Phonological effects in visual word recognition: investigating the impact of feedback activation. J. Exp. Psychol. Learn. Memory, Cogn. **28**, 572–84 (2002). https://doi.org/10.1037/0278-7393.28.3.572

Peynirciolu, Z.F.: A feeling-of-recognition without identification. J. Mem. Lang. **29**(4), 493–500 (1990). https://doi.org/10.1016/0749-596X(90)90068-B

Plaut, D.C., McClelland, J.L., Seidenberg, M.S., Patterson, K.: Understanding normal and impaired word reading: computational principles in quasi-regular domains. Psychol. Rev. **103**, 56–115 (1996). https://doi.org/10.1037//0033-295X.103.1.56

Pugh, K.R., Rexer, K., Katz, L.: Evidence of flexible coding in visual word recognition. J. Exp. Psychol. Hum. Percept. Perform. **20**, 807–825 (1994). https://doi.org/10.1037//0096-1523.20.4.807

Rubenstein, H., Lewis, S.S., Rubenstein, M.A.: Evidence for phonemic recoding in visual word recognition. J. Verbal Learn. Verbal Behav. **10**, 645–657 (1971). https://doi.org/10.1016/S0022-5371(71)80071-3

Ryals, A., Cleary, A.: The recognition without cued recall phenomenon: support for a feature-matching theory over a partial recollection account. J. Mem. Lang. **66**, 747–762 (2012). https://doi.org/10.1016/j.jml.2012.01.002

Schoonbaert, S., Grainger, J.: Letter position coding in printed word perception: effects of repeated and transposed letters. Lang. Cogn. Process. **19** , 333–367 (2004). https://doi.org/10.1080/01690960344000198

Seidenberg, M.S., McClelland, J.L.: A distributed, developmental model of word recognition and naming. Psychol. Rev. **96**, 523–568 (1989). https://doi.org/10.1037/0033-295x.96.4.523

Seidenberg, M.S., Waters, G.S., Barnes, M.A., Tanenhaus, M.K.: When does irregular spelling or pronunciation influence word recognition? J. Verbal Learn. Verbal Behav. **23**, 383–404 (1984). https://doi.org/10.1016/S0022-5371(84)90270-6

Stanovich, K.E., West, R.F.: Exposure to print and orthographic processing. Read. Res. Q. **24**, 402–433 (1989). https://doi.org/10.2307/747605

Stone, G.O., Vanhoy, M., Van Orden, G.C.: Perception is a two-way street: feedforward and feedback phonology in visual word recognition. J. Mem. Lang. **36**, 337–359 (1997). https://doi.org/10.1006/jmla.1996.2487

Strain, E., Herdman, C.M.: Imageability effects in word naming: an individual differences analysis. Can. J. Exp. Psychol. **53**, 347–359 (1999). https://doi.org/10.1037/h0087322

Unsworth, S., Pexman, P.: The impact of reader skill on phonological processing in visual word recognition. Q. J. Exp. Psychol. A, Hum. Exp. Psychol. **56**, 63–81 (2003). https://doi.org/10.1080/02724980244000206

Van Orden, G.C.: A rows is a rose: spelling, sound, and reading. Memory Cogn. **15**, 181–198 (1987). https://doi.org/10.3758/BF03197716

Waters, G.S., Seidenberg, M.S.: Spelling-sound effects in reading: time-course and decision criteria. Memory Cogn. **13**, 557–572 (1985). https://doi.org/10.3758/BF03198326

Whitney, C.: How the brain encodes the order of letters in a printed word: the seriol model and selective literature review. Psychon. Bull. Rev. **8**, 221–243 (2001). https://doi.org/10.3758/BF03196158

Rationality and Dual Process

Coherence as Logical Consistency

Alberto Mura[(⊠)]

Università degli Studi di Sassari, Sassari, Italy
ammura@uniss.it

Abstract. De Finetti characterizes the violation of coherence as an "intrinsic contradiction." Neither the Dutch Book argument in the betting context nor its counterpart in the theory of proper scoring rules shows this except by analogy. They show that violating the laws of probability leads to objectively non-optimal decisions. My contribution aims to illustrate better de Finetti's view that coherence is the probabilistic counterpart of logical consistency. In the context of betting, this paper will show how violating the laws of probability leads to a truly alethically impossible state. I will introduce the notion of an "indirect bet". An indirect bet on an event E is a simultaneous finite set of bets whose payoff "simulates" a bet on E. An indirect bet is such that its payoffs are the same as a bet on E. If we define subjective probability in terms of indirect bets, it is logically impossible to violate the laws of finite probability. On the other hand, if we define subjective probability in the traditional terms of "direct" bets, the dispositions contained in a finite probability set of events that violate the laws of probability can always be "contradicted," since they are dispositions that indirectly express a different bet quotient for at least one event. Conversely, direct and indirect bets always yield the same payoff if the bets satisfy the laws of probability.

Keywords: Coherence · Consistency · indirect bets · de Finetti · Ramsey

1 Introduction

The laws of finite probability may generalize the standard logic of sentences in several respects. The idea of a "Logic of the Uncertain" underlies the interpretation of the laws of probability in the view of Bruno de Finetti. According to this view [7], also shared by Frank Plumpton Ramsey, the laws of probability generalize the logic of sentences applied to belief because they extend the notion of consistency to *partial* degrees of belief ([11, pp. 82–83]). As Sentence Logic allows one to be the bearer of consistent *full* beliefs about sentences, the basic laws of probability allow one to be the bearer of every set of consistent partial beliefs about sentences. Degrees of belief are characterized in terms of bets (a special case of decisions under uncertainty). De Finetti-Ramsey's approach appeals to the well-known Principle of Maximization of Expected Utility (PMEU). The Dutch Book Argument is invoked to show the "intrinsic contradiction" in degrees

J. Baratgin et al. (Eds.): HAR 2024, LNCS 15504, pp. 167–184, 2025.
https://doi.org/10.1007/978-3-031-84595-6_11

of belief violating the finite probability theory (see, for example, [6, p. 85]). It involves the disposition to accept a simultaneous set of bets, resulting in a sure loss. This disposition may be considered *prima facie* as a set of preferences at odds with the assumption that every single bet is fair, which seems to exclude a total negative (as well positive) balance, which surely cannot be considered fair. However, this kind of "contradiction" is very loosely related to logical inconsistency, and it is highly problematic that, on its base, one may seriously take the violation of the laws of probability as a generalization of logical consistency. Ultimately, the alleged contradiction is nothing more than the violation of PMEU and the principles governing preferences from which it derives. One may make similar considerations about coherence as defined in terms of *proper scoring rules*.[1] Since one cannot equate violation of such principles with logical consistency, ultimately, the view that the logic of the uncertain is an extension of deductive logic would seem to be based on a vague analogy. What follows, I will show that, through another road, it is possible to define logical consistency about belief (here called B-consistency) so that violating the laws of probability carries a genuine inconsistency to vindicate de Finetti's and Ramsey's theses.

2   Plan of the Paper

In Sect. 3, I will first discuss the notion of consistency in belief. How does Sentence Logic apply to belief? This question seems at odds with logic developments by which belief is considered a modal operator to give rise to modal logic with specific axioms, called 'Doxastic Modal Logic' ('DML' after that). If this is the case, the immediate application of Sentence Logic to belief becomes problematic. Secondly, I will try to clarify in what sense, according to de Finetti, Sentence Logic can be considered the "Logic of the Certain" and, according to Ramsey, the "Logic of Full Belief." The notions of "certain belief" and "full belief" seem equivalent and can be characterized through a special expressible condition in the language of DML. Adding such a condition to DML involves the coming down of such Logic to Sentence Logic, in the sense that for every formula φ such that $\models_B B(\varphi)$ there is a formula ψ devoid of modal symbols such that $\models_B (\varphi \leftrightarrow \psi)$ and $\models_S (\psi)$, where $\models_B$ and $\models_S$ means, respectively validity in DML and the Sentence Logic. Second, I will give a general definition of consistency, including complete and partial beliefs. Third, I will introduce the notion of an "indirect bet", and in terms of it I will show that one can define consistency in such a way that it coincides with the satisfaction of the axioms of finite probability theory.

3   Consistency and Belief

3.1   Sentence Logic Applies to Belief?

As mentioned above, de Finetti interprets Sentence Logic as the "Logic of the Certain." There are two meanings of the word 'certain'. On the one hand, certain

---

[1] For the notion of *proper scoring rule* and its use by de Finetti, see [3–6].

is that which has the guarantee of truth and is, as such, *objectively* certain (i.e. epistemically necessary). In a second meaning, 'certain' stands for the maximal degree of subjective belief, that is, what is expressed by the value 1 of subjective probability and is generally without objective guarantee. From a subjectivist point of view, 'certain' must be understood in a subjective sense as an extreme degenerate case of doubt.[2] According to today's views, belief is a modal notion properly elucidated by a system of doxastic logic, equipped with a modal unary operator 'B' meaning belief. The syntactic formation rule is simple: if φ is a well-formed formula, $B\varphi$ is also well-formed. Unfortunately, there is no standard Doxastic Modal Logic to refer to. In any case, I refer generically to a system that contains the analogue of the necessitation rule: if $\models \varphi$ then $\models B\varphi$ and, as a theorem, the simplification of iterated beliefs: $BB\varphi \rightarrow B\varphi$.

What is a "Full Belief"? When de Finetti speaks of the "Logic of Uncertainty" and Ramsey of the "Logic of Full Belief", they mean a logic that also satisfies the principle (typically repudiated by doxastic logicians) $\neg B\varphi \leftrightarrow B\neg\varphi$. One can easily show that satisfying this condition entails coming down to Sentence Logic. One can clarify this fact in the following way. Given a formula φ, we call the *actualization* of φ the formula $\tau(\varphi)$, where every occurrence of the operator B is dropped. Analogously, if M is a set of sentences, $\tau(M)$ is the set of the actualization of the formulas of M. The following result then applies.

Theorem 1. *For every set of sentences M, $M \models \varphi$ iff $\tau(M) \models \tau(\varphi)$.*

Proof. Omitted.

∎

In light of this result, the B operator becomes superfluous. This result explains why doxastic logicians reject the formula $\neg B\varphi \leftrightarrow B\varphi$. They understand doxastic logic as modal logic. If this formula were valid, doxastic logic would be isomorphic to non-modal Sentence Logic. It would be as trivial as modal logic. Moreover, Ramsey's "logic of full belief" and de Finetti's "logic of certainty" are in fact not intended to be modal logics, but only an interpretation of Sentence Logic as applied to belief. The truth tables of Sentence Logic become functions between "full belief values". Thus it becomes clear how Sentence Logic can be regarded as the "Logic of the Certain" as intended by de Finetti, or as the "Logic of Full Belief" as understood by Ramsey.

In Sect. 3, I will first discuss the notion of B-consistency. I will give a general definition of B-consistency, including full and partial beliefs. Moreover, I will introduce the notion of "indirect bet", and in terms of it I will show that consistency can be defined in such a way that it coincides with the satisfaction of the axioms of finite probability theory.

---

[2] For the reasons why, according to de Finetti, Probability 1 cannot be generally equated with objective certainty, see de Finetti [6, p. 116–127].

3.2   Consistency for Full Belief

How do we characterize inconsistency for full belief? Logicians have provided several co-extensive definitions (syntactic and semantic) of inconsistency (and the related notion of consistency) for Sentence Logic. One of them is as follows:

Definition 1. *A set M of sentences is said to be logically inconsistent if there exists a sentence φ such that both the following metalinguistic relations are satisfied:*

- $M \models \varphi$
- $M \models \neg\varphi$

This definition, in the light of Theorem 1, also holds for full beliefs. However, the meaning of '$\models$' is dramatically different. Its meaning is no longer alethic. $M \models \varphi$ in the standard Logic means that if all the elements of M are true, φ is also true. In the Logic of belief, the meaning of $M \models \varphi$ (which I will denote in the following as '$\models_B$') is that if one fully believes that all the elements of M are true, then one must fully believe that φ is also true. Full belief in φ and $\neg\varphi$ is not *impossible* as a psychological state but simply unacceptable. In the presence of bivalence, the contradiction amounts to fully believing simultaneously that φ is true and false. B-consistency is nothing more than assigning a single belief value to each sentence. Ramsey was clear about this twofold aspect of consistency and that Sentence Logic allows for both an alethic and a non-alethic interpretation. He writes:

> If p and $\tilde{q}$ are inconsistent so that q follows logically from p, that p implies q is what is called by Wittgenstein a 'tautology' and can be regarded as a degenerate case of a true proposition not involving the idea of consistency. This enables us to regard (not altogether correctly) formal Logic including mathematics as an objective science consisting of objectively necessary propositions. It thus gives us not merely the ἀνάγκη λέγειν, that if we assert p we are bound in consistency to assert q also, but also the ἀνάγκη εἶναι, that if p is true, so must q be. [11, p. 83]

3.3   Consistency for Partial Belief: Preliminary Considerations

B-consistency in the extreme case of full belief (which isomorphic to Sentence Logic) is reduced to assigning a single credence value to each sentence. In that case, one can easily extend this notion of consistency to partial belief. It is a matter of generalizing the Definition 3.2. However, while the non-alethic notion of logical consequence for full belief (Ramsey's ἀνάγκη λέγειν) is coextensive with the alethic notion of logical consequence (Ramsey's ἀνάγκη εἶναι) so that the symbol '$\models_B$' can represent the logical consequence between full beliefs. There is no corresponding alethic notion of logical consequence that is coextensive with the notion of logical consequence between partial beliefs, and this was crystal clear to Ramsey, who thus continues the preceding passage:

But when we extend formal logic to include partial beliefs this direct objective interpretation is lost; if we believe pq to the extent of $\frac{1}{3}$, and $p\tilde{q}$ to the extent of $\frac{1}{3}$ we are bound in consistency to believe $\tilde{p}$ also to the extent of $\frac{1}{3}$. This is the ἀνάγκη λέγειν; but we cannot say that if pq is $\frac{1}{3}$ true and $p\tilde{q}$ is $\frac{1}{3}$ true, $\tilde{p}$ also must be $\frac{1}{3}$ true, for such a statement would be sheer nonsense. There is no corresponding ἀνάγκη εἶναι. Hence, unlike the calculus of consistent full belief, the calculus of objective partial belief cannot be immediately interpreted as a body of objective tautology. [11, p. 83]

4   Bets and Inconsistency

Both Ramsey and de Finetti saw decisions under uncertainty as a way to clarify the notion of partial belief. Although Ramsey's theory is more general than de Finetti's, both use the notion of *bet*. Moreover, both rely on the now famous *Dutch Book Argument* for partial belief and for explaining the B-inconsistency, which involves violating the laws of finite probability.[3] The connection between bets and partial degrees of belief has been known since the origins of the calculus of probability and is particularly featured in Thomas Bayes' essay [1]. One may characterize bets by an ideal currency so that the utility value of monetary amounts is linear. In what follows, instead of the phrase 'monetary amount', I shall use the simple term 'amount', assuming that it satisfies linearity.[4] As has been known since the work by Daniell Bernoulli in 1738 [2], this is not the case for real currencies (such as the euro), so the value of 2 euros is not necessarily twice the value of one euro. One can assume that the currency values involved are small enough to make this problem negligible. Alternatively, with an effort at idealization, one can imagine bets made with utility values. This second solution is the one I will adopt. It is not problematic since (unlike de Finetti) we are not interested, in this context, in solving the concrete problem of *measuring* degrees of belief, but rather to *elucidating* the notion of partial degree of belief. Recourse to idealizations does not constitute a relevant difficulty for this goal.

4.1   Bets

The following discussion and definitions will refer to events rather than sentences. Nothing substantial would change by referring to sentences. Although the notion of betting may be well known, it is necessary here to give an exact definition of it in order to be able later to define, in terms of it, other notions. In what follows, I will use the word 'event' to conform to established terminology. In relation to Sentence Logic, an event is the equivalence class of those sentences that are logically equivalent to a given sentence. The logical equivalence relation, the identity of truth conditions, is an equivalence relation (reflexive, symmetrical

---

[3] For a suggestion on the origin of this locution, see [8, p. 45].

[4] In the Ramsey [11] approach, as well as in the Savages' one [12], utility is linear by the way it is defined in terms of preferences. In light of this approach (endorsed by de Finetti), we may consider bets whose payoffs are made in utility values.

and transitive) that induces a partition in the sentences (whose elements generate a so-called Lindenbaum Boolean algebra). The sentences for each element of such a partition are, in each equivalence class, logically equivalent, and each represents the same event.

Some terminology is needed, which I will use in the following definitions.

Definition 2. *Let E be an event and v be a given amount such that $v > 0$. The exchange $\mathcal{B}$ of an amount v' such that $v' \geq 0$ by a person $\mathbf{X}$ (called the* bettor*) with an offer O that entitles the person to have the amount v such that $v \geq v'$ under the condition that E occurs (null otherwise) is called a* bet *about E. The bet is said to be* on E *if $\mathbf{X}$ pays v' to buy O and is said to be* against E *if $\mathbf{X}$ sells O in exchange for the sum v'. The positive amount v is called the* stake *of the bet.*

Definition 3. *Given a bet $\mathcal{B}$ about an event E, where the amount v' is exchanged with the offer of stake v ($v \geq v'$) under the condition that E occurs (null otherwise), the ratio $\frac{v'}{v}$ is called* the betting quotient *of $\mathcal{B}$.*

By Definition 3 the betting quotient of a bet is within the closed interval $[0, 1]$.

Definition 4. *A betting quotient α at which a bettor $\mathbf{X}$ is willing to bet both* on *and* against *an event E is called a* degree of belief *concerning E by $\mathbf{X}$.*

Definition 5. *Let $E_1, \ldots, E_n$ be events. An assignment of betting quotients to $E_1, \ldots, E_n$ by a bettor $\mathbf{X}$ is said to be* prima facie B-consistent *if and only if for each $E_i (1 \leq i \leq n)$ there is a unique degree of belief α concerning E by $\mathbf{X}$.*

Note that the Definition 5 applies to both full beliefs and partial beliefs. In terms of bets, full beliefs have only the values 0 and 1 as bet quotients. An assignment to all the events can be seen as a *valuation*, albeit not of truth values but of full beliefs. Logic texts assume that a valuation assigns a single value to every event. If allowed, a valuation that assigned both truth values to some event would be immediately B-inconsistent, whether the valuation assigns truth values or full belief values. Assignments of degrees of belief should also be unique to be immediately B-consistent. Because this uniqueness condition is typically taken for granted, I will assume that betting quotient assignments by a bettor are always prima facie consistent.

In addition to the idealization constituted by the linearity of the value of monetary amounts, it is important to assume another idealization: that betting quotients are context independent, and in particular that they represent a person's partial degree of belief in certain events *independent of the simultaneous occurrence of other bets accepted by* $\mathbf{X}$. I will call this idealization (which, as will become clear later, plays an essential role) *condition of absoluteness*. Moreover, partial beliefs depend only on the ratio $\frac{v'}{v}$, so that if a person $\mathbf{X}$ is willing to bet both for and against an event E with amounts v' and v, she is assumed to be willing to bet both for and against E with amounts $\alpha v'$ and αv ($\alpha \in \mathbb{R}$).[5]

---

[5] This idealization derives from Bayesian decision theory and those idealizations that this last one, in turn, contains.

4.2   Logical Consequence for Partial Belief

To properly extend the notion of inconsistency defined by Definition 1 for full beliefs to partial beliefs (from the point of view of what Ramsey called the ἀνάγκη λέγειν), we need a notion of logical consequence for partial beliefs that encompasses logical consequence for full beliefs. In other terms, we need to clarify what it means, in a general way, to say that the degrees of belief $\alpha_1, \ldots, \alpha_n$ in the events $E_1, \ldots, E_n$ logically entail the degree of belief α in the event E. In the case of full beliefs, since the logical consequence relation is coextensive with the relation of logical consequence in Sentence Logic (and in the corresponding Lindenbaum algebra of events), to say that full belief in the events $E_1, \ldots E_n$ logically entails full belief in the event E reduces to the statement that if the full beliefs in $E_1, \ldots E_n$ are true the full belief in E is also true. However, since for partial beliefs, this correspondence with the semantics of Sentence Logic (which would also give us what Ramsey called the ἀνάγκη εἶναι), is missing, we need to look for another way. What we need is a metalinguistic relation that is true when the partial beliefs $\alpha_1, \ldots, \alpha_n$ in the respective events $E_1, \ldots, E_n$ carry with them *implicitly* the partial belief α in the event E. I will show we can achieve this through the notion of *indirect bet*. Before providing an exact definition of this key notion, I will try to illustrate it informally.

Indirect Bets. Suppose a person **X** has such betting dispositions about certain events $E_1, \ldots E_n$. In that case, her degrees of partial belief are respectively $\alpha_1, \ldots, \alpha_n$, and, by the *condition of absoluteness*, these degrees are assumed not to be altered if **X** stipulates *simultaneously* n bets about $E_1, \ldots, E_n$ with betting quotients $\alpha_1, \ldots, \alpha_n$. A finite set M of simultaneous bets about events $E_1, \ldots, E_n$ generally results in different payoffs depending on (a) the truth values of the events on which the bets are placed, (b) which bets are *on* and which bets are *against* each of the events $E_1, \ldots, E_n$, (c) the amounts involved in each bet. Now, it may be the case that there exists an event E for which a finite set M of simultaneous bets about $E_1, \ldots, E_n$ has a constant balance γ whenever E is true and a constant balance β whenever E is false with $\gamma < \beta$. In such a case, the balance of M would be the same as a bet on E. M indeed carries the same payoff as a bet *on* E with the betting quotient $\frac{\gamma}{\gamma + \beta}$.

To offer a concrete example, where the amount is assumed to be expressed in euros, assume that a person **X**, in connection with a horse race, wishes to bet on the event that either horse A or B will win a certain race. The bookmaker allows betting on either A's or B's wins but does not accept bets on disjunctive events such as "either horse A will win or horse B will win." However, **X** can make its disjunctive bet by resorting to indirect betting through two simultaneous bets, one on A's win, the other on B's win. Suppose that the bookmaker sets the betting quotient on A's win at $\frac{1}{6}$ (i.e. risking one euro would, in the case of A's win, collect 6 euros, with a net gain of 5 euros, and suppose that the bookmaker sets the betting quotient on B's win at $\frac{1}{2}$ (i.e. risking one euro would collect 2 euros in the case of B's win, with a net gain of one euro (even money bet). If E is the event "victory of A" and E' is the event victory of B, the two simultaneous bets constitute an indirect bet on the event $E \vee E'$. **X** may achieve this goal by betting on A winning 1 euro and betting 3 euros on B winning. The following table shows the situation:

	A wins	B wins	neither A nor B win
Bet on A	5	−1	−1
Bet on B	−3	3	−3
Total Balance	2	2	−4

As can be seen immediately, the set of the two bets is an indirect bet on $E \vee E'$ (whether A wins or B wins) with the betting quotient equal to $\frac{4}{6} = \frac{2}{3}$. Note that the indirect betting quotient is the sum of the individual betting quotients. The indirect betting quotient agrees with the addition principle of probability theory (considering that E and E' are incompatible events). This situation is general, and it is on it that the present discussion is based.

We are now able to give a general definition of indirect bet.

Definition 6. *A set of bets M is called an indirect bet on [against] E at the α betting quotient if and only if there exist nonnegative real numbers a and b such that: (1) the total balance of M is, independently of the truth-value of the events about which the elements of M are, $b\,[\,-\,b]$ if E is true; (2) the total balance of M is, independently of the truth-value of the events about which the elements of M are, $-a\,[a]$ if E is false; (3) $-a \le b$; (4) $a + b > 0$; (5) $\alpha = \frac{a}{a+b}$.*

Logical Consequence for Partial and Full Belief. The notion of indirect bet allows us to reach the appropriate extension of the notion of logical consequence for belief that encompasses both full and partial beliefs. A *sufficient* condition for the degree of belief ρ in the event E to be a logical consequence of the degrees of belief $\rho_1, \ldots, \rho_n$ assigned to the events $E_1, \ldots, E_n$ is the existence of an indirect bet H on or against E at the betting quotient ρ, where the elements of H are bets on events $E_1, \ldots E_n$ whose betting quotients are $\rho_1, \ldots, \rho_n$ respectively. There are, however, special cases in which the degrees of belief assigned to events $E_1, \ldots, E_n$ do not allow for an indirect betting quotient on E consisting of bets on elements of the set $E_1, \ldots, E_n$ (and, possibly, on other events) but the betting quotient of each indirect bet on E depends only on the degrees of belief assigned to the elements of the set $E_1, \ldots, E_n$. In light of these considerations, the following definition provides the appropriate definition of logical consequence for degrees of belief. In what follows I denote generally the degree α assigned by a person **X** to an event E as $B_\alpha^{\mathbf{X}}(E)$. Since, in this approach, a single person **X** is involved, if there is no ambiguity, I will omit the mention of **X**, writing simply $B_\alpha(E)$.

Definition 7. *$B_\alpha(E)$ is a logical consequence of the set of partial beliefs $M = B_{\alpha_1}(E_1), \ldots, B_{\alpha_n}(E_n)$ $(0 \le n)$ (in symbols $M \models_B B_\alpha(E)$ (1) if exists an indirect* bet about E to the quotient α that contains only bets about elements of the set $\{E_1, \ldots, E_n\}$ or (2) there is a consistent set $\{A_1, \ldots, A_m\}$ $(1 \leq m)$ of events, such that all the events A_i $(0 \leq i \leq m)$ are different from the events E_j $(0 \leq i \leq n)$ and there is at least one indirect bet about E containing bets about events $C_{j_1}, \ldots C_{j_k}$ $(1 \leq k \leq n + m)$ of the set $\{E_1, \ldots, E_n, A_1, \ldots, A_m\}$, such that the betting quotient of E is α, however the betting quotients are assigned to each of the events belonging to the set $\{A_1, \ldots, A_m\}$.

One may easily verify that this definition encompasses standard logical consequences between sentences that refer to full beliefs. It is also monotonic: if the assignment of the belief degree α to the event E is a logical consequence of the assignment of the belief degrees $\alpha_1, \ldots, \alpha_n$ to the events $E_1, \ldots, E_n$, the assignment of the degree of belief α to the event E is a logical consequence of the assignment of the degrees of belief $\alpha_1, \ldots, \alpha_n$ to the events $E_1, \ldots, E_n, A_1, \ldots, A_m$.

Consistency with Respect to Belief. Through the notion of logical consequence of beliefs, it is also easy to extend the notion of consistency to degrees of belief. One may achieve this extension by the following definition.

Definition 8. Let $K = \{B_{\alpha_1} E_1, \ldots, B_{\alpha_n} E_n\}$ a set of assignments of partial degrees. K is B-consistent iff for no event E both $K \models_{\mathcal{B}_\alpha} (E)$ and $K \models_{\mathcal{B}_\beta} E$, where $\alpha \neq \beta$.

In the following discussion, I will not consider conditional probability. One may wonder whether there are any particular difficulties in extending the theory of indirect betting to conditional probability and thus in extending the definitions we have given for "absolute" belief to conditional belief. However, I will give an informal explanation of this extension below.

Conditional Degrees of Belief and Indirect Bets. The betting quotient of a conditional bet expresses the conditional degrees of partial beliefs. A bet on an event E conditional on a second event E' has a payoff of v $(v > 0)$ if $E \wedge E'$ turns out to be true, while it has a payoff of v' $(v' << 0)$ if $\neg A \wedge C$ turns out to be true and it is required if E' turns out to be false, so in such a case its payoff is 0. The betting quotient of such a conditional bet is $\frac{v'}{v}$. It is assumed that the betting quotient of E' is not 0.

In a parallel way, an indirect conditional bet on E conditional on a second event E' is a set of bets such that the total balance is a gain v $(v > 0)$ if $E \wedge E'$ turns out to be true, a gain v' $(v' \leq 0)$ if $\neg A \wedge C$ turns out to be true, and is 0 if $\neg C$ turns out to be true.

Extending the results presented in this paper to conditional bets is straightforward. This result is because, as one can prove, any indirect bet simulating a "direct" bet, either absolute or conditional, can be realized by a set of *absolute* bets on the atoms of the finite Boolean algebra generated by the events involved in the "direct" bet. I omit the proof of this fundamental result here for the sake of simplicity. However, I will use the following example to illustrate the notion of a conditional indirect bet and how this result works.

For simplicity, consider two logically independent events, A and C, such that $B_{\frac{3}{10}}(A)$ and $B_{\frac{3}{5}}(C)$. The atoms of the finite Boolean algebra $\mathbb{F}$ generated by A and C are $A \wedge C$, $A \wedge \neg C$, $\neg A \wedge C$, and $\neg A \wedge \neg C$. Suppose $B_{\frac{1}{10}}(A \wedge C)$, $B_{\frac{1}{5}}(A \wedge \neg C)$, $B_{\frac{1}{2}}(\neg A \wedge C)$, and $B_{\frac{1}{5}}(\neg A \wedge \neg C)$. The corresponding bet quotients are $\frac{1}{9}$ for the bet on $A \wedge C$, $\frac{1}{4}$ for the bet on $A \wedge \neg C$, 1 for the bet on $\neg A \wedge C$, and $\frac{1}{4}$ for the bet on $\neg A \wedge \neg C$. To get a conditional indirect bet on A regarding C, one can resort to a system of absolute bets on the atoms of $\mathbb{F}$. The system of bets shown in the following table constitutes an indirect conditional bet on A given C. In this example, the stakes of the bet on the atoms of $\mathbb{F}$ are 2 for the bet on $A \wedge C$ and 1 for all other bets.

	$A \wedge C$	$A \wedge \neg C$	$\neg A \wedge C$	$\neg A \wedge \neg C$
Bet on $A \wedge C$	18	−2	−2	−2
Bet on $A \wedge \neg C$	−1	4	-1	−1
Bet on $\neg A \wedge C$	−1	−1	1	−1
Bet on $\neg A \wedge \neg C$	−1	−1	−1	4
Total Balance	15	0	−3	0

The betting quotient of this indirect bet on A conditional to C is $\frac{1}{6}$. This value is just the ratio of the betting quotient of $A \wedge C$ and the betting quotient of C, according to the rules of probability. This result is a general fact whenever the betting quotient of the "direct" bets obeys the rules of probability.

4.3   Logical Consequence and Consistency with Respect to Belief vs Finite Probability

The problem now arises in studying the relationship between the notions of logical consequence and consistency for beliefs and finite probability theory. In what follows, I will adopt the following equations and inequalities as axioms of finite probability theory. Of course, any other equivalent axiomatization would do equally well. Given certain events $E_1, \ldots, E_n$ ($n \geq 0$), a probability function may be defined on the finite Boolean algebra generated by such events. Therefore, one can always define finite probability for a finite Boolean algebra in this context.

Definition 9. *Given a finite Boolean algebra $\mathcal{A}$ and a real-valued function $\mathbf{P}$ defined on $\mathcal{A}$, $\mathbf{P}$ is said to be a finite probability function if and only if it satisfies the following conditions (where E and E' are elements of $\mathcal{A}$):*

(A_1) $\mathbf{P}(E) \geq 0$
(A_2) $\mathbf{P}(E \wedge \neg E) = 0$
(A_3) $\mathbf{P}(\neg E) = 1 - \mathbf{P}(E)$

(A_4) $\mathbf{P}(E \vee E') = \mathbf{P}(E) + \mathbf{P}(E') - \mathbf{P}(E \wedge E')$

What is the relationship between the notion of logical consequence for belief and the logical consequence of probability values in finite probability theory? The following result answers this question:

Theorem 2. *Let K be the set of axioms of the theory of finite probability enumerated in Definition 9. Let $\mathcal{A}$ be a finite Boolean algebra of events. Let $E_1, \ldots, E_n$ be elements of $\mathcal{A}$ and let $\mathbf{P}$ be a function satisfying axioms in K defined on $\mathcal{A}$. Let M be the set of probabilistic sentences $\mathbf{P}(E_1) = \rho_1, \ldots, \mathbf{P}(E_n) = \rho$ $(n > 0)$. Let S the set $\{B_{\rho_1} E_1, \ldots, B_{\rho_n} E_n\}$. $\mathbf{P}(E) = \rho$ is a logical consequence of the set $K \cup M$ if and only if $S \models_B B_\rho E$.*

Proof. See Appendix (p. 17)

The time has come to clarify in what sense the violation of the axioms of probability is B-inconsistent without being held to be so only by analogy or metaphor. One should remember that the definition of inconsistency we have given above also includes full beliefs; in that case, it is coextensive with Sentence Logic. The following result shows the violation of the laws of finite probability as inconsistency:

Theorem 3. *Let M be a finite set of sentences, and let N be the set of events expressed by the sentences of M. Any assignment of degrees of belief to the elements of M is inconsistent if and only if, whenever it is applied to the corresponding elements of N, at least one of the axioms of the Definition 9 is violated.*

Proof. See Appendix (p. 18)

Theorems 2 and 3 illustrate, taken together, the sense in which the theory of finite probability preserves Ramsey's ἀνάγκη λέγειν: the *non-alethic* necessity of logic as applied to belief. Ramsey's thesis is that there is no sense in which the logic of probability applied to belief contains, at the same time, an alethic necessity (ἀνάγκη εἶναι). Ramsey surely is right when he notes that there are no truth-functional semantics for probability as a degree of belief. However, this does not rule out the possibility that inconsistent degrees of belief (understood as betting quotients) are somehow impossible. This possibility is because there does not seem to be a serious reason for defining degrees of belief in terms of direct bets rather than in terms of the more general notion of indirect bets. One can also extend the requirement to bet both on and against an event at a single betting quotient to indirect bets (which, from the point of view of payoffs, are indistinguishable from direct bets) in the elicitation of betting quotients, referring them to bets in general, both direct and indirect. In such a case, those who would state betting quotients that violate the axioms of finite probability would assert a logically inconsistent set of assignments and, as such, unsatisfiable. Therefore, I tentatively propose to characterize α as the degree of belief in an event E if α is *the only betting quotient α at which one is willing to bet indirectly* *both on and against E.* It is unnecessary to mention simple or "direct" bets since one may regard them, by Definition 6 as a special case of indirect bets.

Theorem 3 also holds, as a special case, for full beliefs. Thus a corollary of Theorem 3 is that an assignment of full beliefs to certain events is B-consistent if and only if it is consistent in the sense of Sentence Logic.[6]

Suppose we thus characterize the rule of correspondence between degrees of belief and behavioural dispositions. In that case, we immediately derive from Theorem 3 the corollary that, under the idealizing assumptions of linearity and absoluteness about the framework of betting behaviour as conditions under which they represent degrees of belief as a basis for deliberation (which may be called *practical belief*) it is *logically impossible* to be the bearer of betting dispositions as corresponding to the respective degrees of belief that violate the axioms of finite probability.

It is not a difficulty of this proposal that a person **X** may believe that the betting quotients she stated are B-consistent when not. In this case, she has an *alethic* inconsistent belief about her betting dispositions, which is perfectly analogous to someone who believes that all the sentences of an inconsistent set of sentences are true. Thus, what a person thinks of her dispositions may be inconsistent, particularly if she thinks they B-consistently represent her *practical* degrees of belief. After all, it makes no sense to ask whether partial degrees of belief, understood as mere mental states independent of decision dispositions under uncertainty, are consistent. As Ramsey observed:

> Suppose, however, I am wrong about this and that we can decide by introspection the nature of belief, and measure its degree; still, I shall argue, the kind of measurement of belief with which probability is concerned is not this kind but is a measurement of belief *qua* basis of action. [11, p. 67]

Those who rely on the Dutch Book argument hardly appreciate, as noted above, except in an analogical or metaphorical way, the logical inconsistency of degrees of belief that violate the laws of probability and tend to see the violation of the consistency condition as *irrational* behaviour from the point of view of decision-making convenience. Accepting bets involving a sure loss is considered a "ruinous" disposition. Hence, the standard interpretation of the laws of probability (and Bayesian decision theory more generally) as *hypothetical imperatives*, violating which leads to harmful consequences. The present analysis shows how one may view the inconsistency of violation of finite probability theorems as a genuine extension of the logical inconsistency applied to belief as *impossibility* of

---

[6] The paper [9, p. 133] proves that "[t]he logical consistency of a collection of assertions about events can be viewed as a special case of coherent probability assessments in the sense of de Finetti." Since B-consistency is coextensive with de Finetti's coherence, the result proved in [9] is simply equivalent to this corollary of our Theorem 3. As interesting as it is, it says nothing about the logical consistency (insofar as it is conceptually distinct from de Finetti's coherence) of *partial* beliefs.

betting dispositions that assign to sentences one and only one degree of belief. Prescriptive interpretation as hypothetical imperatives is no longer necessary.[7]

However, this account provides also Ramsey's ἀνάγκη λέγειν. This fact happens not only if one takes simple "direct" bets as expressing degrees of belief but even if one bases, as here suggested, the definition of degrees of belief in terms of indirect bets. Suppose one does not want to bear inconsistent partial beliefs (i.e., under our assumptions, a belief in a necessarily false statement about one's current indirect betting dispositions). In that case, our results show that finite probability theory helps avoid this. After all, the normative utility of Logic is precisely to point out inconsistencies and logically false beliefs that, on the surface, may be missed so that one may correct them.

5   Conclusion

The Dutch Book argument illustrates the contradiction by showing that the "incoherent" bettor, on the one hand, is indifferent between betting on and against certain events with certain betting quotients while indifferent between the sure loss and the sure win that the Dutch Book theorem entails. Such behaviour is certainly at odds with the general assumptions of Bayesian decision theory, but it is only analogous to the notion of consistency used in Logic. One can make similar considerations about de Finetti's treatment of consistency in terms of proper scoring rules. Again, what de Finetti shows with this approach is that violating the axioms of finite probability only leads to choices for which (assuming the rule of expected utility maximization) there are definitely better alternatives. Again, the connection with the logical ideas of contradiction and consistency seems vague and at best analogical. On the other hand, the present discussion captures inconsistency in the violation of probabilistic rules by tracing them back to the notion of logical consequence between degrees of belief through the idea of indirect bet.

A   Appendix

Lemma 1. *Let A be a finite Boolean algebra with k atoms and let φ be a function that assigns a betting quotient to each element of A. The following two conditions are equivalent:*

1. *φ satisfies the axioms of finite probability (see Definition 9 above)*
2. *The set of the collections of simultaneous bets on the elements of A with the betting quotients provided by the function φ is a $k-1$ dimensional vector space.*

---

[7] According to D. H. Mellor, Ramsey's view was purely descriptive ([10, p. xviii]). The present paper is not the place to discuss this issue.

Proof. Let $\mathfrak{a}_1, \ldots, \mathfrak{a}_k$ be the atoms of $\mathcal{A}$. One may represent each gamble B defined in terms of the elements of $\mathcal{A}$ as a vector $(g_1, \ldots, g_k)$ of $\mathbb{R}^k$ that specifies for each atom $\mathfrak{a}_i$ ($1 \leq i \leq k$) the gain g_i that B would entail if $\mathfrak{a}_i$ were true. Based on the assumption of the linearity of the utility of the amounts, there exists a linear application $f : \mathbb{R}^k \rightarrow \mathbb{R}$ that associates each gamble G with its fair price (such that $f(G) = f(-G) = 0$). f satisfies the condition:

$$\min(x_1, \ldots, x_k) \leq f(x_1, \ldots, x_k) \leq \max(x_1, \ldots, x_k) \tag{1}$$

A fair bet on an event E with stake v and betting quotient α can then be represented by the vector $(g_1, \ldots, g_k)$ of the kernel $\ker(f)$ of the application f (i.e., of the set of vectors $\mathbf{x}$ such that $f(\mathbf{x}) = 0$ where $g_i = (1 - \alpha)v$ if $\mathfrak{a}_i$ logically entails E and $-\alpha v$ if $\mathfrak{a}_i$ logically entails $\neg E$ ($1 \leq i \leq k$). By the inequality 1 the following equation holds: $f(x, \ldots, x) = x$. It follows that f is surjective and that the matrix of f consists of a vector $(p_1, \ldots, p_n)$ such that $\sum_{i=1}^{i=n} p_i = 1$. Since the dimension of the kernel of a surjective linear application is, for vector spaces of finite dimension, equal to the difference between the dimension of the application's defining domain and that of its image, $\ker(f)$ has dimension $k - 1$. One can represent each event E of $\mathcal{A}$ by the vector $(x_1, \ldots, x_k)$ of $\mathbb{R}^k$ such that, for each i ($1 \leq i \leq k$) has components $x_i = 1$ if $\mathfrak{a}_i$ logically entails E and $x_i = 0$ if $\mathfrak{a}_i$ logically entails $\neg E$. Each $x \in \mathbb{R}$ can be associated injectively with the constant vector of $\mathbb{R}^k (x, \ldots x)$. Let d be such an injection and let $h : \mathbb{R}^k \rightarrow \mathbb{R}^k$ be the endomorphism $d \circ h$ and let $\mathbf{A}$ be the matrix associated with h. Consider the transposed endomorphism ${}^t h$. For each linear form l holds: ${}^t h(l) = l \circ h$. In particular, ${}^t h(f) = f \circ h = f \circ d \circ f = f$ holds. It follows that every row of $\mathbf{A}$ is an eigenvector of ${}^t h$. Therefore each row of the matrix $\mathbf{B} = \mathbf{I}_k - {}^t \mathbf{A}$ belongs to both $\ker(f)$ and $\ker(h)$. On the other hand, all rows of $\mathbf{A}$ are equal to the vector $(p_1, \ldots, p_k)$. Thus, for at least one i ($1 \leq i \leq k$), $p_i \neq 0$. Let us now consider the matrix $\mathbf{C}$ obtained from $\mathbf{B}$ by deleting a row in which $k - 1$ times one of p_i other than 0 occurs. Since $\mathbf{C}$ is of rank $k - 1$, it has linearly independent rows (in fact, its minors of order $k - 1$ have determinant equal to $p_i \neq 0$). And since $\ker(f)$ has $k - 1$ dimensions, the rows of $\mathbf{C}$ constitute a basis of $\ker(f)$. Moreover, each row of $\mathbf{C}$ is a bet on a different atom of $\mathcal{A}$. Since every probability function defined on $\mathcal{A}$ is completely characterized by an assignment of probability values to $k - 1$ atoms, there is a biunivocal correspondence between those assignments of quotients to all elements of $\mathcal{A}$ that are probability functions and those linear applications $f : \mathbb{R}^k \rightarrow \mathbb{R}$ that satisfy the condition 1. It follows:

(a) If φ satisfies the axioms of probability, then the set of collections of bets (which actually are linear combinations of bets) with the betting quotients assigned by φ is a $k - 1$-dimensional vector space;

(b) If φ violates the axioms of probability, the set of betting collections is still a vector space that is *not* $k - 1$ dimensional.

∎

Lemma 2. *This result can be split into two points:*

1. *If ρ is the betting quotient of all indirect bets about event E through bets about events $E_1, \ldots, E_n$ (and nothing else) with respective betting quotients $\rho_1, \ldots, \rho_n$, then the vector representing E is linearly dependent on the set of vectors representing the events $E_1, \ldots, E_n, \top$ (where $\top$ represents the necessary event, that is the disjunctions of all the atoms of the Boolean algebra $\mathfrak{A}$).[8]*
2. *Moreover, in that case, in probability theory, the equation $\mathbf{P}(E)$ is a logical consequence of the equations $\mathbf{P}(E_1) = \rho_1, \ldots, \mathbf{P}(E_n) = \rho_n$.*

Proof.

1. Let $\mathcal{A}$ be the Boolean algebra generated by the set of events $E_1, \ldots, E_n$. Let $f : \mathbb{R}^k \to \mathbb{R}$, $h : \mathbb{R}^k \to \mathbb{R}^k$ and the matrix $\mathbf{C}$ be as in Lemma 1. Let $s : \mathbb{R}^k \to \mathbb{R}^{k-1}$ a linear application that associates with each vector representing an atom $\mathfrak{a}$ for which there exists in the matrix $\mathbf{C}$ a vector $\mathbf{y}$ representing a bet on $\mathfrak{a}$, the same $\mathbf{y}$, and further associate with the constant vector of $\mathbf{R}^k$ $(1, \ldots, 1)$ the vector of $\mathbb{R}^k$ $(0, \ldots, 0)$. Since an indirect bet is a linear combination of bets, and every vector representing an event in $\mathbb{R}^k$ is a linear combination of the vectors representing $k - 1$ atoms and the constant vector of $\mathbb{R}^k$ $(1, \ldots, 1)$, the sentence of the Lemma is proved.
2. We distinguish two cases:
   (a) the probabilistic sentences $\mathbf{P}(E_1) = \rho_1, \ldots, \mathbf{P}(E_n) = \rho_n$ are compatible with the axioms of probability;
   (b) the probabilistic sentences $\mathbf{P}(E_1) = \rho_1, \ldots, \mathbf{P}(E_n) = \rho_n$ are not compatible with the axioms of probability;
   Let us prove the two cases separately:
   (a) it is possible to choose f such that the betting quotients of the bets that the linear application associates with the vectors representing $E_1, \ldots, E_n$ are, respectively, $\rho_1, \ldots, \rho_n$. In that case, the betting quotient of E is ρ. Furthermore, since the betting quotients according to f, as shown in Lemma 1, satisfy the axioms of probability, the proof is concluded.
   (b) it suffices to observe that if the probabilistic sentences $\mathbf{P}(E_1) = \rho_1, \ldots, \mathbf{P}(E_n) = \rho_n$ violate the axioms of probability, any sentence can be deduced from them, because they, in conjunction with the axioms, result in a logically inconsistent system. In particular, one could deduce $\mathbf{P}(E) = \rho$. ∎

Lemma 3. *Given a finite Boolean algebra $\mathfrak{A}$ and a function φ that assigns a betting quotient to all elements of $\mathfrak{A}$, if φ violates the axioms of probability, then for every real number α it is possible to bet indirectly about every element E of $\mathfrak{A}$ with the betting quotient α.*

---

[8] For the notion of linear dependence between vector representing events, see de Finetti [6, pp. 48–56].

Proof. Under Lemma 1, the set of bets on the elements of $\mathfrak{A}$ generates an $n-1$ dimensional vector space $\mathfrak{B}$ (n being the number of atoms of $\mathfrak{A}$) if and only if the betting quotients satisfy the axioms of probability. If the function φ violates the axioms of probability, the dimension of $\mathfrak{B}$ cannot be less than n. Bets on $n-1$ atoms of $\mathfrak{A}$ generate an $n-1$ dimensional vector space contained in the space consisting of all linear combinations of bets. However, since it cannot be $n-1$ by Lemma 1, the dimension of $\mathfrak{B}$ cannot be less than n. On the other hand, the dimension of $\mathfrak{B}$ cannot even be greater than the number n of the components of each vector, equal to the number of atoms of $\mathfrak{A}$. It follows that $\mathfrak{B}$ is an n-dimensional vector space and is therefore isomorphic to $\mathbb{R}^n$. Therefore, every vector of $\mathbb{R}^n$—in the vector representation given in Lemma 1—belongs to the class of vectors representing a collection of bets. In particular, for each event E of $\mathfrak{A}$, that class contains all possible bets on E with any betting quotient. ∎

We now proceed to the proof of the Theorems 2 and 3.

Theorem 4. (see p. 12).

(see p. 12).

Proof. Let $\mathfrak{B}$ be a Boolean algebra containing the events $E_1, \ldots, E_n$. Let φ be a function that associates each element of $\mathfrak{B}$ with a degree of belief such that $\varphi(E_1) = \rho_1, \ldots, \varphi(E_n) = \rho_n$. Let $B_1, \ldots, B_m$ be elements of $\mathfrak{B}$ and let $\tau_1, \ldots, \tau_m$ be, respectively, the values $\varphi(B_1), \ldots, \varphi(B_m)$.

Let K be the set $\{B_{\tau_1} C_1, \ldots, B_{\tau_m} C_m\}$. Suppose that $K \models_{\mathcal{B}} B_\alpha E$. If $\alpha = \rho$, the thesis is proved. If $\alpha \neq \rho$ then, by virtue of Lemma 2 from the following of probabilistic sentences:

$$\{\mathbf{P}(C_1) = \tau_1, \ldots, \mathbf{P}(C_m) = \tau_m\}$$

one can deduce the equation $\mathbf{P}(E) = \alpha$ in the presence of the probability axioms. By hypothesis, from the following set of probabilistic sentences:

$$\{\mathbf{P}(E_1) = \rho_1, \ldots, \mathbf{P}(E_n) = \rho_n\},$$

in the presence of the probability axioms, one can deduce the equation $\mathbf{P}(E) = \rho$. It follows that the function φ is not a function and, therefore, does not satisfy the axioms of probability. By Lemma 3, it is possible to bet about any element of $\mathfrak{B}$ with any betting quotient. We observe that since the vector of $\mathfrak{B}$ representing E is a linear combination of vectors of $\mathfrak{B}$, E also belongs to $\mathfrak{B}$. Therefore, in all cases, it is possible to bet about E by betting on elements of $\mathfrak{B}$ with betting quotients provided by the function φ. We conclude that $K \models_{\mathcal{B}_\rho} E$ where K is the set of assignments $B_{\rho_1} E_1, \ldots B_{\rho_n} E_n$. Let now K be the set of assignments $B_{\rho_1} E_1, \ldots B_{\rho_n} E_n$. Suppose that $K \models_{\mathcal{B}_\rho} E$. By this assumption, given a finite Boolean algebra $\mathfrak{B}$ containing the events $E_1, \ldots, E_n$ and an assignment φ of degrees of belief to elements of $\mathfrak{B}$ that assigns $B_{\rho_1} E_1, \ldots B_{\rho_n} E_n$, there exists an indirect bet about E with the betting quotient ρ consisting of bets about elements $B_1, \ldots, B_m$ with the betting quotients $\tau_1, \ldots, \tau_m$, assigned by φ, possibly in conjunction with bets on another consistent set of events with arbitrary betting quotients. By Lemma 2, from the following set of probabilistic sentences:

$$\{\mathbf{P}(B_1) = \tau_1, \ldots, \mathbf{P}(B_m) = \tau_m\},$$

in the presence of the axioms of probability, the equation $\mathbf{E} = \rho$ is deducible. Furthermore, since E is a linear combination of the elements of $\mathfrak{B}$, E belongs to $\mathfrak{B}$. Therefore, for any probability function φ it is true: $\varphi(E) = \rho$. It follows that for every probability function $\mathbf{P}$ that satisfies the following constraint:

$$\mathbf{P}(E_1) = \rho_1, \ldots, \mathbf{P}(E_n) = \rho_n$$

holds: $\mathbf{P}(E) = \rho$.

∎

Theorem 5. (see p. 12).

Proof. If an assignment of degrees of belief $\rho_1, \ldots, \rho_n$ to respective events $E_1, \ldots, E_n$ violates the axioms of probability, then the following set K of probabilistic sentences:

$$\{\mathbf{P}(E_1) = \rho_1, \ldots, \mathbf{P}(E_n) = \rho_n\}$$

is inconsistent. It follows that any probabilistic sentence is deducible from K. In particular, for some event E and real numbers α and β included in the closed interval $[0, 1]$ such that $\alpha \neq \beta$, it will be deducible both that $\mathbf{P}(E) = \alpha$ and $\mathbf{P}(E) = \beta$. Let H be the set of belief assignments $\{B_{\rho_1} E_1, \ldots B_{\rho_n} E_n\}$. From the Theorem 2 it follows both that $H \models_\mathcal{B} B_\alpha E$ and $H \models_\mathcal{B} B_\beta E$. Therefore, H is B-inconsistent.

For the converse, we will proceed by contradiction. Let H be the set of belief assignments $\{B_{\rho_1} E_1, \ldots B_{\rho_n} E_n\}$. Suppose that the elements of H satisfy the axioms of probability while H is B-inconsistent. In that case, the set of betting collections constitutes, under Lemma 1, a $k-1$ dimensional vector space (k being the number of atoms of $\mathfrak{B}$). Let $\mathfrak{a}_1, \ldots, \mathfrak{a}_{k-1}$ be distinct atoms of $\mathfrak{B}$. The vector representing each element of $\mathfrak{B}$ is linearly dependent on the vectors representing $\mathfrak{a}_1, \ldots, \mathfrak{a}_{k-1}, \top$ (where $\top$ is the necessary event, i.e., the disjunction of all atoms of $\mathfrak{B}$). Therefore, by Lemma 2, one could obtain any indirect bet about elements of $\mathfrak{B}$ by bets about only the atoms of $\mathfrak{B}$. It follows that, for each event A, all indirect bets on A (assuming there are any) have the same betting quotient. This result contradicts the hypothesis and concludes the proof.

∎

References

1. Bayes, T.: An essay towards solving a problem in the doctrine of chances. In: Pearson, E.S., Kendall, M.G. (eds.) Studies in the History of Statistics and Probability, vol. 1, pp. 134–153. Charles Griffin, London (1958)
2. Bernoulli, D.: Exposition of a new theory on the measurement of risk. Econometrica **22**, 23–36 (1954)
3. de Finetti, B.: Does it make sense to speak of 'Good Probability Appraisers'? In: Good, I.J., Mayne, A.J., Smith, J.M. (eds.) The Scientist Speculates: An Anthology of Partly-Baked Ideas, pp. 357–364. Heinemann, London (1962)

4. de Finetti, B.: Logical foundations and measurement of subjective probability. Acta Physiol. (Oxf) **34**, 129–145 (1970)
5. de Finetti, B.: The role of 'Dutch Books' and of 'Proper Scoring Rules'. Br. J. Philos. Sci. **32**(1), 55–56 (1981)
6. de Finetti, B.: Theory of Probability: A Critical Introductory Treatment, vol. 1. John Wiley & Sons, Chichester (1990)
7. de Finetti, B.: On the subjective meaning of probability. In: Monari, P., Cocchi, D. (eds.) Bruno de Finetti: Probabilità e induzione (Induction and Probability), pp. 291–321. CLUEB, Bologna (1992)
8. de Finetti, B.: Philosophical Lectures on Probability. A. Mura ed. Springer, Berlin (2008)
9. Dickey, J.M.: De Finetti coherence and logical consistency. Notre Dame J. Formal Logic **50**(2), 133–139 (2009)
10. Mellor, D. H.: Introduction. In: Ramsey, F.P., Mellor, D.H. (eds.) Philosophical Papers, pp. xi–xxiii. Cambridge University Press, Cambridge (1990)
11. Ramsey, P.F.: Truth and probability. In: Ramsey, F.P., Mellor, D.H. (eds.) Philosophical Papers, pp. 52–94. Cambridge University Press, Cambridge (1990)
12. Savage, L.J.: The Foundations of Statistics, 2nd edn. Dover Publications, New York (1972)

Degrees of Intuition: Coherence in the New Dual Process Theory

Maxime Bourlier[1,2], Daniel Lassiter[3], and Jean Baratgin[1,2,4(✉)]

[1] Université Paris 8, 2 rue de la Liberté, 93200 Saint-Denis, France
[2] Laboratoire CHArt-UP8 RNSR 200515259U, 2 rue de la liberté, 93200 Saint-Denis, France
[3] University of Edinburgh, 3 Charles Street, EH8 9AD Edinburgh, UK
[4] Association P-A-R-I-S, 25 Rue Henri Barbusse, 75005 Paris, France
jean.baratgin@paris-reasoning.eu

Abstract. Dual Process theories of reasoning (DPT) hold that we have the ability to think in two ways. The first is called intuitive and is characterised by its speed and automaticity. The second, known as deliberative, is slower and more costly in cognitive resources. However, it has the advantage of being conscious and therefore sometimes prevents us from falling into the traps to which the intuitive process is prone. Debates surround these theories, notably about the supposed exclusivity of the responses produced by each of the two processes, or about the mechanism behind the activation of the deliberative process. De Neys' model attempts to answer these questions by rethinking the interaction between the two processes. This model posits competing intuitions which, when they cannot be separated to provide an answer, call on the deliberative process to determine which one to keep (or propose a new answer). Another recent change of perspective in the field of reasoning is the recognition that we generally reason under uncertainty, and therefore that the formal logic hitherto used as norm of rationality for deductive reasoning must be replaced by a coherence norm, advocated by de Finetti (1937). In this 'New Paradigm' (NP), a proposition is no longer simply true or false, but rather has a degree of belief associated with it, which can be reported as a probability. We propose to show how this new DPT model and NP are complementary.

Keywords: Reasoning · Dual Process Theory · Subjective Bayesianism · Conditional

1 Dual Process Theories of Reasoning's Biggest Flaw

Introduced by Wason & Evans [59] and popularized by Kahneman & Tversky [32], dual-process theories (DPT) distinguish two kinds of reasoning processes. These are often referred to as "System 1" and "System 2," type 1 and type 2, or—in the terminology that we will adopt in this paper—as "intuitive" and "deliberative" processes [14]. The defining features of these processes may vary

J. Baratgin et al. (Eds.): HAR 2024, LNCS 15504, pp. 185–195, 2025.
https://doi.org/10.1007/978-3-031-84595-6_12

from one theory to another, but some key features always remain consistent throughout the literature. Intuitive reasoning is usually described as fast, automatic, and low-effort, while the deliberative process is slow, conscious, and cognitively demanding.

The distinction between two kinds of reasoning was introduced to explain why, in reasoning tasks, when participants responded quickly, they usually give the same incorrect answer, while participants who took more time to respond could usually arrive at the expected correct answer [32]. From this, models of dual-process theory emerged, in which the intuitive process was expected to produce heuristic answers, that is, answers that rely on adaptive shortcuts allowing us to respond quickly to well-known situations. On the other hand, the deliberative process was supposed to correct the intuitive answer when the latter diverged from the correct logical answer that it computed itself.

Two types of models competed with each other for a long time. Sequential models, the first to appear, stated that the intuitive process was the one activated by default and that the deliberative process was only activated to correct the intuitive answer if a conflict between the two was detected. In the second type, parallel models, both processes are activated from the beginning. The deliberative process is shut down if no conflict is detected between the two processes and if there is a conflict, the deliberative finishes to process its answer [20,31,54]. However, these two types of model share a common flaw.

Many dual process theories oppose the two processes: they rely on the presence of a conflict between the answers computed by intuitive and deliberative processes. In some cases, the two types of reasoning are in agreement, but when they conflict the deliberative process needs to correct the intuitive answer [22,23,25,32]. However, many studies have shown that the answer usually associated with the deliberative process can be also given when participants are under time pressure or cognitive load (e.g., when working memory is taxed). This suggests that such responses are associated with intuitive reasoning, and are not exclusive to the deliberative process [1–3,57].

The assumption that deliberative reasoning works mainly as a corrective to intuitive reasoning raises another problem. Pennycook et al. [49] point out that the responses of the two processes cannot be compared until the deliberative process has finished processing its own answer, and thus no conflict can be detected beforehand. This applies to parallel models in which the deliberative process is not fully done by the time we receive the intuitive answer, as well as to sequential models in which the deliberative process is not supposed to have been activated at all yet.

In parallel models, the deliberative process checks for mistakes in the intuitive response, using the output of the latter to decide whether to override it [54]. However, this model implies that the deliberative process must be able to tell a correct answer from a incorrect one before it has finished computing its own response. The intuitive answer can only be compared to what the deliberative process has been able to compute in the time elapsed in computing the intuitive answer [13,17,22].

Recent sequential models now reject the exclusivity of the responses from both processes [21]. However, they are forced to allow that the deliberative process is activated after any intuitive answer has been computed, just to know if it will be able to compute its own response or not (or justify the intuitive answer) given the available resources.

In both approaches, the decision whether to compute the deliberative answer relies on the deliberative process itself [49]. To solve this issue and to account for the existence of "logical intuitions" [13], correct answers usually associated with the deliberative process but computed by the intuitive process, a new model of dual-process theory has recently been developed.

2 A Promising Model of Dual Process Theory: The Hybrid Model

This new hybrid model of DPT [3,14,15] derives its name from being at the intersection of sequential and parallel models. It resembles sequential models in that the intuitive process is activated by default, with the deliberative process being called upon only when necessary to avoid wasting cognitive resources. This is indeed a major criticism of parallel models from an efficiency standpoint [13,17]. On the other hand, it is close to parallel models as it includes some lateral processing, but this time, it occurs between multiple intuitions instead of between the two types of processes. Instead of a single intuitive response, we have multiple intuitions. For example, there could be one for the heuristic answer and one for the alleged deliberative answer but there could be other intuitions.

Each of these intuitions comes with different strengths, and the one with the highest strength is the answer given automatically by the intuitive process. However, if two intuitions have very similar strengths, the deliberative process enters the field to decide which to prefer. Specifically, as the strengths of the two competing intuitions get close to one another, the uncertainty associated with the intuitive process rises. When a certain threshold of uncertainty is met, the deliberative process gets activated to reduce it. It could do so by modifying the strength of the competing intuitions and sending feedback to the intuitive process or by creating a new answer altogether. When the level of uncertainty returns below the threshold, the deliberative process ends and an answer can be given.

While this model still leaves some questions open (See [16] for more details), it responds to major criticisms of previous DPT models and opens new bridges with other parts of the field of reasoning.

3 From Classical Logic to Bayesian Probability: The New Paradigm of Reasoning

Another branch of the field of reasoning has also have undergone a major change, so much so that this shift has been dubbed "The New Paradigm of reasoning"

(NP) [18,19,46,47]. Conditionals of the form "if p, q" are sometimes considered to be at the heart of reasoning [24]. Propositional logic has long been the norm against which human rationality was tested when interpreting them. This led researchers to doubt human rationality, as they often found that participants' answers did not fit the expectations of classical logic [58]. In propositional logic, conditional sentences translate to "p implies q" ($p \rightarrow q$), where the implication connector is equivalent to the disjunction of q and the negation of p ($q \vee \neg p$). While participants' responses align with this norm when p is true, they diverge when p is false. According to the implication, a conditional is always true when p is false; however, participants often judge the conditional to be irrelevant, rather than true or false, in such cases [5,58]. For example, if I bet that my favorite team will win the tournament if they get to the finals, the bet is in effect as long as my team get to the final. If they don't, the bet is called off, and no one wins or loses.

This discrepancy arises because classical propositional logic only has two possible truth values: true and false. In contrast, humans almost always think in uncertain settings. This realization led many researchers to reconsider the use of propositional logic as the correct norm of rationality to evaluate them, favoring a new probabilistic norm, with Bayesianism as a favored candidate [8,18,44,45,47]. Bayesianism is an epistemology in which we assign degrees of belief to any given event. These degrees of belief can be translated into probabilities and those are updated via Bayes' theorem. Bayes' theorem states that a belief B should be revised on the discovery of new relevant information I by the conditional probability of B on the assumption of I, $P(B|I)$. This conditional probability should be equal to the product of the probability of the new information on the assumption of the belief $P(I|B)$ and the prior probability of the belief $P(B)$ divided by the prior probability of the new information $P(I)$.

$$Bayes' theorem: P(B|I) = \frac{P(I|B) \times P(B)}{P(I)}$$

According to the suppositional theory of conditionals most strongly associated with NP, a conditional "if p, q" is understood as "q supposing p" ($q|p$). *Stalnaker's thesis* [55,56], often called *the Equation*, states that the probability of a conditional is equal to the conditional probability:

$$Stalnaker's \ thesis: P(\text{"If p, q"}) = P(q|p)$$

The Equation seems to be cognitively valid, as participants do indeed judge the probability of a conditional as its conditional probability in most cases [51]. For example, participants presented with the chips in Fig. 1 and asked about the probability of "If the chip is square, then it is black" will generally give the response expected by the suppositional theory, rather than the material conditional: the most frequent response is $P(black|square) = 3/4$, and $P(black \vee \neg square) = 6/7$) is exceedingly rare [48].

Fig. 1. Set of chips from Politzer & al. (2010) used to assess participant's probability judgement of conditionals

Baratgin et al. [5,6,8] also show that the "defective" truth-table of Wason [58] corresponds to de Finetti's trivalent logic [27][1]. The de Finetti logic includes, in addition to true and false, a third value "void" (ø) that serves to represent cases when the truth value is uncertain or simply cannot be assessed (see Fig. 2). The addition of a third value bridges logic and probabilities smoothly, and accounts for many results previously thought to point toward paradox or irrationality [6,36,40,41].

$$
\begin{array}{c|ccc}
A/C & T & \varnothing & F \\
\hline
T & T & \varnothing & F \\
\varnothing & \varnothing & \varnothing & \varnothing \\
F & \varnothing & \varnothing & \varnothing
\end{array}
$$

Fig. 2. De Finetti's trivalent truth table of the conditional. Shaded cell correspond to the "defective" table

4   Coherent is the New Rational

Traditionally, rationality has been defined by conformity to classical logic when evaluating deduction and to frequentist probability (long-run frequency in a reference class) when evaluating induction. Bayesianism revises the standard of rationality to that of coherence of subjective degrees of belief. What matters is that you update your beliefs properly so that they are in accordance with the rest of your beliefs [27,28,52]. This is not to say that any belief is acceptable, but rather that any coherent set of beliefs is possible given one's evidence and prior beliefs. If two people have different beliefs about a given event, their beliefs will eventually converge given enough evidence [9]. However, convergence can be slowed down by the fact that belief revision depends on new information,

---

[1] This logic goes back at least to de Finetti's unpublished work from 1928: see [4].

and new information can be interpreted in different ways by different people. This is one reason why supporters of the Bayesian perspective often emphasize the importance of pragmatics [30] in reasoning tasks: implicit context effects can affect the way that information is taken up [11,38,39,50], as well as other aspects of the language employed in reasoning experiments [37].

Let us now get back to the hybrid model of DPT and see how Subjective Bayesianism could fit inside.

5   Subjective Bayesianism in the Hybrid Dual Process Model

Some defenders of NP do not see a need for DPT, because Bayesian cognitive theories have the resources to explain the distinction between heuristic and logical responses [12,43]. However, we argue that the distinction between automatic and deliberative modes of thinking remains relevant: we all make use of both distinctively in our everyday lives and we can somewhat make the difference from introspection between thoughts that come to our mind by themselves and those we consciously construct. Subjective Bayesianism could improve the hybrid model by providing a better understanding of the nature of intuitions' strength and the mechanism behind their reevaluation by the deliberative process.

5.1   On Intuitions

The concept of intuitions' strength should be aligned with that of degrees of belief. The strength of our intuitions can be seen as a measure of our confidence in them. Therefore, if we assume that beliefs are implicitly conditioned on the subject's total set of beliefs, our intuitions should be condition on them as well. In this case, the probability of conditional events should correspond to their conditional probability directly from intuition. This idea was already present to some extent in de Finetti's work [28]. However, de Finetti held that only one intuition could come out of a particular body of knowledge without additional information; therefore, conflicting intuitions would not be possible. However, he also explained that any additional piece of information gathered would change the initial set of knowledge even slightly and could therefore lead to a different answer. If we think of the mind as being in a perpetual search for information, two answers could arrive very close to one another and be perceived as simultaneous. Moreover, de Finetti's discussion focuses on actual probabilities. It is plausible that our minds do not compute quantitative probabilities intuitively, but rather qualitative comparative probabilities [7]. This would make it possible to rank events as more or less likely [26,29,35,42,53] and therefore leave room for potential indecisiveness between two answers.

In ongoing research we manipulated cognitive load using a dual-task paradigm and asked participants about the comparative probability of conditional events in the set presented in Fig. 1 from Politzer & al. (2010). Preliminary results support the hypothesis that we intuitively compute qualitative conditional probabilities [10].

5.2   On Deliberation

The deliberative process could be a way to consciously distinguish what information in our pool of knowledge is relevant and what is not, or at least to consciously decide what weight each piece of information should have. It could also make us consciously explore our environment for new information to help us decide. De Finetti emphasizes that having an initial opinion based on limited information is reasonable, and that we can look for further information only when we feel it is required [28]. Such a deliberative process would be compatible with the two roles that have been assigned to it in other models of DPT in the past: 1. To correct the answer of the intuitive process. Here, deliberation would correct the strength of competing intuitions rather than the intuitive answer itself. 2. To rationalize the intuitive answer, that is, to find justification in its favor. In the hybrid model, we would choose which intuition to defend from among the competing intuitions. Moreover, it would explain both rationalization and correction with the same probability revision mechanism, differing only in what the values of the priors are changed to and what the output of the revision process would be.

5.3   Future Work

The next steps of our work will be 1. to see if changing the probabilities of events after participant have developed intuitions about them could lead to conflicting intuitions. 2. to see if using the deliberative process leads to the strength of intuitions being modified according to the rules of probabilities.

Our main challenge is to determine the probability a participant assigns to an event and the strength of their intuition about it to see if they match. To address this, we will create artificial scenarios with two possible outcomes and ask participants to predict the outcome over multiple trials. Initially, their guesses will be random, but over time, they will become more accurate as subtle cues in the scenarios are linked to specific outcomes with certain probabilities. We aim for participants to develop an intuition about the outcomes. Afterward, we will change the probabilities associated with the cues. We expect this to create a second intuition that conflicts with the first, making it difficult for participants to rely on their intuition to answer.

We also plan on manipulating the probabilities of conditionals using linguistic devices such as modality and discourse coherence [33,34] in judgement tasks adapted to allow participants to answer intuitively without the need to compute actual probabilities.

6   Conclusion

De Finetti, the founding father of Subjective Bayesianism, has provided a revolutionary new conception of what probabilities are, and his conclusion is... They do not exist! Probabilities are not tied to events, but to the mind that conceives them. This reconceived notion of probability as a mental construct has allowed psychologists to better grasp human inference production. The hybrid model of DPT completes the picture by giving us an explanation of why we sometimes have to think harder to make those inferences. In this model, the deliberative process is a fail-safe ensuring that we maintain coherence in face of conflicting intuitions that would prevent us to choose an answer.

Disclosure of Interests The authors have no competing interests to declare that are relevant to the content of this article.

References

1. Bago, B., De Neys, W.: Fast logic?: examining the time course assumption of dual process theory. Cognition **158**, 90–109 (2017)
2. Bago, B., De Neys, W.: The smart system 1: evidence for the intuitive nature of correct responding on the bat-and-ball problem. Think. Reason. **25**(3), 257–299 (2019)
3. Bago, B., Neys, W.D.: Advancing the specification of dual process models of higher cognition: a critical test of the hybrid model view. Think. Reason. **26**(1), 1–30 (2020). https://doi.org/10.1080/13546783.2018.1552194
4. Baratgin, J.: Discovering early de finetti's writings on trivalent theory of conditionals. Argumanta **6**(2), 267–291 (2021). https://doi.org/10.14275/2465-2334/202112.bar
5. Baratgin, J., Over, D.E., Politzer, G.: Uncertainty and the de finetti tables. Think. Reason. **19**(3–4), 308–328 (2013). https://doi.org/10.1080/13546783.2013.809018
6. Baratgin, J., Over, D.E., Politzer, G.: New psychological paradigm for conditionals and general de Finetti tables. Mind Lang. **29**(1), 73–84 (2014). https://doi.org/10.1111/mila.12042
7. Baratgin, J., Politzer, G.: Logic, probability and inference: a methodology for a new paradigm. In: Macchi, L., Bagassi, M., Viale, R. (eds.) Cognitive Unconscious and Human Rationality. pp. 119–142. MIT Press, Cambridge (2016). https://doi.org/10.7551/mitpress/10100.003.0010
8. Baratgin, J., Politzer, G., Over, D.E., Takahashi, T.: The psychology of uncertainty and three-valued truth tables. Front. Psychol. **9**, 1479 (2018). https://doi.org/10.3389/fpsyg.2018.01479
9. Blackwell, D., Dubins, L.: Merging of opinions with increasing information. Ann. Math. Stat. **33**(3), 882–886 (1962)
10. Bourlier, M., Doutrebente, F., Lassiter, D., Baratgin, J.: Are we intuitively bayesian ? a dual process approach to the probabilistic theory of reasoning. In: Presented at the International Conference on Thinking 2024 in Milan, Italy (2024)
11. Bourlier, M., Jacquet, B., Lassiter, D., Baratgin, J.: Coherence, not conditional meaning, accounts for the relevance effect. Front. Psychol. **14** (2023). https://doi.org/10.3389/fpsyg.2023.1150550
12. Chater, N., Zhu, J.Q., Spicer, J., Sundh, J., León-Villagrá, P., Sanborn, A.: Probabilistic biases meet the bayesian brain. Curr. Dir. Psychol. Sci. **29**(5), 506–512 (2020)
13. De Neys, W.: Bias and conflict: a case for logical intuitions. Perspect. Psychol. Sci. **7**(1), 28–38 (2012)

14. De Neys, W.: Dual Process Theory 2.0. Routledge, Abingdon (2017)
15. De Neys, W.: Advancing theorizing about fast-and-slow thinking. Behav. Brain Sci., 1–68 (2022). https://doi.org/10.1017/S0140525X2200142X
16. De Neys, W.: Further advancing fast-and-slow theorizing. Behav. Brain Sci. **46**, e146 (2023)
17. De Neys, W., Glumicic, T.: Conflict monitoring in dual process theories of thinking. Cognition **106**(3), 1248–1299 (2008). https://doi.org/10.1016/j.cognition.2007.06.002
18. Elqayam, S.: The new paradigm in psychology of reasoning. In: Linden, J.B., Thompson, V.A. (eds.) The Routledge International Handbook of Thinking and Reasoning, pp. 130–150. Routledge (2017)
19. Elqayam, S., Over, D.E.: New paradigm psychology of reasoning: an introduction to the special issue edited by Elqayam, Bonnefon, and Over. Think. Reason. **34**, 249–265 (2013). https://doi.org/10.1080/13546783.2013.841591
20. Epstein, S.: Integration of the cognitive and the psychodynamic unconscious. Am. Psychol. **49**(8), 709 (1994)
21. Evans, J.S.B.T.: Reflections on reflection: the nature and function of type 2 processes in dual-process theories of reasoning. Think. Reason. **25**(4), 383–415 (2019). https://doi.org/10.1080/13546783.2019.1623071
22. Evans, J.S.B.: On the resolution of conflict in dual process theories of reasoning. Think. Reason. **13**(4), 321–339 (2007)
23. Evans, J.S.B.: Dual-process theories of reasoning: contemporary issues and developmental applications. Dev. Rev. **31**(2–3), 86–102 (2011)
24. Evans, J.S.B., Over, D.E.: If: Supposition, Pragmatics, and Dual Processes. Oxford University Press, New York (2004)
25. Evans, J.S.B., Stanovich, K.E.: Dual-process theories of higher cognition: advancing the debate. Perspect. Psychol. Sci. **8**(3), 223–241 (2013)
26. Fine, T.L.: Theories of Probability: An Examination of Foundations. Academic press, Cambridge (2014)
27. de Finetti, B.: La logique de la probabilité. In: Actes du congrès international de philosophie scientifique, vol. 4, pp. 1–9. Hermann Editeurs Paris (1936)
28. de Finetti, B.: Introductory remarks to discussion on statistical methods and inference (1968)
29. Fishburn, P.C.: The axioms of subjective probability. Stat. Sci. **1**(3), 335–345 (1986)
30. Grice, P.: Studies in the Way of Words. Harvard University Press, Cambridge (1989)
31. Handley, S.J., Trippas, D.: Dual processes and the interplay between knowledge and structure: a new parallel processing model. In: Psychology of Learning and Motivation, vol. 62, pp. 33–58. Elsevier (2015)
32. Kahneman, D.: Thinking, Fast and Slow. Farrar, Straus and Giroux, New York (2011)
33. Kehler, A.: Coherence, reference, and the theory of grammar. No. 104 in CSLI lecture notes, CSLI Publications, Stanford (2002)
34. Kehler, A.: Discourse coherence. In: Horn, L.R., Ward, G. (eds.) The Handbook of Pragmatics, pp. 241–265. Blackwell Publishing (2006)
35. Krantz, D., Luce, D., Suppes, P., Tversky, A. (eds.): Foundations of Measurement, vol. I: Additive and Polynomial Representations. Academic Press, New York (1971)

36. Lassiter, D.: What we can learn from how trivalent conditionals avoid triviality. Inquiry **63**(9–10), 1087–1114 (2020). https://doi.org/10.1080/0020174X.2019.1698457

37. Lassiter, D.: The crucial role of compositional semantics in the study of reasoning. In: Proceedings of the 2nd International Conference on Human and Artificial Rationalities (HAR 2023) (2023)

38. Macchi, L., Poli, F., Caravona, L., Vezzoli, M., Franchella, M.A., Bagassi, M.: How to get rid of the belief bias: boosting analytical thinking via pragmatics. Eur. J. Psychol. **15**(3), 595 (2019)

39. Mosconi, G., Macchi, L.: The role of pragmatic rules in the conjunction fallacy. Mind Soc. **2**, 31–57 (2001)

40. Mura, A.: Probability and the logic of de finetti's trievents. In: Galavotti, M.C. (ed.) Bruno de Finetti Radical Probabilist, pp. 201–242. College Publications, London (2009)

41. Mura, A.: Bypassing lewis' triviality results. a kripke-style partial semantics for compounds of adams' conditionals. Argumanta **6**(2), 293–354 (2021). https://doi.org/10.14275/2465-2334/202112.mur

42. Narens, L.: Theories of Probability: An Examination of Logical and Qualitative Foundations, vol. 2. World Scientific, Singapore (2007)

43. Oaksford, M.: Could bayesian cognitive science undermine dual-process theories of reasoning? Behav. Brain Sci. **46** (2023)

44. Oaksford, M., Chater, N.: The probabilistic approach to human reasoning. Trends Cogn. Sci. **5**(8), 349–357 (2001). https://doi.org/10.1016/S1364-6613(00)01699-5

45. Oaksford, M., Chater, N.: Bayesian Rationality: The Probabilistic Approach to Human Reasoning. Oxford University Press, Oxford (2007)

46. Over, D.E.: New paradigm psychology of reasoning. Think. Reason. **15**(4), 431–438 (2009). https://doi.org/10.1080/13546780903266188

47. Over, D.E.: The development of the new paradigm in the psychology of reasonings. In: Elqayam, S., Douven, I., Evans, J.S.B.T., Cruz, N. (eds.) Logic and Uncertainty in the Human Mind A Tribute to David E. Over, pp. 161–177. Routledge (2020). https://doi.org/10.4324/9781315111902-15

48. Over, D.E., Evans, J.S.B.T.: The probability of conditionals: the psychological evidence. Mind Lang. **18**(4), 340–358 (2003). https://doi.org/10.1111/1468-0017.00231

49. Pennycook, G., Fugelsang, J.A., Koehler, D.J.: What makes us think? a three-stage dual-process model of analytic engagement. Cogn. Psychol. **80**, 34–72 (2015). https://doi.org/10.1016/j.cogpsych.2015.05.001

50. Politzer, G., Macchi, L.: Reasoning and pragmatics. Mind Soc. **1**, 73–93 (2000)

51. Politzer, G., Over, D.E., Baratgin, J.: Betting on conditionals. Think. Reason. **16**(3), 172–197 (2010). https://doi.org/10.1080/13546783.2010.504581

52. Ramsey, F.P.: Truth and probability. In: Mellor, D. (ed.) Philosophical Papers, pp. 52–94. Cambridge University Press, Cambridge (1926/1990)

53. Savage, L.J.: The Foundations of Statistics. Courier Corporation, Chelmsford (1972)

54. Sloman, S.A.: The empirical case for two systems of reasoning. Psychol. Bull. **119**(1), 3 (1996)

55. Stalnaker, R.C.: A theory of conditional. In: Rescher, N. (ed.) Studies in Logical Theory, pp. 98–112. Basil Blackwell, Oxford (1968)

56. Stalnaker, R.C.: Probability and conditionals. Phil. Sci. **37**(1), 64–80 (1970)

57. Trémolière, B., Bonnefon, J.F.: Efficient kill-save ratios ease up the cognitive demands on counterintuitive moral utilitarianism. Pers. Soc. Psychol. Bull. **40**(7), 923–930 (2014)
58. Wason, P.C.: Reasoning en, b. foss (comp.). In: New Horizons in Psychology, pp. 135–151 (1966)
59. Wason, P.C., Evans, J.S.B.: Dual processes in reasoning? Cognition **3**(2), 141–154 (1974)

A Tribute to Kahneman and Tversky in the Context of Mathematics Education

Ulrich Hoffrage[1] and Laura Martignon[2]([✉])

[1] Faculty of Business and Economics (HEC Lausanne), University of Lausanne, Lausanne, Switzerland
[2] Institute of Mathematics, Ludwigsburg University of Education, Ludwigsburg, Germany
`martignon@ph-ludwigsburg.de`

Abstract. Daniel Kahneman and Amos Tversky have changed the way we think about human rationality. Based on their contributions, the Enlightenment notion of rational humans thinking according to the laws of logic and probability had to be replaced by a more realistic conceptualisation of human thinking. However, their experiments on the so-called "errors" of the human mind were countered by a plethora of other experiments aimed at identifying the conditions that reduce or even eliminate these errors. Both the discoveries of Kahneman and Tversky, and the reactions they provoked, also had an impact on the way we teach critical thinking and probability in school. Maths educators have been prompted by these discoveries to find new, more adaptable ways to address reasoning and decision making under uncertainty in the classroom. This paper presents tools and methods that have been developed to promote logical reasoning in tasks on the so-called conjunction rule and tasks on strictly probabilistic reasoning. These tools use adequate and dynamic representations partly inspired by Barbara Tversky's work on dynamic visualizations and the way the human mind constructs thoughts. The two studies reported here were carried out in schools and demonstrate the effectiveness of these tools.

Keywords: Conjunction Fallacy · Intensional · Extensional · Icon Arrays

1 Introduction

A famous minister from the German region of Bavaria has repeatedly insisted on giving a good reason for forbidding marijuana. According to his reasoning, marijuana is a gateway drug to Cocaine. He has cited that more than 70% of cocaine users have their drug initiation with marijuana. Without pretending to defend marijuana, we do claim that there is a problem with his reasoning. What matters here is not the conditional probability that someone who makes use of cocaine today began by using marijuana in the past; the relevant conditional probability is that someone who starts with marijuana ends up using cocaine. This number is much smaller.

There are endless examples of wrong conditioning in reasoning by politicians, but also influencers and marketing agents. Recent advertisement posters in French subway

© The Author(s), under exclusive license to Springer Nature Switzerland AG 2025
J. Baratgin et al. (Eds.): HAR 2024, LNCS 15504, pp. 196–207, 2025.
https://doi.org/10.1007/978-3-031-84595-6_13

stations suggest that young men should try to drive like women do, because 88% of young people's mortal car accidents are caused by young men. Here again what matters for the assertion is the reverse conditional probability, namely the proportion of young men who have severe accidents, compared to that of women and adjusted for driven kilometer. This proportion could easily be well below one per thousand, and if men drove more than 88% of the kilometers, they would drive even more safely than women. This is just one of the many flaws human reasoning can exhibit, as has been empirically demonstrated in the famous collection of human fallacies in reasoning – not just probabilistic – discovered by Kahneman and Tversky and their close colleagues during the second half of the last century. According to the discoveries they made (see, for instance, Eddy 1982), people are bad at estimating the predictive value of a feature, like the result of a test for detecting a disease. People even make mistakes when assessing the probability of a conjunction of events, committing the so-called *conjunction fallacy*. Understanding how and why all those so-called flaws come about is a first step towards eliminating them and towards critical thinking.

2   How Can School Foster Critical Thinking?

The OECD has declared Critical Thinking as one of the relevant competencies for the 21st century. School is the main institution that can foster critical thinking of all individuals in modern times, as education should be granted every citizen of a civilized society. Education stands, in more than one sense, as opposed to Nudges. While nudges only change features of the decision-making environment, education boosts citizens' conscious judgment and decision-making capabilities. Education should therefore be more robust when it comes to initiating behavioral change across different environments (Hertwig and Grüne-Yanoff 2017).

To make it more visible and tangible to practitioners, the OECD has worked with networks of schools and teachers in 11 countries to develop and test a set of pedagogical resources that exemplify what it means to teach, learn and make progress in critical thinking in primary and secondary education. Our intervention programs have been inspired by the OECD practices. Our intention has also been to implement experience-based approaches based on activities supported by digital plugins. We begin by describing our intervention studies on the conjunction problem.

3   Extension Versus Intension in the Conjunction Fallacy

The original, by now classical, paper by Tversky and Kahneman (1983) on the conjunction fallacy was entitled "Extensional vs. Intuitive Reasoning: The Conjunction Fallacy in Probability Judgment". From our perspective, deeply influenced by the "Logique de Port Royal", that paper illustrated the importance of the intensional vs. extensional dichotomy in the context of reasoning and decision (see Stenning et al. 2017, for a detailed treatment). Let us briefly recall the meaning of the terms "intensional" and "extensional": whereas the intensional meaning (or the intension of a term) refers to the qualities or attributes that the term connotes, the extensional meaning (or the extension of a term) consists of the list of all items or individuals the term denotes. The pair of terms originated from the environment of Aristotelian logic and was established as 'epitome of the term' ("comprehension de l'idée") and 'scope of the term' ("étendue de l'idee") by the Port-Royal's logic. In Mathematics, the distinction between intensional meaning and extensional meaning was of great importance when, at the end of the 19th century, the fundamentals of the discipline were written down by scientists like Russell or by Frege. A set can namely be described intensionally or extensionally. Take, for instance, the set of even numbers beteen 1 and 10, formally written as {x| x is a natural number divisible by 2, larger than 1 and less than or equal to 10}, but it can also be written as {2, 4, 6, 8, 10}. The first formal expression describes the set intensionally, while the second expression is purely extensional. To put it somewhat casually, intensional descriptions of meaning are composed of attributes (or predicates) of the described item, while extensional descriptions are lists of the objects with those attributes.

Going back to the conjunction error, we want to stress that it was conceived extensionally: the probability of a conjunction should be rated as smaller than that of each of its components because the set that corresponds to the conjunction is contained in each of the sets of its components (for the first full and strictly extensional treatment of the fallacy, see Hertwig 1994).

Let us remember what the conjunction fallacy is about. Tversky and Kahneman (1983) demonstrated this fallacy with a short vignette about Linda:

Linda is 31 years old, single, outspoken, and very bright. She majored in philosophy. As a student, she was deeply concerned withissues of discrimination and social justice, and also participated in anti-nuclear demonstrations.

Subjects are then given some statements about Linda that they should rank according to their probability of being true, among them:

- Linda is a bank teller
- Linda is a bank teller and a feminist.

The Linda sketch naturally and strongly evokes an intensional interpretation as a story, enunciating qualities of Linda (i.e., a thumbnail biography). However, when ranking the probabilities that Linda is a bank teller, or that she is both a bank teller and a feminist, a switch to an extensional interpretation is triggered. Thus, while a detective searching for Linda among bank tellers would certainly look for a woman with political interests – and why not even a feminist – the probabilist knows that there are more bank tellers in the world than bank tellers who are also feminists. The detective thinks intensionally, while the probabilist measures sets extensionally.

Conjunctions and their probabilities are usually taught in schools based on Venn diagrams, which are ovals representing sets. This representation expresses sizes of sets but not of populations. We posit that an even more extensional approach based on representations like icon arrays foster the understanding of probabilities to an even higher degree than working with Venn diagrams. In the next section we describe how an extensional approach can be fostered by creating conjunctions in icon arrays.

4   Creating Your Own Conjunctions

An experience-based approach to learning about extensional aspects of conjunctions can be implemented by playing with the following dynamical website created by the second author and Tim Erickson: https://www.eeps.com/projects/wwg/wwg-en.html. The third section on that website is called the Conjunction Playground. Here the user can, in a playful manner, create her/his conjunctions in icon arrays. The user can choose between two task contents, namely, Soccer and Chess (one with 25 and another one with 100 icons), or Computers and Gardening (with 36 icons) and will subsequently see the icon array displayed in Fig. 1 (left panel) or that of Fig. 2 (left panel), respectively. The user plays by dragging the labels he/she finds at the bottom of the icon array, like a football or a chess-token for the icon array depicted in Fig. 1, and a plant or a computer for the icon array depicted in Fig. 2. After the user has played with the content, the icon array may look like the ones displayed in the right panels of Fig. 1 and Fig. 2, respectively.

Fig. 1. The left panel shows the screen after the user had clicked on Conjunction Playground and chose Soccer and Chess as task content. Having then selected icons from the top menu, 25 boys are displayed. At the bottom left side, the user can tip on the soccer-ball icon and drag it on those boys who, in his or her imagination, are good at soccer. In addition, the knight can be dragged on all those boys who are considered to be good at chess. After having dragged the labels on the individual icons, the array may look like that displayed in the right panel.

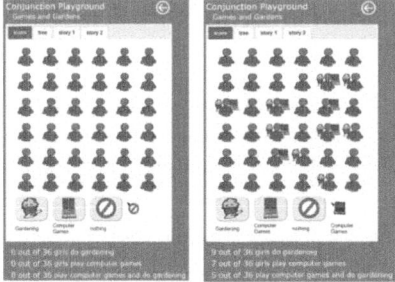

Fig. 2. For the task content Computers and Gardening, the icon array displays 36 icons of girls. Like for the content displayed in Fig. 1, the user can decide which girls do gardening and which play computer games. The left panel shows the starting point, the right panel a possible result produced by a given user.

After the user has played around in these icon arrays, clicking on icons and dragging features on them, the user can go to the menu at the top of the array. He/she can now click on a button labeled tree, and on buttons labeled Story 1 and Story 2. To start with the former, the button tree leads to a double tree like those illustrated in Fig. 3. These double trees exhibit the different four conjunction nodes in the middle row: in the situation shown in Fig. 3 (left panel), there two boys who are good both at soccer and at chess, 11 boys who are good at soccer but not at chess, etc. In the right panel of Fig. 3 we see the conjunction nodes in the middle row indicating that 5 girls play computer games and do gardening, another 4 who love gardening but do not play computer games, etc. Note that the numbers in the double tree count all possible combinations of the features (and their margins), while the numbers that are displayed below the icon array (see Figs. 1 and 2) only include some of them (albeit the crucial ones). Both the numbers below the icon arrays (Figs. 1 and 2), and the numbers in the double tree (Fig. 3) are automatically updated with every drag-and-drop of any of the features.

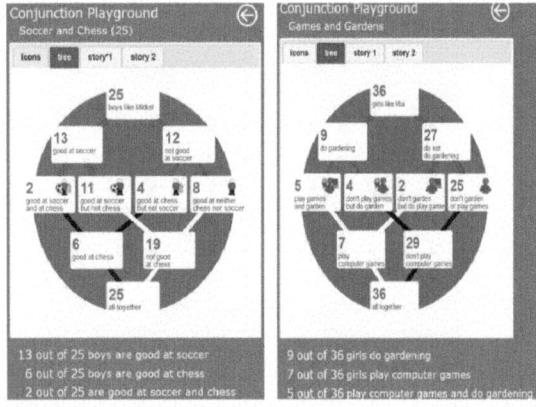

Fig. 3. Double trees that are displayed when the user clicks on trees after having generated the icon arrays shown in the right panels of Fig. 1 and Fig. 2.

Next to icons and trees, the top menu also offers the possibility to open story 1 and story 2. While Story 1 is strictly extensional, Story 2 is intensional. In the case of Soccer and Chess they read as follows:

Story 1 (extensional): Mikkel loves Sudoku and mathematics. He does lots of puzzles and has memorized pi to 50 places. Imagine 100 boys like Mikkel. Are there more boys who are good at soccer, or more boys that are good at soccer and chess?

Story 2 (intensional): Mikkel loves Sudoku and mathematics. He does lots of puzzles and has memorized pi to 50 places. What would you bet: That Mikkel is good at soccer, or that Mikkel is good at soccer and chess?

In the case of Computers and Gardening the two cover stories are:

Story 1 (extensional): Mia is a vegetarian. She eats organic yogurt and is the president of the Ecology Club. Imagine 36 girls like Mia. Are there more girls who play computer games, or more girls that play computer games and do gardening?

Story 2 (intensional): Mia is a vegetarian. She eats organic yogurt and is the president of the Ecology Club. What would you bet: That Mia plays computer games, or that Mia plays computer games and does gardening?

The intervention study we implemented (Study 1) required from students, each one operating from his/her laptop or ipad, that they freely created their own conjunctions enactively. They tended to place the icons for soccer and chess such that many boys appear to be good at soccer while only a few appear to be good at chess. An example for this kind of assignment is illustrated in Fig. 1, where 13 boys are good at soccer but only 6 are good at chess. The conjunction of being good at soccer and chess consists of 2 icons only. Students can easily see that there are more boys who are good at chess than boys who are good at both. Note that this insight conflicts with thinking along Kahneman and Tversky's representativeness heuristic. This heuristic suggests that (1) being good at soccer is typical for boys and that, hence, (2) the conjunction (which explicitly contains the term soccer) is more likely than the single event being good at chess (which does not contain the typical event).

We propose that the experience gained by playing around on that website leads to an embodied understanding of extensional conjunctions. Stated as a hypothesis:

Hypothesis: Gaining experience by creating conjunctions in icon arrays will foster extensional reasoning, thereby protecting from committing the conjunction fallacy.

To test whether this hypothesis can be empirically supported, we have implemented two intervention studies to demonstrate the effectiveness of such an experience-based approach.

5   Study 1: Testing the Facilitating Effect of Enacting Visual Representations in Conjunction Tasks

Both intervention studies were based on comparisons between two school classes (one with 28 and the other with 26 school students as participants) who were put in the treatment condition and two school classes (one with 26 and the other with 27 school students as participants) who served as a control condition. Our participants were 15 years old and in the 9th class of a German Gymnasium. All classes had the same pre-test with questions about probability and one conjunction task with the cover story about Mikkel in Story 2. The treatment consisted of two school hours (each 45 min) of experience-based work with the conjunction playground concluded by diagrams on the black board under the title Conjunction of events and a conceptual explanation. The control classes had 2 h of exercises on solving equations. Four weeks after the intervention treatment, all four classes were subject to the same post-test. The post-test contained two tasks that corresponded to the Linda task both in the cover story and in the question to be answered. This means that the cover stories were intensional descriptions, like in Story 2 both in the case of Mikkel and of Mia. The wordings were different from those in the Mikkel and Mia tasks, but referred again to persons or animals, with two different predicates.

5.1 Results of Pre- and Post-tests

The results are displayed in Table 1. There was practically no difference between the two conditions in the pre-test. Almost all students committed the conjunction fallacy, as predicted and repeatedly shown by Kahneman and Tversky. In the treatment condition, only 11.1% of the students (6 of 54) made correct extensional choices for conjunctions in both conjunction tasks the pre-test contained, and in the control condition, only 1.9% (1 of 53) made correct extensional choices. The pattern changed, however, in the post-test. In the treatment group, after having accumulated experience on the website, 85.2% of the students (46 of 54) provided correct choices for the two conjunctions in the post-test. Not surprisingly, there was no such improvement in the control condition: students' error rate stayed on the same high level as it was for the pre-test. We could, hence, observe a significant effect of the experience-based training.

Table 1. Numbers (and proportions) of students who provided correct answers (i.e., extensional choices) in the pre- and post-tests in the treatment and control condition in Study 1 (on conjunction tasks).

	Treatment condition	Control condition
Pre-test	6 (of 54)	1 (of 53)
Post-test	46 (of 54)	1 (of 53)

Following the teachings of Gerd Gigerenzer and his school, who have criticized the use of statistical rituals in cases like this, where the results speak for themselves, to the naked eye, we do not report results of inferential statistic tests here (and also not in Study 2 below). In this case, in which the studies were conducted in one school without randomizations, Gigerenzer would even speak of "mindless statistics" (see Gigerenzer 1993 and 2004).

The website where we find the Conjunction Playground used in our intervention studies offers another scenario, namely that of conditioning an event upon another event, with the usual components. These components are again an icon array and buttons leading to double trees for representation and two stories, a fully extensional one and an intensional one. The conjunction of two events is the basic element for measuring the dependence of one event upon the other. If the event S represents, for instance, *is good at soccer* in the small population of boys we treated above, and C represents the event *is good at chess*, then the conditional probability of a boy being good at chess given that he is also good at soccer is measured by the proportion of those who are good at both among those who are good at soccer. The two events are called independent if this proportion is the same as the proportion of boys who are good at chess in the whole population. As we described in our introduction, learning to deal with conditional probabilities is fundamental, and is often not easy for humans. To help students and adults overcoming such problems, we have designed interventions that foster conditional reasoning.

Under similar conditions as those described in Sect. 4, we performed another intervention study, Study 2, this time on probabilistic conditioning and Bayesian inference.

Participants were again pupils of the same four 9th classes we had for Study 1 described above. Study 2 was conducted three weeks after Study 1. Those youngsters who were in the treatment group for the conjunction tasks in Study 1 were also the ones for the treatment group for the intervention on probabilistic conditioning and Bayesian inferences in Study 2.

Instead of working with a pre-test we could base our assessment of students' knowledge and understanding of Bayesian reasoning on a test given by the teachers after having completed their teaching on probabilities. The test contained one task based on the following contingency table for an Elisa test for HIV in a certain population of 101.000 people (Table 2).

Table 2. Statistical properties of the Elisa Test, that were used in the HIV task in the pre-test of Study 2

Elisa Test	Disease present	Disease absent	Total
positive	995	500	1.495
negative	5	99.500	99.505
Total	1.000	100.000	101.000

The questions in the task were devoted to finding probabilities of patients satisfying different conditions:

a) That a patient with the disease tests positive
b) That a patient without the disease tests negative
c) That a patient with the disease tests negative
d) That a patient without the disease tests positive
e) That a patient who tests positive does have the disease

Our intervention in Study 2 consisted of two hours of experience-based learning working with the dynamic page https://www.eeps.com/projects/wwg/wwg-en.html (specifically, its first section dubbed "The explanatory power of features"). One of the contents in this first section is a disease in a certain population (with a high base rate), with the icon array displayed in Fig. 4.

Going from Fig. 4, left panel, to Fig. 4, right panel, visualizes the effect of *sorting*, which facilitates the counting. Students experience this action as the first "statistical" move, or even "the beginning of statistics".

The next statistical step is to add classification to sorting. Classification is based on forming classes according to features. Handy representations can be tables, or, if a hierarchy is desired, trees and double trees, as in Fig. 5.

The tool and the visualizations behind "The explanatory power of features" is designed to support children and adults becoming informed and competent when:

• understanding base rates (of having the disease and of not having the disease),
• dealing both with the sensitivity of a test (i.e., the probability of testing positive when having the disease) or its specificity (i.e., the probability of testing negative when not having the disease),

- inferring the positive predictive value of a test (PPV, i.e., the probability of having the disease when testing positive) and its negative predictive value (NPV, i.e., the probability of having the disease when testing negative),
- understanding how the PPV and the NPV hinges on sensitivity, specificity, and base rates.

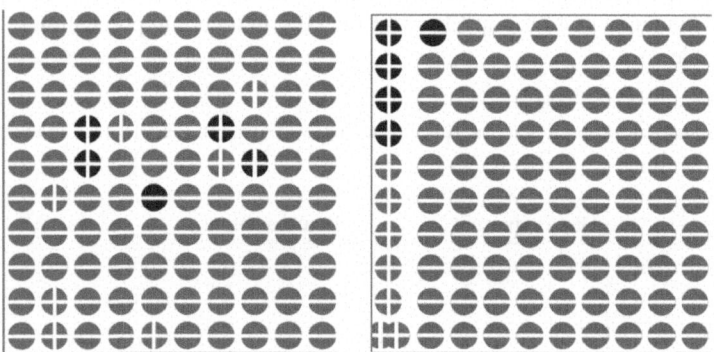

Fig. 4. An icon array for 100 patients who may or not have HIV (left panel: unsorted, right panel: sorted). Legend: ⬤ means "disease is absent and test is negative", ⬤ means "disease is absent and test is positive", ⬤ means "disease is present and test is negative", and ⬤ means "disease is present and test is positive".

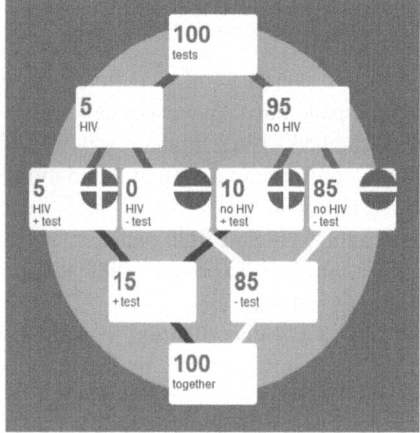

Fig. 5. The double tree arranges counts of the margins of the features "HIV" and "not HIV", those of the margins of the features "positive test" and "negative test", and, in the middle row, counts of the conjunctions of these features. The counts in Fig. 5 correspond to those in the icon array displayed in Fig. 4. The double tree facilitates Bayesian reasoning.

The treatment group in our Study 2 had only two school hours of experience-based work with this webpage and a brief conceptualization. The control group had a class on solving equations in correspondence with the schedule. Three weeks later both groups were tested with two tasks with other cover stories on medical tests but equivalent situations and questions. Prior to our intervention, that is, in the pre-test, the two conditions did not differ (Table 3 shows the number of students who answered questions (a) or (e) (or both) correctly, and Table 4 shows the number of students who answered all questions correctly). In contrast, there was a big difference between the two conditions in the post-test, after treatment: While the results in the post-test were excellent for the treatment group (45 school students answered all questions correctly; Table 3), and there were 51 out of 53 who answered either (a) or (e) (or both) correctly (Table 4), there was practically no improvement in the control group, both when considering only questions (a) and (e) (Table 3) and when considering all five test questions (Table 4).

Table 3. Results of pre-tests and post-tests in the treatment and control condition in Study 2 (on probabilistic conditioning and Bayesian inference). The numbers denote how many students answered either question (a) or (e) (or both) correctly

	Treatment condition Questions a) or e)	Control condition Questions a) or e)
Pre-test	21 (of 54)	21 (of 53)
Post-test	52 (of 54)	23 (of 53)

Table 4. Results of pre-test and post-tests in the treatment and control condition in Study 2 (on probabilistic conditioning and Bayesian inference). The numbers denote how many students answered all questions (a) - (e) correctly

	Treatment condition	Control condition
Pre-test	20 (of 54)	17 (of 53)
Post-test	45 (of 54)	23 (of 53)

Our results clearly show that, at least for these students, using this dynamical website with the icon arrays, sorted and unsorted, and the corresponding double trees is a successful method to train extensional reasoning both for assessing conjunctions (Study 1) and for conditional reasoning (Study 2).

6 Conclusion

Daniel Kahneman and Amos Tversky (and their close colleagues) introduced and used extremely interesting tasks that school students can work on and should be able to cope with from 9th class on in order to acquire the competency of critical thinking as a basis for sound decision making. For some of these tasks, for instance, Bayesian inference tasks, Gerd Gigerenzer and his group developed instruments that could improve adults' performance (like the so-called "natural frequencies"; Gigerenzer and Hoffrage 1995; Hoffrage et al. 2000). Likewise, Gigerenzer and his group introduced icon arrays to communicate medical information to adults (Galesic et al. 2009). Brase (2008) analyzed different representations for conditional and Bayesian reasoning as well, and his results also confirmed the advantages of using such icon arrays. Icon arrays are an excellent representation of extensional information. Our work consists in making use of these ideas in the context of math education, adding the dynamical tools for experience-based learning. We have also been inspired by the work of Gage and Spiegelhalter (2016), who, on their turn, introduced natural frequencies for teaching probability in English schools.

The webpages presented here were inspired by the studies on the advantages of icon arrays as representation of information in probabilistic reasoning tasks mentioned above, adding the component of dynamical, interactive and embodied treatment. Our main contribution is the enactive character of these webpages, which allow for an experience-based learning for probabilistic reasoning (see Martignon and Hoffrage 2019) and a movement based thinking, as Barbara Tversky has consistently proposed (Tversky 2019).

Schools in Germany are slowly beginning to implement such practices and it has been our purpose to implement and test such procedures in order to exhibit their didactical power.

Disclosure of Interests.   We declare that we have no competing interest of any kind, related to the work presented here.

References

Brase, G.L.: The power of representation and interpretation: doubling statistical reasoning performance with icons and frequentist interpretations of ambiguous numbers. J. Cogn. Psychol. **26**(1), 81–97 (2014)

Eddy, D.: Probabilistic reasoning in clinical medicine: problems and opportunities. In: Tversky, A., Kahneman, D., Slovic, P. (eds.) Judgment Under Uncertainty: Heuristics and Biases, pp. 249–267. Cambridge University Press, Cambridge (1982)

Frege, G.: Die Grundlagen der Arithmetik: Eine logisch-mathematische Untersuchung über den Begriff der Zahl. Verlag von Wilhelm Koebner, Breslau (1960). (In English)

Frege, G.: The Foundations of Arithmetic: A Logico-Mathematical Enquiry into the Concept of Number. Translated by Austin, J. L., 2nd edn. Northwestern University Press, Evanston (1884)

Gage, J., Spiegelhalter, D.: Teaching Probability. Cambridge University Press, Cambridge (2016)

Galesic, M., Garcia-Retamero, R., Gigerenzer, G.: Using icon arrays to communicate medical risks to low-numeracy people. Health Psychol. **28**, 210–216 (2009)

Garcia-Retamero, R., Galesic, M., Gigerenzer, G.: Do icon arrays help reduce denominator neglect? Med. Decis. Mak. **30**, 672–684 (2010). http://mdm.sagepub.com/content/30/6/672

Gigerenzer, G.: The superego, the ego, and the Id in statistical reasoning. In: Keren, G., Lewis, C. (eds.), A Handbook for Data Analysis in the Behavioral Sciences: Methodological Issues, pp. 313–339. Erlbaum, Hillsdale (1993)

Gigerenzer, G.: Mindless statistics. J. Socio-Econ. **33**, 587–606 (2004)

Gigerenzer, G., Hoffrage, U.: How to improve Bayesian reasoning without instruction: frequency formats. Psychol. Rev. **102**, 684–704 (1995)

Hertwig, R.: Why dr. gould's homunculus doesn't like to think like dr. gould: the conjunction fallacy reconsidered. Dissertation at the University of Konstanz (1994)

Hertwig, R., Grüne-Yanoff, T.: Nudging and boosting: steering or empowering good decisions. Perspect. Psychol. Sci. **12**(6), 973–986 (2017)

Hoffrage, U., Lindsey, S., Hertwig, R., Gigerenzer, G.: Communicating statistical information. Sci. **290**, 2261–2262 (2000)

Martignon, L., Hoffrage, U.: Wer wagt, gewinnt? Wie Sie die Risikokompetenz von Kindern und Jugendlichen fördern können. Hogrefe, Göttingen (2019)

Russell, B.: The Principles of Mathematics. Cambridge University Press, Cambridge (1903)

Stenning, K., Martignon, L., Varga, A.: Probability-free judgment: integrating fast and frugal heuristics with a logic of interpretation. Decision **4**(3), 136–158 (2017)

Tversky, A., Kahneman, D.: Extensional vs. intuitive reasoning: the conjunction fallacy in probability judgment. Psychol. Rev. **90**, 293–315 (1983)

Tversky, B.: Mind in Motion: How Action Shapes Thought. Basic Books, New York (2019)

Effect of Brief Mindfulness Training on Cognitive Rigidity

Léa Lachaud[1,2](✉) ⓘ and Jérémy Louis[1,2] ⓘ

[1] Univ Paris Est Créteil, CHArt, 94380 Bonneuil, France
{lea.lachaud,jeremy.louis}@u-pec.fr
[2] Lutin Userlab, CHArt, Cité des Sciences et de l'Industrie, Paris, France

Abstract. The practice of mindfulness involves the training of the processes involved in the self-regulation of attention. Attentional flexibility, as opposed to mental rigidity, is one of the main cognitive skills enhanced by the practice of mindfulness. In psychology, the *Einstellung* effect refers to the fact that individuals automatically apply old, known strategies when solving a problem, even though new strategies might enable them to solve it more efficiently. This phenomenon reflected mental rigidity that could be measured by the Water-Jar Task (WJT) [13]. Previous experiments have shown that meditators who have practiced mindfulness regularly for several years were less susceptible to this phenomenon, revealing less mental rigidity [5, 15]. Following these experiments, we investigate the effect of brief mindfulness training (10 min) on mental rigidity, by replicating the WJT protocol used by Greenberg et al. [5] on 62 participants split into two groups. The mindful group was trained in mindfulness meditation and the control group listened to a podcast on the topic of meditation just before the task. Given that brief mindfulness training improves insight problem-solving, we can speculate that it might also reduce cognitive rigidity via increased attentional flexibility. So, we hypothesize that the mindful group will obtain a lower score of cognitive rigidity than the control group. The results showed that brief mindfulness training and trait mindfulness reduce the cognitive rigidity score on the WJT. These findings support the idea that mindfulness could contribute to de-automating certain cognitive processes and increase creativity.

Keywords: Mindfulness · Cognitive Rigidity · *Einstellung* effect · Water-Jar Task

1 Introduction

When facing problems in everyday life, humans tend to adopt already-known strategies to solve them. This automatic, rapid, economical, and useful mechanism for survival is referred to as heuristic [12, 20]. However, those strategies are sometimes applied to problems that could be more easily solved by using new strategies [7]. This tendency to quickly resolve problems is associated with cognitive rigidity, which can be described as the inclination to persist in the use and formulation of a strategy, idea, or behavior [18]. In contrast, cognitive flexibility is an executive function that promotes the development

© The Author(s), under exclusive license to Springer Nature Switzerland AG 2025
J. Baratgin et al. (Eds.): HAR 2024, LNCS 15504, pp. 208–216, 2025.
https://doi.org/10.1007/978-3-031-84595-6_14

of new perspectives on a situation or object. This flexibility allows for adjusting one's response to environmental stimuli [18]. More precisely, Scott [18] indicates that "flexibility consists in the ready alteration of images, by selectively changing the attributes assigned to them; alternatively viewed, it consists in ready alteration of the relations among attributes, so that they can intersect the set of object-images in new ways. By contrast, cognitive rigidity consists both in maintaining fixed images of objects and in maintaining constant correlations among the attributes conceived in the cognitive domain" (p. 412). A lack of cognitive flexibility is associated with difficulty in changing mental and behavioural responses, which can exacerbate pathologies like depression and anxiety, when ruminating for example [9]. However, skills associated with the process of cognitive flexibility, such as coping to changing environmental demands, reconfiguring mental resources, or changing perspectives [9], are trained through the practice of mindfulness [8].

The Water-Jar Task (WJT) is a test assessing cognitive rigidity [13]. This task showed that, after being accustomed to solving a problem using a complex procedure, participants are unable to resolve the same kind using a much simpler procedure. This is the *Einstellung* effect [13]. About the Luchins's experiments (1942): "In the original experiment [13], participants had to resolve five mathematics problems with the same complex procedure (B-A-2C) followed by five similar problems with a simpler procedure (A + C or A−C). The first experiment was conducted on 222 college students. Just before the WJT, half of the participants received the instruction to be awake and attentive while they solved each problem ("Don't Be Blind" (DDB) group) and others were instructed to solve problems without this hint (Plain group). Results showed that the DDB group was more successful in changing their procedure when the problems were similar. This effect was also observed through other experiments [13] in which adults and children (9 to 14 years) were participating. Adding a warning to the participants drew their attention to the problems. Increased awareness during problem solving could therefore reduce cognitive rigidity. Yet, mindfulness is known to increase self-regulation and awareness [1]. Thus, it could be replacing the DDB effect acting as a warning. Using a similar protocol (see the Method part, where the replicated paradigm is described), Greenberg et al. [5] have shown in two experiments that mindfulness can help reduce cognitive rigidity by acting on the *Einstellung* effect. In these experiments, experienced mindfulness meditators (14 participants in experiment 1) and participants in a Mindfulness-Based Cognitive Therapy program [19] (38 participants in experiment 2) scored lower on the WJT than non-meditators (21 in experiment 1 and 38 in experiment 2). In the two experiments, non-meditators were on the waiting list to receive the mindfulness training later. The author explained this result by the fact that mindfulness could immunize being "blind" to past experiences thanks to the adoption of the "beginner's mind" and by the fact that it focuses attention on the present moment [1]. Mindfulness could thus improve procedural change in the context of a problem involving rigid, repetitive resolution patterns, by promoting the generation of new solutions. In addition, in two studies (N =

157), Ostafin and Kassman [15] showed that a brief 10-min mindfulness training session just before solving insight problems of a similar nature[1] could enhance their resolution (compared to a control group who listened to an audio text on natural history). They also highlighted the positive correlation between mindfulness trait (measured by the score on the *Mindfulness Attention Awareness Scale* (MAAS) [3]), mindfulness state, and the resolution of these problems. The mindfulness trait is stable and persisting in individuals, whereas the mindfulness state refers to the state of mind the meditator is in when engaged in meditation [2, 4].

This current study aims to investigate if 10 min of mindfulness training is enough to help to reduce cognitive rigidity through the WJT, as for insight problems [15]. Given that Greenberg et al. [5] have shown an effect of six weeks of training on reducing cognitive rigidity, we replicate their WJT protocol to investigate whether brief mindfulness training helps reduce cognitive rigidity score too. Following the results of Ostafin and Kassman on insight problems [15], we hypothesize that a brief 10-min mindfulness training session would decrease the cognitive rigidity score on the WJT. We also hypothesize that the cognitive rigidity score may be correlated with scores obtained on the MAAS scale [3] and with the mindfulness state, as was the case for insight problems [15]. Therefore, we hypothesize that brief mindfulness training could increase the use of unusual simple alternative strategies rather than complex habitual strategies to solve problems.

2 Method

2.1 Participants

Participants were recruited from visitors to the Museum of Cité des Sciences et de l'Industrie in Paris (France). They were all novices in mindfulness practice and were unfamiliar with the Water Jar Task (WJT). A G*power analysis showed that 24 to 109 participants were needed to replicate the results of Greenberg et al. [5], but they only had 27 and 51 subjects (respectively experiments 1 and 2). To account for the rather small effects in the 2012 study, we recruited 81 participants. Of those participants, 29 were excluded from the study. Exclusion criteria included: (i) obtaining fewer than six correct trials out of the first ten presented problems to ensure proper habituation to the application of the complex strategy, (ii) using fractions instead of the number of jars, and (iii) making calculation errors on the critical and extinction trials. Finally, 62 participants were retained (36 females, 26 males, aged M = 25.9; SD = 8). Participants were randomly assigned to one of two groups: the mindful group or the control group. The mindful group consisted of 32 participants (19 females, 13 males, aged M = 26.03; SD = 7.81), and the control group consisted of 30 participants (17 females, 13 males, aged M = 25.77; SD = 8,34).

---

[1] Insight problems' solutions have the particularity of arising suddenly into the consciousness, without conventional logical rules assisting in their resolution. In this type of problem, the use of procedures or strategies from past problem-solving leads to a dead end [15]. In this study [15], participants had to resolve three insight problems including the prisoner's rope problem, the antique coin problem and the inverted steel pyramid problem, and two non-insight problems including card problem and the criminal problem [16]. Mindfulness has helped to resolve only the insight problems.

2.2 Material

Water Jar Task

The WJT was assessed in its computerized version, which was adapted and translated from Greenberg et al. [5] using the software Soscisurvey. The problems were adapted from the two original studies [13, 17] (see Table 1).

Table 1. Water Jar task's items (replicated from Greenberg et al. [5])

Trial type	Jar A	Jar B	Jar C	Goal to obtain	Shortest Solution
Example	29	3	0	20	A-3B
Set	31	61	12	6	B-A-2C
Set	22	57	10	15	B-A-2C
Set	18	59	16	9	B-A-2C
Set	20	67	13	21	B-A-2C
Set	22	57	10	15	B-A-2C
Set	21	127	3	100	B-A-2C
Set	18	43	10	5	B-A-2C
Set	24	52	3	22	B-A-2C
Set	19	42	3	17	B-A-2C
Set	14	163	25	99	B-A-2C
Critical	18	48	4	22	A + C
Critical	15	39	3	18	A + C
Critical	23	49	3	20	A-C
Critical	7	16	2	5	A-C
Extinction	14	39	8	6	A-C
Extinction	13	37	5	18	A + C

The participants viewed a screen on which three jars labeled with the letters A, B, and C were represented. A number under each jar indicated its size (unit). A bucket on the right side of the screen, also accompanied by a number representing its size, symbolized the target to achieve. On the left, a water fountain indicated to the participants that it was possible to fill the jars at will. Participants entered their solutions by indicating in the designated boxes how many jars A, B, or C they wanted to add or subtract (see Fig. 1).

The participants were allowed to use a notepad and a pen to resolve calculations. After ensuring that the participants had understood the instructions and navigation through the interface by providing an example, they proceeded with the experiment by themselves. The first ten trials were to be solved using the complex strategy (B-A-2C). The objective of these initial ten trials was to familiarize the participants with applying a specific strategy to solve this type of problem. Subsequently, four critical trials were presented. These trials could either be solved using the complex method (B-A-2C) or

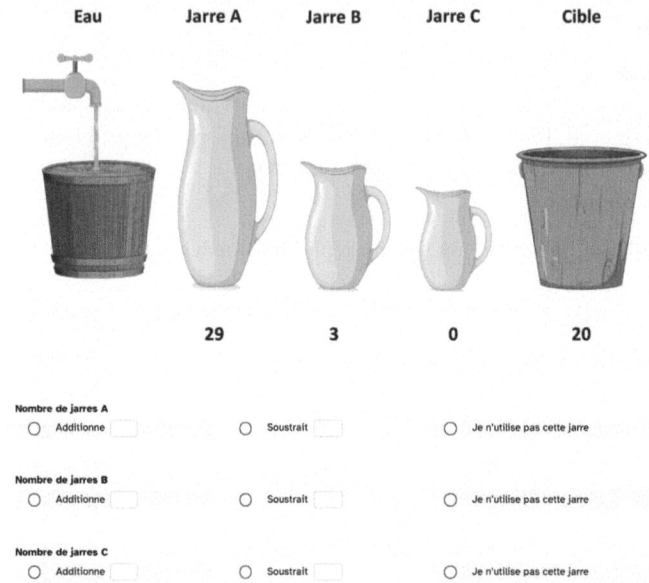

Fig. 1. Illustration of computer display on example trial (French version).

a simpler method (A + C or A-C). Finally, two extinction trials were presented. These trials could only be solved using a simple formula (A + C or A-C). Participants were instructed not to spend more than 5 min per problem. A rigidity score was calculated by adding one point for each critical or extinction trial solved with the complex formula, and for each extinction trial solved in more than 60 s.

Mindfulness Attention Awareness Scale (MAAS)
To measure participants' mindfulness trait, the Mindfulness Attention Awareness Scale (MAAS) [3] was utilized in its French version [6]. This questionnaire consists of 15 items to which participants respond using a 6-point Likert scale (1 = almost always; 6 = almost never, reverse-scored). The mean of the 15 items provides the participants' mindfulness trait score (higher scores indicating higher levels of mindfulness trait). Additionally, Item 8 of the MAAS[2] was used to measure the mindfulness state (as a continuous measure) immediately after listening to the sound induction, as did Ostafin and Kassman [15].

Audio Induction
Just like Ostafin and Kassman [15], who aimed to compare the scores of two groups of meditators and non-meditators on insight problems, we used either a breath and body-focused meditation or a podcast on the topic of meditation as auditory inductions. The two auditory inductions lasted 10 min.

---

[2] Item 8 of the MAAS: "At this moment (right now) I feel like I will rush through activities without being really attentive to them" [3] adapted by Ostafin and Kassman [15].

2.3 Procedure

The participants entered a room equipped with a computer and headphones. After signing a consent form, they indicated whether they were French native speakers or not and completed the MAAS. Next, the mindfulness group listened to the meditation auditory induction while the control group listened to the podcast, both using the headphones. Following this, they responded to the question measuring mindfulness state and indicated if they were disturbed during their listening. Then, they performed the WJT. At the end of the experiment, the participants completed a demographic survey which included questions about their gender, age, education level, study type, occupation, and experience with meditation. At the end of the experiment, participants were debriefed.

3 Results

The analyses were conducted as presented in the original study we replicated [5] with a one-tailed t-test, followed by an ANOVA adding age as a covariate. Since there is no equivalent to the Psychometric Entrance Test (PET) score[3] in France, the number of higher education years was tested but showed no significant effect.

By the first hypothesis, following the brief mindfulness training, the mindful group (M = 2.69, SD = 2.62) received significantly lower rigidity scores than the control group (M = 3.77, SD = 2.08), t(51) = 1.79, p = .039, d = .455 (see Fig. 2). This effect slightly increased in a one-way ANOVA in which Age was added as covariate to the analysis, F(1,60) = 4.34, p = .041, d = .27. Without the covariant, the analysis was only marginally significant.

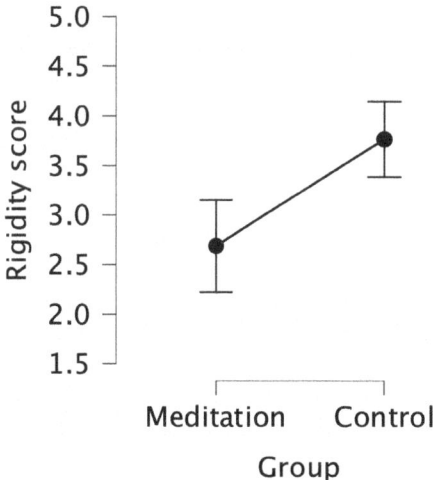

Fig. 2. Mean rigidity score (SD as error bar) depending to audio induction

---

[3] PET is a test used in Israel to measure academic abilities. The score on this test served as a control variable in the original experiment [5].

According to our second hypothesis, the rigidity score was negatively related to the mindfulness trait, $t(60) = -.3$, $p = .018$, $d = .04$; and was not significantly related to either the mindfulness state, the age or the educational level. The only other significative correlation was the positive relation between the mindfulness state and trait, $t(60) = .51$, $p < .001$, $d = .07$.

4 Discussion

The results obtained in this study showed that not only did a brief 10-min mindfulness training reduce the cognitive rigidity score on the WJT, but also that the MAAS score was negatively correlated with the rigidity score. The higher the mindfulness trait, the lower the rigidity score. These findings corroborate those obtained by Greenberg et al. [5] as well as Ostafin and Kassman [15], who hypothesized that mindfulness lessens the influence of past habits while enabling the generation of new solutions to problems. Our study completes the results obtained by Ostafin and Kassman [15], showing that 10 min of mindfulness training and high mindfulness trait could help resolving insight and *Einstellung* problems.

While Greenberg et al. [5] demonstrated a correlation between long-term regular mindfulness training (meditative retreat and 6-week mindfulness program) and the reduction of cognitive rigidity, our study suggests that brief mindfulness practice (10 min) just before the WJT already induces this benefit. The potential of such brief mindfulness training has already been explored in similar problem-solving contexts [15]. This result could be explained by the enhancement of the cognitive flexibility induced by mindfulness training as well as the reduction of the automatic mechanisms' influences [8]. These abilities would enable reasoners to more easily adopt a "beginner's mind", which promotes a sidestep that allows for approaching problems from new perspectives [1]. According to Malinowski [14], focused attention meditation is a mental training that mobilizes three major processes: emotional flexibility, attention, and cognitive flexibility. It engages different attentional processes associated with brain networks (sustained attention, distraction, monitoring, disengagement, and shifting attention). This training could increase the capacity of individuals to act with awareness, flexibility, and autonomy, which seems involved in the WJT.

The mindfulness trait, measured by the MAAS [3], was also associated with the reduction of cognitive rigidity. As Ostafin and Kassman [15] have demonstrated, the mindfulness trait could contribute to the improvement of the ability to solve problems requiring unusual and creative responses. It could do so by increasing attention to the present moment and reducing ruminations and anticipations [8].

Finally, although the mindfulness state was correlated with the mindfulness trait, it was not associated with the reduction of cognitive rigidity. This observation contradicts Ostafin and Kassman's results [15]. This could be explained by the fact that the assessment of the mindfulness state was conducted using a tool that was too flimsy and was not scientifically validated. We solely used item 8 of the MAAS questionnaire to assess this trait, as the methodology developed by Ostafin and Kassman [15], but this measure may be lacking. However, for the same reason, the correlation between trait and state measures appears to confirm the stability of the MAAS tool.

The correlation between mindfulness practice and the reduction of certain cognitive automatisms has already been demonstrated through other protocols [10, 11]. This study thus strengthens the link between mindfulness training and de-automatization or the ability to abandon old problem-solving strategies in favor of more adaptive ones.

5   Conclusion

This study highlights the relationship between mindfulness and cognitive rigidity. The results suggested that brief mindfulness training and a strong mindfulness trait can help reduce cognitive rigidity in the context of the Water Jet Task. These findings support the notion that mindfulness could contribute to de-automating certain cognitive processes and increase creativity. The ability to step aside, inherent to mindfulness, activated by increased attention to the present moment, leads to increased mental flexibility [21]. Consequently, it could be the cause of the effect observed in the present study.

Acknowledgments. Thank you to the Lutin-Userlab Laboratory for providing the experimental facilities and to the P-A-R-I-S Reasoning team for their collaboration on this project.

Disclosure of Interests. The authors have no competing interests to declare that are relevant to the content of this article.

References

1. Bishop, S.R., et al.: Mindfulness: a proposed operational definition. Clin. Psychol. Sci. Pract. **11**(3), 230 (2004). https://doi.org/10.1093/clipsy.bph077
2. Brown, K.W., Ryan, R.M.: The benefits of being present: mindfulness and its role in psychological well-being. J. Pers. Soc. Psychol. **84**(4), 822 (2003). https://doi.org/10.1037/0022-3514.84.4.822
3. Brown, K.W., Ryan, R.M.: Mindful attention awareness scale. J. Pers. Soc. Psychol. (2003) https://doi.org/10.1037/t04259-000
4. Cahn, B.R., Polich, J.: Meditation states and traits: EEG, ERP, and neuroimaging studies. Psychol. Bull. **132**(2), 180 (2006). https://doi.org/10.1037/0033-2909.132.2.180
5. Greenberg, J., Reiner, K., Meiran, N.: Mind the Trap»: mindfulness practice reduces cognitive rigidity. PLoS ONE **7**(5), e36206 (2012). https://doi.org/10.1371/journal.pone.0036206
6. Jermann, F., et al.: Mindful Attention Awareness Scale (MAAS): psychometric properties of the French translation and exploration of its relations with emotion regulation strategies. Psychol. Assess. **21**(4), 506–514 (2009). https://doi.org/10.1037/a0017032
7. Kahneman, D.: Thinking, Fast and Slow. Macmillan, New York (2011)
8. Kang, Y., Gruber, J., Gray, J.R.: Mindfulness and de-automatization. Emot. Rev. **5**(2), 192–201 (2013). https://doi.org/10.1177/1754073912451629
9. Kashdan, T.B., Rottenberg, J.: Psychological flexibility as a fundamental aspect of health. Clin. Psychol. Rev. **30**(7), 865–878 (2010). https://doi.org/10.1016/j.cpr.2010.03.001
10. Lachaud, L., Jacquet, B., Baratgin, J.: Reducing choice-blindness? an experimental study comparing experienced meditators to non-meditators. Eur. J. of Invest. Health Psychol. Educ **12**(11), 1607–1620 (2022). https://doi.org/10.3390/ejihpe12110113

11. Lachaud, L., Jacquet, B., Bourlier, M., Baratgin, J.: Mindfulness-based stress reduction is linked with an improved Cognitive Reflection Test score. Front. Psychol. **14**, 1272324 (2023). https://doi.org/10.3389/fpsyg.2023.1272324

12. Lindsay, P. H., Norman, D.A.: Traitement de l'information et comportement humain: une introduction à la psychologie. Editions Etudes vivantes (1980)

13. Luchins, A.S.: Mechanization in problem solving: the effect of Einstellung. Psychol. Monogr. **54**(6), i–95 (1942). https://doi.org/10.1037/h0093502

14. Malinowski, P.: Neural mechanisms of attentional control in mindfulness meditation. Front. Neurosci. **7**, 8 (2013). https://doi.org/10.3389/fnins.2013.00008

15. Ostafin, B.D., Kassman, K.T.: Stepping out of history: mindfulness improves insight problem solving. Conscious. Cogn. **21**(2), 1031–1036 (2012). https://doi.org/10.1016/j.concog.2012.02.014

16. Schooler, J.W., Ohlsson, S., Brooks, K.: Thoughts beyond words: when language overshadows insight. J. Exp. Psychol. Gen. **122**, 166–183 (1993)

17. Schultz, P., Searleman, A.: Rigidity of thought and behavior: 100 years of research. Genet. Soc. Gen. Psychol. Monogr. **128**(2), 165–207 (2002)

18. Scott, W.A.: Cognitive complexity and cognitive flexibility. Sociometry **25**(4), 405–414 (1962). https://doi.org/10.2307/2785779

19. Segal, Z.V., Williams, J.M.G., Teasdale, J.D.: Mindfulness-Based Cognitive Therapy for Depression: A New Approach to Preventing Relapse. The Guilford Press, New York (2002)

20. Simon, H.A.: A behavioral model of rational choice. Q. J. Econ. Econ. **69**(1), 99–118 (1955). https://doi.org/10.2307/1884852

21. Vago, D.R., Silbersweig, D.A.: Self-awareness, self-regulation, and self-transcendence (S-ART): a framework for understanding the neurobiological mechanisms of mindfulness. Front. Hum. Neurosci. **6**, 296 (2012). https://doi.org/10.3389/fnhum.2012.00296

Reasoning and Special Needs

Research Trends in Information Accessibility to Web Content for Adults with Developmental Disabilities: A Literature Review

Kai Seino[✉] [iD]

National Rehabilitation Center for Persons with Disabilities, Namiki 4-1, Tokorozawa, Japan
seino-kai@rehab.go.jp

Abstract. Introduction: Information accessibility to Web content is becoming increasingly important for people with developmental disabilities, who often face unique challenges in acquiring information due to cognitive impairments. Despite the establishment of guidelines like the Web Content Accessibility Guidelines (WCAG) 2.1, there is limited research on effective strategies to enhance Web accessibility for this population. Methods: We conducted a comprehensive review of research published in the last decade (2014–2024) using the PubMed and Google Scholar databases. The search included terms related to developmental disabilities and information accessibility in the adult population. In total, 247 articles were identified, with six meeting the criteria for detailed review. Results: Several key themes emerged from the review. Simplified formats, using plain language and visual aids, such as bar graphs and simple icons, and a user-centered design are crucial for ensuring accessibility and comprehension in individuals with intellectual and developmental disabilities. In addition, user participation in the design process can significantly enhance the usability of digital health solutions, and assistive technologies can support information accessibility. Dis-cussion: This review suggests that while the WCAG 2.1 guidelines provide a foundation, there are gaps in addressing the specific needs of people with developmental disabilities. Thus, there is a need for customized solutions that consider the diverse range of cognitive impairments and incorporate evolving technologies. Training and support, tailored to individual needs, will be critical for improving accessibility. Future research should focus on understanding the interactions between developmental disabilities and digital content processing.

Keywords: Developmental Disabilities · Cognitive Dysfunction · Cognitive · Information Accessibility · Assistive Technology

1 Introduction

Information accessibility to Web content for people with disabilities is becoming increasingly important. According to the World Health Organization [1], approximately 15% of the world's population have some form of disability. However, it has been reported that people with disabilities are often excluded from access to knowledge [2]. In recent years,

J. Baratgin et al. (Eds.): HAR 2024, LNCS 15504, pp. 219–233, 2025.
https://doi.org/10.1007/978-3-031-84595-6_15

information and communication devices such as personal computers and smartphones have become widely available and can be used by almost anyone in their daily lives. This has made it essential for many people, including those with disabilities, to acquire information from Web content to access the services they need today. In response, the World Wide Web Consortium (W3C) established the Web Content Accessibility Guidelines (WCAG) 2.1 as an international standard to enhance information accessibility for people with disabilities [3]; Other web-related ones besides W3C that deal with disability in general include library policies [4–7] and guidelines on web accessibility and web readability [8, 9]. However, despite these guidelines, people with different types of disabilities face a variety of unique challenges that require individualized adjustments and interventions.

The aims of this study were to identify research trends related to improving information accessibility to Web content for adults with developmental disabilities, and to examine the interaction between people with disabilities and their Web content. Adults in this study were defined as individuals aged $\geq$ 18 years.

2 Background

2.1 Information Accessibility for People with Disabilities

Information accessibility for people with disabilities is a process and practice aimed at ensuring the confidentiality, integrity, and availability of information and information systems while taking into account the specific needs and requirements of people with disabilities. Information accessibility includes the following aspects [10, 11]:

1) Accessibility: designing and developing information and systems so that they are accessible to people with different types of disabilities including visual, hearing, cognitive, and physical disabilities.
2) Usability: ensuring that information and systems are user-friendly and can be used effectively by people with disabilities, taking into account the specific needs and abilities of people with disabilities.
3) Privacy and security: ensuring that systems are in place to protect the privacy and personal information of individuals with disabilities and ensure the security of information systems and data from unauthorized access, modification, or destruction.
4) Compliance: complying with laws, regulations, and standards related to accessibility, privacy, and security for people with disabilities, including the Americans with Disabilities Act (ADA) [12] and WCAG 2.1 [3].

Ensuring information accessibility for people with disabilities is important for the following reasons [10]:

1) Equal access: promoting equal access to information and services, enabling people with disabilities to fully participate in various aspects of life, including education, employment, and social activities.
2) Independence and empowerment: by providing accessible and easy-to-use information systems, people with disabilities can become more independent, empowered, less dependent on others, and more autonomous.

3) Inclusion and diversity: addressing information assurance for people with disabilities fosters an inclusive and diverse society in which everyone can contribute and benefit from technological advances.

However, achieving information accessibility for people with disabilities can be challenging because the needs and requirements of individuals with disabilities vary widely by type and severity of disability, making it difficult to develop solutions that work for everyone [10, 11]. In addition, keeping up with rapid technological advances and ensuring that people with disabilities have access to new technology and are safe are ongoing challenges.

Organizations may face resource constraints or lack the expertise needed to effectively implement and maintain information assurance measures for people with disabilities. To address these challenges, it is essential that individuals with disabilities, disability advocates, technology developers, policymakers, and information assurance professionals work together to ensure that information systems are designed and implemented with accessibility, usability, privacy, and security in mind.

2.2 WCAG

W3C is currently recommended that WCAG standards are followed when developing Web content to guarantee information accessibility for people with disabilities [13–16]. WCAG 2.1 is based on four principles that provide the foundation for Web accessibility [14, 15]:

1) Perceivable: information and user interface components must be presented to users in ways they can perceive.
2) Operable: user interface components and navigation must be operable.
3) Understandable: information and operation of the user interface must be understandable.
4) Robust: content must be sufficiently robust that it can be interpreted reliably by a wide variety of user agents, including assistive technologies.

Under each principle, there are guidelines that provide basic goals to make content more accessible. For example, under the "Perceivable" principle, one guideline is "Provide text alternatives for non-text content [14, 15]." Other guidelines cover topics such as ensuring compatibility with assistive technologies, making content easily navigable and readable, and maximizing compatibility across devices and browsers [14–18]. WCAG 2.1 also provides advisory techniques for meeting the success criteria, such as examples, code samples, and tests.

Although WCAG provides a valuable foundation for making Web content more accessible to people with disabilities, there are three major limitations and challenges to its implementation:

1) Technical limitations: implementing WCAG guidelines for e-learning resources and other applications creates additional challenges beyond general Web accessibility, such as taking into account different learning styles and educational considerations [19, 20]; WCAG has conformance checking tools, but current tools can only detect

approximately 30% of WCAG Level A and AA compliance issues, and automated testing alone does not detect many accessibility barriers [19].

2) Implementation challenges: although WCAG is the de facto legal standard for Web accessibility, designing for a specific disability may not always create accessible Webpages for people with other disabilities [17].

3) Need for a holistic approach: for example, effective e-learning accessibility requires a holistic framework that considers usability, pedagogy, learning styles, technical aspects, and resource issues, in addition to adhering to the WCAG guidelines.

2.3 Information Accessibility Practices for People with Disabilities

General Disability. Existing research has underscored the importance of information accessibility for people with disabilities. People with disabilities, especially those with visual impairments, are at risk of being excluded from Web-based information resources because of access barriers [21]. People with visual, hearing, and intellectual disabilities often face challenges owing to a lack of accessible Web design and con-tent [22]. While current technologies such as screen readers and subtitles have improved accessibility, they do not address all types of disabilities. Furthermore, there is growing recognition that adaptive technology and individualized support will be key to effectively address these accessibility challenges [23]. Indeed, it has been shown that information accessibility can be improved and social participation and activity can be facilitated when appropriate environments and support are provided. For ex-ample, improving accessibility to information and communication technology pro-grams by providing computers, assistive technology, tutoring, and Internet access has been shown to greatly improve computer work performance, skills, and social participation among youth with severe disabilities [24].

Developmental Disabilities. Developmental disorders, also called developmental neu-ropathies, are a group of conditions characterized by impairments or delays in one or more developmental domains. According to the International Classification of Dis-eases Working Group of the World Health Organization, developmental disorders are defined as "a group of developmental conditions characterized by significant impairment of cog-nitive functions, which are associated with limitations of learning, adaptive behavior, and skills" [25, 26]. Developmental disabilities include a variety of conditions such as autism spectrum disorder (ASD), attention deficit hyperactivity disorder (ADHD), learn-ing disabilities (such as dyslexia), intellectual disabilities, and communication disorders [25, 26]. Compared to physical disabilities, there is limited research on information accessibility that addresses cognitive disabilities. These disabilities are often accom-panied by mental and developmental disabilities such as thinking and theory-of-mind disorders, which pose additional challenges to information accessibility [27]. A study of cognitive determinants in individuals with developmental cognitive disabilities found that although they could access the Web, they had difficulty using W3C accessibility-compliant Websites [28]. Thus, it is likely that the current Web accessibility guidelines do not adequately address the needs of people with cognitive disabilities.

2.4   Cognitive Function and Information Processing in Persons with Developmental Disabilities

When considering information accessibility for individuals with developmental disabilities, thought should be given to the various cognitive functions such as information processing, sensory processing, attention, and visuospatial ability because these functions differ depending on the type of developmental disability. For example, people with ASD often have difficulty in processing sensory information [29]. Impairments may also appear in visuospatial processing, particularly the integration of local and global visual information [28]. In addition, 10%–33% of adults with ASD use only simple phrases, have verbal and nonverbal intelligence quotients in the intellectual disability range, and need support [30]. Whereas in children with ADHD, sensory processing difficulties are common and can cause problems with attention, behavior, and work productivity [31]. Such individuals may struggle to coordinate and interpret sensory inputs properly and have difficulty in processing and understanding information [32].

These differences in cognitive dysfunction by type of developmental disability mean that the most effective means of presenting information is likely different for the different disabilities. To examine this in relation to text-based materials, Seino et al. [33] conducted a study in Japanese high school students with learning disabilities, ADHD, and ASD. They found that for individuals with learning disabilities, illustrations, pictures, appropriately sized text, and concise and grammatically correct sentences were important; for those with ADHD, short sentences, bold text, bullet points, and concisely organized information were important; and for those with ASD, pictures, flowcharts, concise and clear language, and positive expressions were important. Thus, when considering information accessibility, it is important to consider the information processing characteristics of the various developmental disability types. However, because people's information processing and comprehension differ when presented in physical paper media and digital formats such as Web content [34], it is likely that different considerations are also needed when developing Web content versus physical media.

3   Method

A literature review was conducted in two stages. First, the PubMed and Google Scholar databases were searched using the statements shown in Table 1; the search was limited to papers published in the last 10 years (2014–2024) and was conducted on May 30, 2024. Then, the reference lists of the papers extracted from the databases were reviewed for any relevant papers that the database search had missed; these papers were then retrieved from their sources.

Table 1. Search statements.

Database	Search statement
PubMed	((Developmental Disabilities) OR (Developmental Disability) OR (Developmental Disorder) OR (Autism) OR (Attention Deficit Hyperactivity Disorder) OR (ADHD) OR (Learning Disabilities) OR (Learning Disability)) AND ((Information Accessibility) OR (Information Access)) AND (Adult)
Google Scholar	("Developmental Disabilities" OR "Developmental Disability" OR "Developmental Disorder" OR "Autism" OR "Attention Deficit Hyperactivity Disorder" OR "ADHD" OR "Learning Disabilities" OR "Learning Disability") AND ("Information Accessibility" OR "Information Access") AND "Adults"

4 Results

4.1 Number of Papers Identified

The database search yielded 247 articles (PubMed = 62; Google Scholar = 185). I Inclusion criteria for articles were that they study relevant elements of information accessibility for people with developmental disabilities. The extracted papers were reviewed by the authors by reading the title, abstract, and text to ensure that they met the inclusion criteria. The extracted articles that met the objectives of this study were thereby selected for review, and a total of four articles (PubMed = 2, Google Scholar = 2) were identified. Two additional articles were added from the search of printed literature for a final total of six articles for detailed review.

4.2 Research Trends in Information Accessibility for Developmental Disabilities

Factors to Consider for Improving Information Accessibility

Information Behavior. There is a paucity of research on developmental disabilities with regard to information behavior, including information seeking and retrieval. The term "information behavior" here is defined as "the totality of human behavior in relation to sources and channels of information" [35]. Information seeking is primari-ly concerned with the different methods people use to find and access information sources. Information retrieval deals with the interactions between computer-based information systems and users. Berget et al. [35] onducted a literature review on the impact of various disabilities on information seeking and retrieval and found that while visual, hearing, and reading impairments, have been extensively studied, very little or no research has been conducted on autism or on specific cognitive impair-ments. Because information behavior may be relevant to accessing Web content and retrieving information, accumulating research on information behavior in individuals with developmental disabilities will be useful for improving information accessibility.

Effective Visual Design. Wu et al. [36] examined how visual designs can help people with developmental disabilities with co-occurring intellectual disabilities (intellectual and developmental disabilities; IDD). The method was a Web-based experiment measuring the accuracy and efficiency of interpreting data visualizations. The results showed that bar graphs were more accessible than line graphs for time-series data, and that stacked bar graphs were more accessible than pie charts for percentage data. For graph decoration, simple icons improved response time but not necessarily accu-racy. For data continuity, discrete representations were more effective for IDD users who preferred visualizations with real-world metaphors and simple images. However, because Wu et al.'s [36] study focused on IDD, separate consideration of the two types of disabilities is necessary when applying the outcomes of the study to one or the other alone. In addition, there are likely various other designs that also facilitate information processing and information acquisition from Web content; therefore, graph designs other than those examined in Wu et al. [36] should be explored.

User Participation in Development. Research on accessibility, ease of use, and patici-patory design has been conducted. Henni et al. [13] conducted a scoping review of the experiences, needs, and barriers to the use of digital health solutions by people with disabilities in general and found that people with disabilities highly valued digital health solutions designed to address specific challenges related to their disabilities. From the review results, many studies have reported the different types of apps avail-able for smartphones and tablets. Other studies have examined the use of conversational agents, games, social media, robotic technology, Websites, interactive maps, augmented reality, augmentation aids for low-vision users, and strategies on how to design digital health solutions. In general, regardless of the type of disability, digital health solutions need to be easy to use and intuitive. Most studies have emphasized the importance of involving people with disabilities from the earliest stages of the development process to improve the accessibility and usability of the final product [13].

Research Related to Web Content

Mobile Application Rationale and Interfaces. Păsărelu et al. [37] conducted a system-atic review of mobile applications as Internet interventions for individuals with ADHD. A total of 109 apps were included in the study, most of which were designed for the treat-ment of ADHD or for symptom assessment. There were also apps related to behavioral therapy–based intervention; time and task management; and the moni-toring and record-ing of ADHD symptoms. The most popular apps were, in descend-ing order, (1) those assessing ADHD symptoms; (2) ADHD support and management tools; (3) those track-ing symptoms, medications, and other health-related indicators; and (4) games designed to help improve concentration and attention. However, none of the apps reviewed reported any content rationale; information on the development process or interface design; or empirical evidence of their effectiveness or efficacy; in light of Henni et al.'s [13] report, involving people with disabilities as users in the development process may help improve information accessibility. Furthermore, con-sidering the reports by Seino et al. [33] and Wu et al. [36], it will be important to ex-amine the level and amount of content and the effectiveness of visual presentation methods.

Assistive Technology. Shattuck et al. [38] conducted a scoping review of various services for adults with ASD. Nine studies were reported that evaluated the impact of interventions, including assistive technology (AT) as it relates to information accessibility; six studied the use of AT to improve employment outcomes [39–44]; one stud-ied the use of a computer-based intervention for college students [45]; one studied the use of AT to support social skills development [46]; one studied the use of AT to teach community shopping skills [47].

Regarding employment, Bross et al. [39] reported that a video modeling intervention increased the use of customer service phrases by a person with autism working as a cashier in a retail store. Gentry et al. [40] studied the effects of using mobile devices with task management tool capabilities as an occupational aid for workers with ASD. With appropriate training, the intervention resulted in a decrease in the amount of support time the workers needed on the job. Smith et al. [42] evaluated the feasibility and effectiveness of "virtual reality job interview training" for people with ASD. The intervention was a computerized virtual reality training simulation that was used via computer software or the internet. The evaluation results showed that participants that received the intervention showed improvement in their job interview role-play performance.

Regarding computer-based interventions, White et al. [46] reported the feasibility of a computer-based psychosocial intervention for college students with ASD aged 18 or older. The study participants were assigned to either a virtual reality–based brain–computer interface or the colleges' Life Success Program. The brain–computer interface used electroencephalography in a virtual reality environment to facilitate emotional recognition and practice social interaction skills.

Regarding skill acquisition, Burckley et al. [47] evaluated the use of a mobile device to provide visual tips and video prompts to teach community shopping skills to young adults with ASD and intellectual disabilities. As a result of the intervention, the participants' shopping skills improved significantly, subsequent gains in skills were maintained, and shopping skills were generalized to those that were not taught. Such an approach may be effective for teaching independent shopping skills in the community.

Overall, these studies show that Web content, including computers, mobile devices, and videos, can be effective interventions for improving information accessibility in individuals with disabilities.

Telework Support for Workers with Developmental Disabilities. Seino [48] reviewed the literature on teleworking for people with disabilities during the pandemic. They found that teleworking presents some unique challenges. Two of the main challenges of teleworking include 1) the digital divide, where access to the technology and high-speed Internet needed for teleworking may be limited; and 2) the need for teleworkers with disabilities to be able to access appropriate technology and a comfortable work-space. There are three influencing factors for the two points:

(1) Health and well-being: physical and mental health affect work experience.
(2) Work skills and values: personal work skills and values, as well as the ability to manage time and tasks, are important.

(3) Work environment: the quality of the work environment, including ergonomic considerations and organizational support, is critical to the success of teleworking for people with disabilities.

Seino recommended the need for customized telework practices that address the specific needs of people with disabilities, as well as training and ongoing support to adapt to a remote work environment that uses computers and information and communication devices. It is possible that training and support may also be effective for information accessibility to address the individual characteristics and needs of people with developmental disabilities, as well as for accessing Web content and information behavior.

4.3 Elements for Improving Information Accessibility for People with Developmental Disabilities

Based on the literature review, the following elements were considered to improve information accessibility.

Visual Design. Recent research suggests the importance of visual design in improving Web accessibility for people with developmental disabilities. Wu et al. [36] showed how different types of graphs, icons, and visual aids can improve the understanding and manipulation of Web content for people with IDD.

Simplified Information Formats. It is important to provide information in a simplified format that is easy to understand. This includes using plain language, visual aids such as images and symbols, text simplification, incorporation of images and symbols, and audio/video descriptions to make Web content accessible to people with cognitive disabilities [49, 50].

User-Centered Design and Participation. Involving users with developmental disabilities in the design process is increasingly being recognized as important for improving information accessibility. Participatory design approaches may help ensure that Web content is tailored to meet the needs of the developmentally disabled population, leading to more effective and inclusive results [13, 36].

Assistive Technology. The integration of assistive technologies such as screen readers, speech recognition software, and adaptive input devices plays an important role in improving the accessibility of Web content [33, 38].

Compliance with Accessibility Guidelines. It is important to adhere to established accessibility guidelines such as WCAG 2.1. These guidelines provide a framework for making Web content recognizable, operable, understandable, and robust, thereby improving accessibility to people with developmental disabilities [13].

5 Consideration

5.1 Summary of Results

This study aimed to identify research trends in information accessibility for adults with developmental disabilities when acquiring information from Web content. Six relevant studies were identified through an extensive literature review. Key themes included the need for a user-friendly visual design, the importance of simplified information formats, the role of assistive technologies, and the importance of user involvement in the content development process. These themes suggest multifaceted challenges and potential solutions for improving the accessibility of Web content for people with developmental disabilities. Of the six studies identified, four (13, 37, 38, and 48) relate to information accessibility for people with developmental disabilities, but do not deal directly with taking it on. Thus, the study indicates that there is little research on information accessibility for people with developmental disabilities.A previous study has reported a similar lack of studies on information behavior in individuals with developmental disabilities [35].

To the best of our knowledge, this is the first review of the literature across various aspects of Web content accessibility for adults with developmental disabilities. Investigating information accessibility for people with developmental disabilities, which is under-researched, will make it easier for them to connect to the information and services they need, which will promote health and social participation and improve their quality of life.

5.2 Comparison and Relation to Previous Studies

The results of the present review suggest that people with developmental disabilities have challenges working in digital environments, such as while teleworking, and require environmental adjustments and individualized support and training [33, 38]. A variety of assistive and digital devices have been successfully used as interventions in people with developmental disabilities [38]. Thus, as already noted [23], technology and individualized support are necessary for improving information accessibility.

It has been suggested that individuals with developmental and intellectual disabilities are unique in their acquisition of visual information [36]. Examples include impairments in visual information processing in ASD [28] and the characteristics of information processing in individuals with developmental disabilities when working with paper media [33]. Future studies examining developmental disabilities alone and identifying differences between physical media and digital formats will be useful for understanding more about the information processing of Web content in individuals with developmental disabilities.

If the present study leads to improved information accessibility for people with developmental disabilities, it will have contributed to improving the independence, empowerment, and social inclusion of these individuals, which are important topics identified in previous study [10]. However, the actual situation of people with developmental disabilities with regard to their exclusion from access [2, 21, 27] remains largely unknown.

It is also unclear how the cognitive impairments found in individuals with developmental disabilities noted in previous studies affect information acquisition [27]. Therefore, further studies are warranted.

5.3 Study Limitations and Future Prospects

The study has several limitations. First, owing to the small number of applicable references, this review likely does not capture the full range of realities and issues in this field. This could be due to a lack of actual research or a lack of direct research on information accessibility. Increasing the number of studies conducted in this area is needed. Second, because the types and definitions of developmental disabilities covered in the literature varied and included those with comorbid disorders, we were unable to examine the relationship between cognitive dysfunction and information processing. Future surveys and experiments that consider cognitive dysfunction will be needed to clarify the relationship between the cognitive characteristics of individuals with developmental disabilities and information processing. Finally, because we did not specify the Web content, it was difficult to clarify the actual status of information accessibility. In the future, by focusing on specific information such as public medical services, it should be possible to clarify the actual status of information accessibility and the necessary requirements to improve accessibility for such services based on previous studies.

5.4 Examination of Human Interactions with Digital Media

This study investigated the acquisition of Web content by people with developmental disabilities. The results suggest that people with developmental disabilities require simplified, visual information that is tailored to the characteristics of their disability. However, it remains unclear how the cognitive characteristics of the various developmental disabilities affect the processing of digital devices and digital data. Further research is needed to understand more about the interactions between people with developmental disabilities and digital information.

6 Proposals and Solutions

The present review highlights not only the lack of research on information accessibility for adults with developmental disabilities but also the paucity of information on which we can derive specific suggestions and solutions. Based on the present literature review, we offer several proposals and solutions. Comprehensive consideration of these solutions and measures tailored to the characteristics of individual developmental disabilities is expected to lead to improved information accessibility.

6.1 Proposals

The following four specific suggestions have the potential to enhance information accessibility for adults with developmental disabilities:

1) Accumulate research on information accessibility for people with developmental disabilities.
2) Use visual design and easy-to-understand formats tailored to the information provided.
3) Use assistive technologies.
4) Promote development with user participation.

6.2 Solutions

There are five practical solutions to information accessibility challenges for adults with developmental disabilities:

1) Use of visual design: visual design can be effective for people with developmental disabilities in improving comprehension and the usability of Web content. Therefore, effective use of graphs, icons, and visual aids is important.
2) Simplified information provision: rather than presenting complex information as is, plain language, simplified text, and visual aids should be incorporated to make it easier for people with developmental disabilities to understand. Audio and video explanations can also be effective tools.
3) User-centered design: involving users with developmental disabilities themselves in the design process is important for creating Web content that meets their needs. Reflecting users' opinions will help achieve more effective and inclusive information accessibility.
4) Use of assistive technologies: assistive technologies such as screen readers, voice recognition software, and adaptive input devices can play an important role in helping people with developmental disabilities access Web content.
5) Adherence to accessibility guidelines: adherence to existing guidelines, such as WCAG 2.1, can make Web content more accessible to a wider audience, including people with developmental disabilities.

7 Conclusion

This study reviewed research trends in information accessibility to Web content among adults with developmental disabilities. Visual design, understandable and simplified information formats, user involvement, and assistive technologies were found to be important. However, little research has been conducted on information accessibility for people with developmental disabilities, meaning the actual conditions and effective factors remain largely unknown. Despite this lack of understanding, the present study confirms that there are several elements that impact the information accessibility of people with developmental disabilities. In the future, it will be necessary to clarify these elements and resolve issues so that individuals with developmental disabilities can effectively access the information and services they need.

Acknowledgments. This work was supported by JSPS KAKENHI Grant Number JP 24K05493.

References

1. World Health Organization. World report on disability. Geneva (2011)
2. Garbutt, M., Kyobe, M.: Knowledge practices of people with disabilities and the role of ICT. In: Proceedings of the Fourth International Conference on Information and Communication Technology and Accessibility (ICTA), pp. 1–6. IEEE, New York (2013)
3. World Wide Web Consortium (W3C). Web Content Accessibility Guidelines (WCAG) 2.1 (2023). https://www.w3.org/TR/WCAG21/. Accessed 04 July 2024
4. Australian Library and Information Association. Guidelines on library standards for people with disabilities (1998). https://www.alia.org.au/about-alia/policies-and-guidelines/alia-pol icies/guidelines-library-standards-people-disabilities. Accessed 30 May 2024
5. Burgstahler, S.: Equal Access: Universal Design of Libraries. University of Washington, Seattle (2018)
6. Canadian Library Association. Guidelines on library and information services for people with disabilities (2016). http://cfla-fcab.ca/en/guidelines-and-position-papers/guidelines-on-library-and-information-services-for-people-with-disabilities/. Accessed 30 May 2024
7. Irvall, B., Nielsen, G.S.: Access To Libraries For Persons With Disabilities, vol. 89 Checklist. IFLA, The Hague (2005)
8. Miniukovich, A., Angeli, A.D., Sulpizio, S., Venuti, P.: Design guidelines for Web readability. In: Proceedings of the 2017 Conference on Designing Interactive Systems, Edinburgh, UK (2017)
9. Venturini, G., Gena, C.: Testing Web-based solutions for improving reading tasks in students with dyslexia. In: Paternò, F., Spano, L.D. (eds.) Proceedings of the 12th Biannual Conference on Italian SIGCHI Chapter, pp. 1–6. ACM, New York (2017)
10. Shree, P., Desai, J., Pandey, A.: Universal design of digital learning environments. Procedia Comput. Sci. **78**, 239–247 (2016). https://doi.org/10.1016/j.procs.2016.02.044
11. Willman, A.: Accessibility in Higher Education: Policies and Practice. Routledge, New York (2019)
12. Civil Rights Division, US Department of Justice. The Americans with Disabilities Act (ADA). https://www.ada.gov/law-and-regs/ada/. Accessed 04 July 2024
13. Henni, S.H., Maurud, S., Fuglerud, K.S., et al.: The experiences, needs, and barriers of people with impairments related to usability and accessibility of digital health solutions, levels of involvement in the design process, and strategies for participatory and universal design: a scoping review. BMC Public Health **22**, 35 (2022). https://doi.org/10.1186/s12889-021-123 93-1
14. Almourad, M.B., Hussein, M.J., Kamoun, F., Wattar, Z.: Analysis of WCAG 2.0 data accessibility success criterion of e-government Websites. Period. Eng. Nat. Sci. (PEN) **7**, 496–503 (2019)
15. Casaro, D.E., Alfonzo, P.L., Mariño, S.I., Godoy, M.V.: Applying WCAG 2.0 guidelines in online banking services-an empirical case study in Argentina (2015)
16. Chawla, M.N., Rana, P.: A practitioner's approach to assess the WCAG 2.0 website accessibility challenges. In: Proceedings of the 2019 Amity International Conference on Artificial Intelligence (AICAI), pp. 958–966 (2019)
17. Currier, B.: Comparing dyslexia and visual impairments under W3C's WCAG: a legal standard for web design? (2015)
18. Patra, M.R., Dash, A.R.: Accessibility of Indian government web portals with respect to WCAG 2.0 and GIGW guidelines. In: Proceedings of the 11th International Conference on Theory and Practice of Electronic Governance (2018)
19. Kelly, R., Qian, Y., Lo, L.: Digital accessibility and usability: practical considerations for web developers (2019)

20. Crawford, M.J., Killaspy, H., Barnes, T.R.E., Barrett, B., Byford, S., Clayton, K., et al.: Group art therapy as an adjunctive treatment for people with schizophrenia: multicentre pragmatic randomised trial. BMJ **346** (2013)
21. Brophy, P., Craven, J.: Web accessibility. Libr. Trends **55**, 950–972 (2007). https://doi.org/10.1007/978-1-4471-7440-0
22. Jaeger, P.T.: Disability and the Internet: Confronting a Digital Divide. Lynne Rienner Publishers, Boulder (2012)
23. Lazar, J., Goldstein, D.F., Taylor, A.: Ensuring Digital Accessibility through Process and Policy. Morgan Kaufmann, Burlington (2015)
24. Annable, G., Goggin, G., Stienstra, D.: Accessibility, disability and inclusion in information technologies: introduction. Inf. Soc. **23**, 145–147 (2007). https://doi.org/10.1080/01972240701323523
25. Santos, M.D.: Intellectual developmental disorders: towards a new name, definition and framework for mental retardation/intellectual disability in ICD-11. Nascer e Crescer. **21**, 248–249 (2012)
26. Salvador-Carulla, L., Reed, G.M., Vaez-Azizi, L.M., Cooper, S.A., Martinez-Leal, R., Bertelli, M., et al.: Intellectual developmental disorders: towards a new name, definition and framework for "mental retardation/intellectual disability" in ICD-11. World Psychiat. **10**(3), 175–180 (2011). https://doi.org/10.1002/j.2051-5545.2011.tb00045.x. PMID: 21991267; PMCID: PMC3188762
27. Small, J., Schallau, P., Brown, K., Appleyard, R.: Web accessibility for people with cognitive disabilities. In: Proceedings of the CHI '05 Extended Abstracts on Human Factors in Computing Systems (2005). https://doi.org/10.1145/1056808.1057024
28. Cardillo, R., Vio, C., Mammarella, I.C.: A comparison of local-global visuospatial processing in autism spectrum disorder, nonverbal learning disability, ADHD, and typical development. Res. Dev. Disabil. **103**, 103682 (2020). https://doi.org/10.1016/j.ridd.2020.103682. PMID: 32442872
29. Happé, F.G., Mansour, H., Barrett, P., Brown, T., Abbott, P., Charlton, R.A.: Demographic and cognitive profile of individuals seeking a diagnosis of autism spectrum disorder in adulthood. J. Autism Dev. Disord. **46**, 3469–3480 (2016)
30. Developmental disabilities monitoring network surveillance year 2010 principal investigators; centers for disease control and prevention. prevalence of autism spectrum disorder among children aged 8 years—autism and developmental disabilities monitoring network, 11 sites, United States, 2010. MMWR Surveill Summ. **63**, 1–21 (2014)
31. Borden, K.: Sensory strategies appear to improve classroom behaviours and attention in children with psychiatric disabilities (2005)
32. Ghosh, P., Ghosh, S., Mondal, S., Moulik, S.: Assessing sensory processing disorders in a child guidance clinic with focus on ADHD. Eastern J. Psychiat. (2019)
33. Seino, K., Enomoto, Y., Miyazawa, S.: Textual requirements for high school students with developmental disabilities: text analysis about assessing educators and employment support professionals in Japan. J. Dev. Disabil. Res. **1**, 1–17 (2023)
34. Hui-pin, Q.: Multimedia information and digital image processing technology (2007)
35. Berget, G., MacFarlane, A.: What Is known about the impact of impairments on information seeking and searching? J. Am. Soc. Inf. Sci. **71**, 596–611 (2020). https://doi.org/10.1002/asi.24256
36. Wu, K., Petersen, E., Ahmad, T., Burlinson, D., Tanis, S., Szafir, D.: Understanding data accessibility for people with intellectual and developmental disabilities. In: Proceedings of the 2021 CHI Conference on Human Factors in Computing Systems (2021). https://doi.org/10.1145/3411764.3445743

37. Păsărelu, C.R., Andersson, G., Dobrean, A.: Attention-deficit/hyperactivity disorder mobile apps: a systematic review. Int. J. Med. Inf. **138**, 104133 (2020). https://doi.org/10.1016/j.ijm edinf.2020.104133. PMID: 32283479

38. Shattuck, P.T., et al.: Services for adults with autism spectrum disorder: a systems perspective. Curr. Psychiat. Rep. **22**(3), 13 (2020). https://doi.org/10.1007/s11920-020-1136-7. PMID: 32026004; PMCID: PMC7002329

39. Bross, L.A., Travers, J.C., Munandar, V.D., Morningstar, M.: Video modeling to improve customer service skills of an employed young adult with autism. Focus Autism Other Dev. Disabil. **34**(4), 226–235 (2018)

40. Gentry, T., Kriner, R., Sima, A., McDonough, J., Wehman, P.: Reducing the need for personal supports among workers with autism using an iPod touch as an assistive technology: delayed randomized control trial. J. Autism Dev. Disord. **45**(3), 669–684 (2015)

41. Hill, D.A., Belcher, L., Brigman, H.E., Renner, S., Stephens, B.: The Apple iPad™ as an innovative employment support for young adults with autism spectrum disorder and other developmental disabilities. J. Appl. Rehabil. Couns. **44**(1), 28–37 (2013)

42. Smith, M.J., et al.: Virtual reality job interview training in adults with autism spectrum disorder. J. Autism Dev. Disord. **44**(10), 2450–2463 (2014)

43. Smith, M.J., et al.: Brief report: vocational outcomes for young adults with autism spectrum disorders at six months after virtual reality job interview training. J. Autism Dev. Disord. **45**(10), 3364–3369 (2015)

44. Wehman, P., Brooke, V., Brooke, A.M., Ham, W., Schall, C., McDonough, J., et al.: Employment for adults with autism spectrum disorders: a retrospective review of a customized employment approach. Res. Dev. Disabl. **53**, 61–72 (2016)

45. White, S.W., Richey, J.A., Gracanin, D., Coffman, M., Elias, R., LaConte, S., et al.: Psychosocial and computer-assisted intervention for college students with autism spectrum disorder: preliminary support for feasibility. Educ. Train Autism Dev. Disabil. **51**(3), 307–317 (2016)

46. Kandalaft, M.R., Didehbani, N., Krawczyk, D.C., Allen, T.T., Chapman, S.B.: Virtual reality social cognition training for young adults with high-functioning autism. J. Autism Dev. Disord. **43**(1), 34–44 (2013). https://doi.org/10.1007/s10803-012-1544-6

47. Burckley, E., Tincani, M., Guld Fisher, A.: An iPad™-based picture and video activity schedule increases community shopping skills of a young adult with autism spectrum disorder and intellectual disability. Dev. Neurorehabil. **18**(2), 131–136 (2015). https://doi.org/10.3109/175 18423.2014.945045. PMID: 25084013

48. Seino, K.: Diversity and inclusion in telework for persons with disabilities after COVID-19: a literature review. Effect. Hum. Resour. Manag. Multigenerat. Workplace, 102–114 (2024)https://doi.org/10.4018/979-8-3693-2173-7.ch005

49. Murthy, S., Parker Harris, S., Hsieh, K.: Information needs of caregivers of adults with intellectual and/or developmental disabilities in India. J. Intellect. Disabil., 17446295241254933 (2024). https://doi.org/10.1177/17446295241254933. PMID: 38749503

50. Benson-Goldberg, S., Geist, L., Erickson, K.: Simplified COVID-19 guidance for adults with intellectual and developmental disabilities. J. Appl. Res. Intellect Disabil. **37**(3) (2024). https://doi.org/10.1111/jar.13222. PMID: 38494739

NAO Robot to Help People with Alzheimer's Manage the Recall of Activities in Prospective Memory

Kerem Tahan[1,2]([⊠]) and Bernard N'kaoua[1]

[1] Université de Bordeaux, Bordeaux Population Health U-1219, 33000 Bordeaux, France
k.tahan@colisee.fr, bernard.nkaoua@u-bordeaux.fr
[2] Colisée Group, 33300 Bordeaux, France

Abstract. This study aims to evaluate the ability of an assistive robot to help caregivers manage Prospective Memory (PM) activities in people with Alzheimer's disease (AD) living in an institution. The Prospective Memory tasks corresponded to daily life activities to be carried out 15 min later (time-based PM). Fourteen people with AD diagnosed by a professional and assessed using the MMSE and the BDAE participated in the study. Each participant underwent 2 individual sessions, one managed by the NAO robot and the other by a caregiver. PM tasks included bringing (after a delay) a new person into the office (a staff member or another resident), an object (such as a glass of water or a newspaper), or performing a specific action (like sending an envelope or having a snack). The results show that MP performance was identical when the tasks were performed with the robot or with the caregiver. Additionally, engagement and attention levels were higher with the robot than with the caregiver. These results were discussed in terms of the robots' capabilities to provide valuable assistance in the management of the daily activities of institutionalized patients.

Keywords: Aging · Dementia · Prospective Memory · Assistive Robot · NAO

1 Introduction

Prospective memory (PM) plays a crucial role in autonomy and management of daily life activities. This form of memory refers to the ability to plan and carry out deferred intentions in the future, such as remembering to take medication at a certain time of day or to go to an appointment, etc. (Einstein and McDaniel 1990). This capacity is essential for maintaining the autonomy and functional independence of elderly people and in particular people with Alzheimer's Disease (AD), and maintaining this skill constitutes an element key to the quality of life of these persons.

In this context, studies have shown a significant decline in prospective memory with age, particularly in patients with AD (Hering et al. 2018). This leads to difficulties in remembering future intentions, frequent forgetting, and errors in managing daily activities, which can significantly impact their safety and well-being. Supporting the PM

J. Baratgin et al. (Eds.): HAR 2024, LNCS 15504, pp. 234–249, 2025.
https://doi.org/10.1007/978-3-031-84595-6_16

of people with AD by implementing procedures to help them not forget the activities to be carried out in the future is therefore a major health issue.

In recent years, companion robots have largely demonstrated their effectiveness in helping elderly people manage their daily activities. In addition to providing assistance in certain activities, companion robots have been shown to improve the pleasure, joy, and well-being of people with AD (Tulsulkar et al. 2021; Yu et al. 2022). In this context, the objective of this work was to evaluate the ability of the NAO robot to help people with AD living in institutions to plan and remember the PM tasks that they are required to carry out.

2 Problem

According to the World Health Organization (WHO), dementia affects more than 9.9 million new people each year, with Alzheimer's disease accounting for 60% to 70% of dementia cases. This pathology is therefore a major public health issue, requiring effective interventions to improve the quality of life of patients and their families. Alzheimer's disease (AD) is characterized by memory loss, word-finding difficulties, executive dysfunction, and confusion in unfamiliar environments with secondary impairment of mood and quality of life (Wilson et al. 2011; Wilson et al. 1999). On the behavioral level, the patient's communication ability deteriorates, leading to psychological and relational disorders such as apathy, depressive symptoms, anxiety, agitation, mood disturbances, sleep and appetite disorders, as well as episodes of disinhibition, hallucinations, and delusions (Balestreri et al. 2000). Although studies of memory impairment in aging have primarily focused on retrospective memory, prospective memory difficulties have been shown to account for a large proportion of memory difficulties reported by older adults (Kliegel et al. 2003). Prospective memory (PM) describes the ability to successfully plan and execute delayed intentions into the future (e.g., Einstein and McDaniel 1990). PM tasks are traditionally divided into two main types based on the triggering cue: event-based or time-based (McDaniel and Einstein 2007; Kvavilashvili and Ellis 1996). Event-based PM tasks require elderly people to remember to execute a planned action when a specific event occurs (ie. Don't forget to turn off the oven when an alarm sounds). In contrast, time-based PM tasks are triggered by a temporal cue indicating that a given action needs to be performed (ie. Don't forget to turn off the oven in 15 min).

Among the cognitive impairments associated with AD, PM occupies a central place as this form of memory plays a crucial role in performing daily activities, especially in people at risk of developing AD (Huppert et al. 2000; Zöllig et al. 2010).

Therefore, PM disorders are central to cognitive decline and loss of autonomy in the daily activities of aging people (Hering et al. 2018). Providing assistance to elderly people, particularly those with AD (living at home or in institutions), in managing PM is crucial for maintaining their functional independence as long as possible.

In recent years, companion robots have been considered a promising solution to improve the quality of life for elderly people, especially those with dementia or cognitive impairments. Designed to mimic the behaviors of pets or to interact autonomously, these robots aim to reduce social isolation, improve mood, and promote social interactions. Studies highlight the positive impact of companion robots on the well-being of elderly people, both in institutional settings and at home (Mordoch et al. 2013; Wada et al. 2005; Saito et al. 2003).

Pineau et al. (2003) noted that assistive robots in nursing homes help elderly people with mild cognitive or physical impairments by providing essential assistance to caregiving staff. Mobile robots, capable of moving autonomously, providing reminders, and monitoring restricted areas, contribute to the overall well-being of residents. These results have encouraged studies investigating the ability of assistive robots to help the caregiver in many activities of daily living.

For example, Pou-Prom et al. (2020) examined the utility of robots in helping residents with AD maintain conversational abilities. In their experiment, patients saw images and were asked to describe them or answer questions asked by a human or a robot. The results showed that the conversations were generally better with human interlocutors (longer, with more engagement and fewer misunderstandings), but that the robot was generally well liked by the participants and was able to capture their interest. Another study focused on the use of a robot to help the caregiver offer psychomotor activities to residents. Rouaix et al. (2017) compared a session carried out by a therapist to another carried out by both the therapist and the NAO robot. For both groups, patients were asked to do physical and breathing exercises and talk about their bodies to increase their body awareness and verbal skills. The results showed that participants responded more emotionally when interacting with the robot, although there was no difference in the level of engagement between the 2 groups. Additionally, having a robot increased participants' well-being and satisfaction.

While companion robots have already been shown to be useful in helping people with AD living in institutions manage daily activities (Abdollahi et al. 2017), no studies have clearly focused on the robot's ability to help residents manage daily PM tasks in real-world contexts. Indeed, elderly people with AD must plan and carry out numerous PM tasks (not forgetting an appointment at a given time, not forgetting meal times, taking medication, etc.) and forgetting to carry out these activities constitutes a significant obstacle to autonomy. Better understanding how the NAO robot can be useful to help them carry out these tasks is therefore an important issue. An experimental study has already shown the relevance of companion robots in helping patients with AD remember activities to be performed in the future (Tahan et al. 2024). However, this study was conducted in an experimental setting and not in the daily living environment of patients living in institutions.

In this context, the present study aims to compare the NAO companion robot and the caregiver in planning and reminding daily tasks that require PM in institutionalized AD patients. Different PM tasks were proposed and cued (if the person forgot to carry them out) either by a caregiver or by a robot. These tasks corresponded to daily life activities (such as picking up a person and bringing them back to the office, or getting a glass of water or a newspaper, etc.). The instruction was given to the participant (for example, go get a glass of water in 15 min) and a cue (with a weak cue, then a strong cue) was given after 15 min if the person had forgotten to carry out this task. The instructions (tasks to be carried out) and cues (if the task was not carried out) were given either by the robot or by the caregiver and the performances of the participants in these 2 conditions were compared. If the performances of participants with AD (ability to complete the task within the allotted time) are identical in both conditions (not significantly different), it would mean that the robot is as effective as the caregiver in indicating PM tasks (e.g., not forgetting a doctor's appointment) and providing reminders if the task is not completed on time. In this case, this would mean that the robot could help people with AD to effectively plan and carry out PM activities of daily life.

Additionally, Mabire et al. (2016) developed an ethogram to evaluate and quantify the social interactions of elderly people with AD. For our study, we used the two most relevant categories: social interactions with the caregiver or the robot (interactive behaviors) and self-centered behaviors (facial expressions, gazes, etc.). The category "resident-resident social interactions" was not included since the experiment took place in individual sessions (human or robot with participant), which did not allow resident-resident interactions to be possible.

Therefore, another complementary objective of the present study will be to compare the quality of social interactions of residents when performing PM tasks managed either by the robot or by the caregiver. As we have seen, the use of an assistance robot is associated with positive behaviors (joy, smile, reduction in anxiety, well-being, etc.) among residents. But on the other hand, interactions, particularly verbal, can sometimes be difficult between residents and the robot due to technological constraints (Tahan and N'Kaoua 2023).

Taking all these elements into account, our general hypotheses were:

- PM tasks presented and cued by people with AD will be as successful in the caregiver condition (tasks presented and cued by the caregiver) as in the robot condition (tasks presented and cued by the robot).
- The participants' behaviors will be more positive (joy, smile, etc.) with the robot compared to the caregiver, but there will be more interactions in the situation managed by the caregiver than in the situation managed by the robot.

The operational hypotheses were therefore:

- The scores on the PM tasks in the 2 conditions (tasks managed by the robot and tasks managed by the caregiver) will not be significantly different;
- In the indicators of the SOBRI grid, positive behaviors (joy, laughter, etc.) will be more numerous in the robot condition compared to the caregiver condition. On the other hand, interaction behaviors (participant interaction with the robot or the caregiver) will be more numerous with the caregiver than with the robot.

3 Method

3.1 Participants

Fourteen institutionalized persons with Alzheimer disease participated in the study. The clinical diagnosis of Alzheimer's disease was made according to ICD-10 criteria 10 by a senior consultant doctor in conjunction with a multidisciplinary team assessment.

The participants consisted of 12 women (86%) and 2 men (14%), with ages ranging from 79 to 99 years. To characterize the participants, we used the Mini-Mental State Examination (MMSE) (Folstein et al. 1975), the State Anxiety Inventory by Spielberger

1971, and the "Commands" (BDAE 1) and "Logical and Reasoning, Complex Intellectual Operations" (BDAE 2) subtests of the BDAE (Goodglass and Kaplan 1972). The inclusion criteria were residents diagnosed with Alzheimer's Type Dementia and having an MMSE score greater than 14. Twenty participants were willing to participate but 6 participants were excluded because they presented excessive sensory (hearing or visual), cognitive or behavior disorders. Before the intervention, the study was presented to all participants so that they could decide if they wanted to participate. A consent form was completed by all persons wishing to participate in the study.

Table 1. Descriptive characteristics of the participants. BDAE 1 corresponds to the "Commands" subtest and BDAE 2 to the "Logic and reasoning, complex intellectual operations" subtest.

	Age	MMSE	BDAE 1	BDAE 2
N	14	14	14	14
Average	90.6	22.4	14.1	8.36
Median	92.0	22.5	15.0	9.00
Standard deviation	5.81	3.75	1.64	2.84
Minimum	79	14	10	3
Maximum	99	27	15	12

Table 1 indicates that our participant sample is elderly (Mean 90.6; SD: 5.51) and has an MMSE greater than 14 (indicating moderate to mild dementia). For the BDAE 1 ("commands" subtest), the average of our sample is 14.1. An average greater than 10 indicates that the instructions are understood and carried out by the participant. For the BDEA 2 "Logic and reasoning, complex intellectual operations" the average of our sample is 8.36. An average above 5 indicates good logic and reasoning skills.

These different scores, as well as the observations made during the tests and interviews, indicate that the participants were able to understand the instructions and implement them.

3.2 Material

NAO Robot
We used the NAO robot, recognized as one of the most widespread humanoid robots globally. It is widely used in many sectors such as research, healthcare services, and education. Its financial accessibility, compared to other robots on the market, makes it a relevant choice for institutions seeking innovation in their activities.

The NAO robot measures 58 cm in height and weighs 4.3 kg. It has 25 degrees of freedom, allowing for great agility in its movements and smooth adaptation to its environment. Its two 2D cameras, seven sensory touch sensors, as well as its four microphones and speakers, enable it to perceive and interact with its environment in a sophisticated manner. These combined features make the NAO robot an ideal choice for our study on assisting caregivers in managing the prospective memory tasks of elderly people.

Nao is a programmable robot. The questions asked to the participants were implemented (using a user-friendly interface) and the questions and more generally the verbal interactions were strictly the same as for human interlocutors (human condition). The NAO robot is capable of moving and performing different gestures with its upper limbs and head. In our experiment, we did not program any locomotor movements since the robot was seated facing the participant. We focused on arm and hand gestures, which are already implemented and correspond to gestures usually used when speaking. NAO blinks (LEDs that turn on and off) and is able to follow a person in front of him with his gaze.

Prospective Memory Tasks

Each participant in the experiment completed two individual sessions: one with the caregiver and another with the robot (controlled by a human but without interference with the experience).

In both cases, time-based PM tasks were proposed to them (either by the caregiver or by the robot).

The tasks were randomly chosen from a panel of 7 different tasks. They could be asked to bring a person to an office (a staff member or another resident), an object (such as a glass of water or a newspaper), or to perform a specific action (like mailing an envelope or having a snack). We had previously verified that all these tasks were feasible by the participants in the study and that they corresponded to tasks they might be expected to perform in their daily lives.

For each condition (caregiver vs. robot), the procedure was entirely managed either by the robot or by the caregiver. The session began with the presentation of the prospective memory task instruction (before the experiment, the experimenter ensured that the participants understood the types of instructions used). In all conditions, the task had to be performed 15 min later (Rabin et al. 2014; Tierney et al. 2016).

After receiving the instruction, the participant had to complete a quiz and could choose the theme from general culture, music, cinema, literature, wildlife and flora, history/geography, or riddles. Classically, time-based PM tasks involve the following sequence: the instruction (do something in "X" minutes); an interfering task (activity which avoids mental repetition of the instructions) and finally the completion of the task by the subject. This sequence aims to get closer to daily life in which interference is often present between planning the task to be carried out and the moment when the task must be carried out. In our study we chose to offer participants general knowledge type quizzes as interfering tasks. The quizzes were of the same nature whatever the theme and the same themes were proposed either for the robot condition or for the caregiver condition in order to avoid a different interfering activity between the 2 conditions. The quizzes were in the form of multiple choice questions (3 possible answers for each question). For example, for a musical quiz, an excerpt of a song was followed by the question who is the performer; for a general culture quiz, a question was which of these 3 countries is not in Europe, etc."

Once the 15-min period had elapsed, the participant was expected to remember the previously given task. If the participant did not perform the task, a reminder was provided with a weak cue, followed by a strong cue if the task was still not performed. Four types of responses were possible, with an overall score ranging from 0 to 3 for each:

- A correct response where the participant executes the intention stated in the instruction within the allotted time, earning a score of 3;
- A correct response after prompting with a weak cue, only if the participant does not initially perform the intention. The experimenter asks, "Don't you have something to do?" This earns a score of 2;
- A correct response after a strong informational cue, if the participant still has not performed the intention with the weak cue. The experimenter asks, "What were you supposed to remind me of after 15 min?" This earns a score of 1;
- No response or an incorrect response, where the participant does not perform the intention despite the provided cues, earning no points.

The experiment was conducted by a psychologist assisted by two psychology students. The manipulation and programming of the robot are well known and mastered by the experimenters. In particular, the psychologist working in the institution has already used the robot for different activities with residents. The PM tasks that the residents had to carry out were developed by the psychologist who works daily with people with Alzheimer's Disease and who has good knowledge of the activities usually carried out by residents.

Quantification of Interactions
To quantify the interactions of the residents, we used the Social Observation Behaviors Residents Index (SOBRI) observation grid proposed by Mabire et al. (2016). This grid lists 126 behaviors divided into four categories. For our study, we used the two most relevant categories: social interactions with the caregiver or the robot (facial expressions, gazes, verbal interactions, quasi-linguistic interactions, and interactive behaviors), and self-centered behaviors (facial expressions, gazes, undirected verbalizations, sadness behaviors, stereotyped behaviors, comfort behaviors, inactivity behaviors, waiting behaviors, and movements). As indicated above, the category "resident-resident social interactions" was not included since the experiment took place in individual sessions (human or robot with participant), which did not allow resident-resident interactions to be possible.

During the sessions with the caregiver or the robot, a strategically positioned camera covered all interactions of the participant. The video recordings were reviewed on a computer and analyzed manually. Whenever a behavior listed in the grid was observed, the evaluator paused the video and recorded it on a copy of the ethogram.

Anxiety Assessment
Finally, the sessions concluded with the Spielberger questionnaire (STAI 2018), which is an anxiety scale assessing the apprehension, tension, nervousness, and worry that the subject feels at the given moment. It is described as an "indicator of transient changes in anxiety caused by aversive or therapeutic situations." Responses are rated on a scale from 1 to 4 (Not at all, A little, Moderately, and Very much). Total scores range from 20 to 80.

3.3 Procedure

Each participant carries out 2 PM tasks, one managed by the caregiver and the other managed by the robot. For each participant, the 2 tasks are drawn at random from the 7 possible tasks. The same task can be drawn at random either in the robot condition or in the caregiver condition (the 7 tasks were implemented so that the robot could execute them).

Before the experience with the robot, the robot was presented to the resident for approximately 30 min. During this time, the resident saw the robot move, and a quiz was taken to check the resident's ability to hear and answer the questions. No residents had previously been exposed to interactions with the robot. The order of the sessions (human vs. robot) was counterbalanced and a period of 4 days was left between the 2 conditions. At the end of the second task, the participant's anxiety is assessed using the STAI.

4   Research Procedure

To analyse our results, we carried out various statistical analyses.

Prospective Memory Tasks
We compared the scores on the PM task (dependent variable) in the 2 conditions (human vs. robot). The "conditions" variable therefore constitutes our independent variable with 2 modalities (human vs. robot). A non-parametric test (taking into account the size of the samples) comparing 2 means for a paired sample was used (Wilcoxon test). To complete this analysis, we also carried out a chi2 test of association by creating a contigance table distributing the "conditions" variable (to 2 modalities) to and the "performance score" variable (to 4 modalities: no cue, low cue, strong cue, no answer).

Anxiety Assessment
We compared anxiety scores (dependent variable) on the STAI in the 2 conditions using a Wilcoxon test. Again, the "conditions" variable constitutes our independent variable with 2 modalities (human vs. robot).

Quantification of Interactions
We compared our 2 conditions (human vs. robot) for each indicator of the SOBRI grid by first indicating the results for the indicators corresponding to self-centered compromises then for the indicators corresponding to interaction behaviors.

5   Results

Prospective Memory Tasks
As indicated above, the results of the experiment were analyzed using a non-parametric Wilcoxon signed-rank test, as all participants were assessed in both conditions (activity management with a robot and with a human). This analysis consisted of comparing the scores on MP tasks (material section: Prospective Memory Tasks) obtained in the human condition to those obtained in the robot condition.

The analyses did not reveal a significant difference in performance between the robot and human conditions ($p = 0.301$). The mean scores in the "robot" condition were slightly higher ($M = 1$, $SD = 1.109$) compared to the "human" condition ($M = 0.64$, $SD = 0.08$), but this difference is not significant.

The results were also analysed using a χ^2 association test (Table 2), focusing on the frequencies (contingency table) distributed according to the level of responses (4 modalities: no cue, weak cue, strong cue, no response) and the condition (robot vs. human). The analysis did not reveal a significant relationship between the condition and the level of response [χ^2 (3) = 4.46; p = 0.463].

Table 2. Contingency table of performance scores according to the condition (robot vs human)

	No cue	Low cue	Strong cue	No answer	Total
Human	0 (0%)	3 (21.4%)	3 (21.4%)	8 (57.1%)	14 (100%)
Robot	1 (7.1%)	5 (35.7%)	1 (7.1%)	7 (50%)	14 (100%)
Total	1 (3,6%)	8 (28.6%)	4 (14.3%)	15 (53.6%)	28 (100%)

This result indicates that the responses are distributed in the same way for both conditions (human vs. robot). The 'no response' modality is the most represented (about 50%). Very few participants complete the task without a cue (3.6%). The other participants improve their performance with a low cue (28%) and then a strong cue (14%).

Anxiety Assessment
Regarding the participants' anxiety, no significant difference (p = 0.16) was observed between the robot (M = 28.8, SD = 4.76) and human (M = 32.2, SD = 9.70) conditions on the STAI tests (Wilcoxon test), indicating that the condition had no impact on the induced anxiety levels.

Quantification of Interactions
For self-centered behaviors (Table 3; only significant results or results showing a tendency are presented), significant differences between the human and robot conditions were observed on two indicators. Participants 'ramble' (incomprehensible verbalization) less often in the robot condition (p = 0.01) than in the human condition. We also observe the same result with the "repositioning behavior" indicator (p = 0.01), meaning that participants reset less frequently in the robot condition compared to the human condition. Repositioning behavior involves moving back into one's seat and its presence is a sign of discomfort and inattentiveness.

We also observe a trend for 4 indicators:

– 'Surprise' (positive facial expression) (p = 0.06): participants are more surprised in the human condition compared to the robot condition.
– 'Distant gaze' (p = 0.06): participants are more attentive in the robot condition compared to the human condition.
– 'Grimace' (negative facial expression) (p = 0.07): participants grimace less in the robot condition than in the human condition.
– 'Rolling eyes' (negative facial expression) (p = 0.07): participants roll their eyes less often in the robot condition compared to the human condition."

Table 3. Self-centered behaviors. For each behavior, the average observation of this behavior (with the human or with the robot), the value of the Wilcoxon W statistic, and the p value is indicated

indicator	Human average	Robot average	statistics	p-value
Surprise	1,71	0,50	32	0,06
Grimace	1,64	0,64	38	0,07
Roll eyes	1	0,29	25	0,07
Distant gaze	11,2	7,64	83	0,06
Ramble	1	0,21	10	0,01
repositioning	1,07	0,29	30	0,01

For interaction indicators (Resident-human/social robot interactions), differences between the two conditions (human and robot) were observed for the following indicators: smiling (p = 0.02), laughing (p = 0.02), looking at the animator (p = 0.01), asking questions (p = 0.02), responding (p = 0.004), and making jokes (p = 0.01). These results show that participants exhibited more joyful behaviors with human and had a higher level of requests, likely due to difficulties in understanding the human's diction (Table 4; only significant results or results showing a tendency are presented). We also observed a trend for the indicator 'reaching out to others' (p = 0.05). In all cases, the human condition is associated with higher scores than the robot condition.

Table 4. Resident-human/robot social interactions. For each behavior, the average observation of this behavior (with the human or with the robot), the value of the Wilcoxon W statistic, and the p value is indicated

indicator	Human average	Robot average	statistics	p-value
Smile	5,57	3,14	59	0,02
Laugh	5	1,36	58,50	0,02
Towards animator	13,2	3,64	94	0,01
Ask	10,1	8,14	77	0,03
Make a joke	2,71	0,71	60	0,02
Reach out to others	1,64	0,36	32	0,06

6 Discussion

The objective of our study was to evaluate the ability of the NAO robot to assist elderly people with dementia living in institutions in the planning and recall of time-based PM activities. To our knowledge, no previous research has been conducted on the use of social assistance robots for task reminders requiring prospective memory. However, there are numerous digital tools designed to aid elderly individuals with dementia in their daily activities. In this perspective of supporting the autonomy of the elderly, the study by El Haj et al. (2021) showed that using a digital calendar on a smartphone led to fewer omissions of prospective events among individuals with Alzheimer's disease compared to using a paper calendar. Hodges et al. (2006) demonstrated that a camera worn around the neck (which records situations and compares them to past situations) was associated with positive effects on the memory of individuals with cognitive impairments by providing visual reminders that facilitated social interactions and the well-being of participants. Finally, reminders on smartphones can also provide assistance to individuals with dementia and constitute an effective, accessible, and non-stigmatizing tool (Ferguson et al. 2015). Similarly, Scullin et al. (2022) investigated the feasibility and effectiveness of smartphone-based strategies for prospective memory in people with cognitive impairment. They show that older adults with cognitive impairment can learn smartphone-based memory strategies, thereby improving recall of tasks in PM and improving people's autonomy.

The objective of our study was to assist residents in recalling prospective tasks using the NAO robot. Our results show that the performance of time-based task recall does not differ significantly between the two conditions (Robot or Human). The robot therefore does not differ significantly from human in the management and recall of activities. Other studies have also demonstrated the effectiveness of social assistance robots. In the study by Pou-Prom et al. (2020), the use of a chatbot was beneficial for elderly individuals with dementia. Participants often preferred discussions with humans due to their engagement and longer duration. However, the robot was well-received and sparked considerable interest despite some communication issues. This finding is also supported by the article by Valentí-Soler et al. (2015), which shows that robots can be effectively used for cognitive exercises aimed at maintaining autonomy and cognitive function in elderly individuals with mild cognitive impairment (MCI). The authors propose training methods tailored to the individual's needs and condition. Additionally, the study by Moro et al. (2019) demonstrates that robots, especially those with dynamic social features such as facial expressions and gestures, can be as effective as humans in assisting individuals with dementia in daily activities. The results highlight the appeal and usefulness of the robot Casper compared to other devices like the robot Ed or a tablet. Participants showed strong engagement during interactions with Casper due to its dynamic social characteristics, human-like appearance, and expressive face (which distinguishes it from the robot Ed or a tablet). Another study focused on the use of a robot to assist caregivers in providing psychomotor activities to residents (Rouaix et al. 2017). The authors compared a session conducted solely by a therapist to another where the therapist was assisted by the NAO robot. In both groups, participants performed physical and breathing exercises while discussing their bodies to improve body awareness and verbal skills. The results showed a more pronounced emotional reaction from participants during interactions with the robot, although the level of engagement between the two groups did not show differences. Additionally, the presence of the robot was associated with improved well-being and satisfaction of the participants. Wang et al. (2017) evaluated the help that a robot could provide in managing the daily activities of elderly people with mild to moderate AD. Elderly people were encouraged by a remote-controlled robot to wash their hands in the bathroom and make a cup of tea in the kitchen. Semi-structured interviews were then conducted with the elderly but also with caregivers. The analysis of the responses indicates that older people expressed the possibility that robots could help them with their daily activities, but did not want a robot. Caregivers identified many opportunities and were more open to robots. The authors concluded that positive consequences of robots in caregiving scenarios could include reduced frustration, stress, and relationship tensions, as well as increased social interaction via the robot. A negative consequence could be a reduction in interactions with caregivers.

In our study, we also analyzed the distribution of response modalities to the PM task (response without any help, response with a low cue, response with a high cue, and no response) depending on the condition (task managed by the robot or by the caregiver). The analysis did not reveal a significant link between the condition (robot vs. human) and the response modality. On the other hand, the help provided by the cues (weak then strong) is effective (increase in the number of residents carrying out the tasks thanks to the aids) and this improvement linked to the aids is no different depending on whether the help is provided by the robot or the caregiver (absence of significant link between the human vs. robot condition and the type of response to the chi2 test).

Moreover, it should be noted that almost none of the participants responded without any help (approximately 3%) and more than half were unable to respond even with all the provided cues (approximately 53%). This result aligns with numerous studies showing the difficulties of time-based PM in Alzheimer's patients, which are related to the low cognitive resources available (time-based PM tasks are cognitively demanding) as well as disruptions in internal time processing (Martin et al. 2001; Jàger and Kliegel 2008).

Another result of our study is that the anxiety scores measured by the STAI questionnaire showed no significant difference between the robot and human conditions, indicating the absence of impact of the condition on the level of anxiety induced. The score on this scale ranges from 20 to 80 and the higher the score, the greater the anxiety. In our study, the scores did not differ depending on the condition (robot vs. human) and the anxiety scores were low in both conditions (28.8 for the robot condition and 32.2 for the human condition). The study by Pino et al. (2019) also highlights this point by showing that the presence of the NAO robot helped reduce anxiety by reinforcing therapeutic behavior and adherence to treatments among participants during interactions with the psychologist.

Regarding the observation of behaviors, we used the SOBRI grid to identify self-centered behaviors and interaction behaviors depending on the condition (robot or human). For self-centered behaviors, our results show more positive behaviors in the robot condition than in the human condition (with the exception of surprise behavior which is more important in the human condition than in the robot condition). Indeed, participants exhibited less behaviors such as "wandering" and "readjusting" with a robot compared to the human. They also tended to make less faces and roll their eyes with a robot than with the human. Our results are consistent with numerous studies showing that participants feel increased well-being and satisfaction in the presence of a robot, highlighting a reduction in agitation or anxiety behaviors (Wu et al. 2014; Kidd et al. 2006; Saito et al. 2003).

The analysis of interaction behaviors (resident vs. human or robot), on the other hand, shows significantly more positive behaviors directed towards the human than towards the robot. Some studies have shown that robots can play a facilitative role in social interactions (Moyle et al. 2017; Hung et al. 2021), but other studies have highlighted the difficulties faced by older people and people with Alzheimer's disease to interact with robots, particularly due to comprehension problems. For example, in our study, behaviors such as "smiling" and "laughing" are more common in the human condition. These positive facial expressions indicate that residents find interactions with a human more enjoyable or pleasant. This is consistent with findings by Broekens et al. (2009) showing that social interactions are often richer and more varied with humans, which may explain the increase in positive facial expressions. In our study, residents were more inclined to "make jokes" with a human host than with the robot. This trend can be explained by the fact that human interactions are often more spontaneous and varied, allowing for more playful and humorous exchanges. Kidd et al. (2006) and Wada et al. (2006) also conclude that interactions with social robots may lack the spontaneity necessary to elicit humorous interactions.

7 Limitations

Future research should continue with a larger sample size and evaluate the long-term effects of robot use on the autonomy and quality of life of both residents and healthcare staff. Another limitation of our study was that the sample was not gender-balanced (14 people, including 2 men). Furthermore, the participants did not all perform the same tasks, but the tasks were chosen at random and the same tasks were offered either in the robot condition or in the human condition. Finally, using a more sophisticated robot with enhanced verbal and non-verbal communication capabilities would be beneficial.

8 Conclusion

To conclude, our study represents a significant advance in the field of gerontology by comparing the NAO robot to a human in helping residents with Alzheimer's disease perform tasks requiring prospective memory.

Our results indicate that the help provided by cues (weak and strong) improves residents' performance in PM tasks and this improvement does not differ significantly depending on whether the help is provided by the robot or the caregiver (it should be noted, however, that the statistical test does not allow us to conclude that the 2 situations are equivalent). Additionally, positive self-centered behaviors are higher with the robot than with the human. Conversely, interaction behaviors are more positive with the human than with the robot.

In this context, robots equipped with artificial intelligence are extremely promising, providing responses tailored to the interlocutor without the need for an experimenter, as in our study, to manage the procedures implemented by the robot. A more advanced robot in terms of autonomy and interaction capabilities could offer more effective solutions for both self-centered behaviors and interaction behaviors with the robot. Thus providing more effective assistance to both the caregiver and to the resident.

Acknowledgement. This study is supported by the Colisée Group and more particularly by Doctor Vincent KLOTZ (Chief Medical Officer of Colisée Group) and Doctor Monique GIRARD (Medical Director and Geriatric Consultant of Colisée Group France).

Disclosure Statement. The author reported no potential conflict of interest.

References

1. Balestreri, L., Grossberg, A., Grossberg, G.T.: Behavioral and psychological symptoms of dementia as a risk factor for nursing home placement. Int. Psychogeriatr.Psychogeriatr. **12**(S1), 59–62 (2000)
2. Broekens, J., Heerink, M., Rosendal, H.: Assistive social robots in elderly care: a review. Gerontechnology **8**(2), 94–103 (2009)
3. Einstein, G.O., McDaniel, M.A.: Normal aging and prospective memory. J. Exp. Psychol. Learn. Mem. Cogn.Cogn. **16**(4), 717 (1990)
4. El Haj, M., Moustafa, A.A., Gallouj, K., Allain, P.: Cuing prospective memory with smartphone-based calendars in Alzheimer's disease. Arch. Clin. Neuropsychol. Clin. Neuropsychol. **36**(3), 316–321 (2021)
5. Ferguson, S., Friedland, D., Woodberry, E.: Smartphone technology: gentle reminders of everyday tasks for those with prospective memory difficulties post-brain injury. Brain Inj. **29**(5), 583–591 (2015)
6. Folstein, M.F., Folstein, S.E., McHugh, P.R.: "Mini-mental state": a practical method for grading the cognitive state of patients for the clinician. J. Psychiatr. Res.Psychiatr. Res. **12**(3), 189–198 (1975). https://doi.org/10.1016/0022-3956(75)90026-6
7. Goodglass, H., Kaplan, E.: HDAE (BDAE) Boston Diagnostic Aphasia Examination. Issy-les-Moulineaux: Editions du Centre de Psychologie Appliquée (1972)
8. Hering, A., Kliegel, M., Rendell, P.G., Craik, F.I., Rose, N.S.: Prospective memory is a key predictor of functional independence in older adults. J. Int. Neuropsychol. Soc. Neuropsychol. Soc. **24**(6), 640–645 (2018)
9. Hodges, S., et al.: SenseCam: a retrospective memory aid. In: UbiComp 2006: Ubiquitous Computing: 8th International Conference, UbiComp 2006 Orange County, CA, USA, 17–21 September 2006, Proceedings, vol. 8, pp. 177–193). Springer, Heidelberg (2006)
10. Huppert, F.A., Johnson, T., Nickson, J.: High prevalence of prospective memory impairment in the elderly and in early-stage dementia: Findings from a population-based study. Appl. Cogn. Psychol. **14**(7), S63–S81 (2000). https://doi.org/10.1002/acp.771
11. Hung, L., et al.: Exploring the perceptions of people with dementia about the social robot PARO in a hospital setting. Dementia **20**(2), 485–504 (2021)
12. Jager, T., Kliegel, M.: Time-based and event-based prospective memory across adulthood: Underlying mechanisms and differential costs on the ongoing task. J. Gen. Psychol. **135**(1), 4–22 (2008)
13. Kidd, C D., Taggart, W., Turkle, S.: A sociable robot to encourage social interaction among the elderly. In: Proceedings 2006 IEEE International Conference on Robotics and Automation, 2006, pp. 3972–3976. ICRA. IEEE (2006)
14. Kliegel, M., et al.: Prospective memory and ageing: is task importance relevant? Int. J. Psychol. **38**(4), 207–214 (2003). https://doi.org/10.1080/00207590344000132
15. Kvavilashvili, L., Ellis, J.A: Varieties of intention: some distinctions and classifications. In: Brandimonte, M., Einstein, G.O., McDaniel, M.A. (eds.) Prospective Memory: Theory and Applications, pp. 23–51. Lawrence Erlbaum Associates Publishers (1996). https://psycnet.apa.org/record/2002-02930-002

16. Mabire, J.B., Gay, M.C., Vrignaud, P., Garitte, C., Vernooij-Dassen, M.: Social interactions between people with dementia: pilot evaluation of an observational instrument in a nursing home. Int. Psychogeriatr. **28**(6), 1005–1015 (2016)

17. Martin, M., Schumann-Hengsteler, R.: How task demands influence time-based prospective memory performance in young and older adults. Int. J. Behav. Dev. **25**(4), 386–391 (2001)

18. McDaniel, M.A., Einstein, G.O.: Prospective memory: an overview and synthesis of an emerging field (2007)

19. Mordoch, E., Osterreicher, A., Guse, L., Roger, K., Thompson, G.: Use of social commitment robots in the care of elderly people with dementia: a literature review. Maturitas **74**(1), 14–20 (2013)

20. Moro, C., Lin, S., Nejat, G., Mihailidis, A.: Social robots and seniors: a comparative study on the influence of dynamic social features on human–robot interaction. Int. J. Soc. Robot. **11**, 5–24 (2019)

21. Moyle, W., et al.: Use of a robotic seal as a therapeutic tool to improve dementia symptoms: a cluster-randomized controlled trial. J. Am. Med. Dir. Assoc. **18**(9), 766–773 (2017)

22. Pineau, J., Montemerlo, M., Pollack, M., Roy, N., Thrun, S.: Towards robotic assistants in nursing homes: challenges and results. Robot. Auton. Syst. **42**(3–4), 271–281 (2003)

23. Pino, O., Palestra, G., Trevino, R., De Carolis, B.: The humanoid robot NAO as trainer in a memory program for elderly people with mild cognitive impairment. Int. J. Soc. Robot. **12**, 21–33 (2020)

24. Pou-Prom, C., Raimondo, S., Rudzicz, F.: A conversational robot for older adults with alzheimer's disease. ACM Trans Hum-Robot. Interact. (THRI) **9**(3), 1–25 (2020)

25. Rabin, L.A., Chi, S.Y., Wang, C., Fogel, J., Kann, S.J., Aronov, A.: Prospective memory on a novel clinical task in older adults with mild cognitive impairment and subjective cognitive decline. Neuropsychol. Rehabil. **24**(6), 868–893 (2014)

26. Rouaix, N., Retru-Chavastel, L., Rigaud, A., Monnet, C., Lenoir, H., Pino, M.: Affective and engagement issues in the conception and assessment of a robot-assisted psychomotor therapy for persons with dementia. Front. Psychol. **8**, 950 (2017). https://doi.org/10.3389/fpsyg.2017.00950

27. Saito, T., Shibata, T., Wada, K., Tanie, K.: Relationship between interaction with the mental commit robot and change of stress reaction of the elderly. In: Proceedings 2003 IEEE International Symposium on Computational Intelligence in Robotics and Automation. Computational Intelligence in Robotics and Automation for the New Millennium (Cat. No. 03EX694), vol. 1, pp. 119–124. IEEE (2003)

28. Scullin, M.K., et al.: Using smartphone technology to improve prospective memory functioning: a randomized controlled trial. J. Am. Geriatr. Soc.Geriatr. Soc. **70**(2), 459–469 (2022)

29. Spielberger, C.D., Gonzalez-Reigosa, F., Martinez-Urrutia, A., Natalicio, L.F., Natalicio, D.S.: The state-trait anxiety inventory. Revista Interamericana de Psicologia/Interamerican J. Psychol. **5**(3 & 4) (1971)

30. Tahan, K., Cayrier, A., Baratgin, J., N'kaoua, B.: ZORA robot to assist a caregiver in prospective memory tasks: a preliminary study: prospective memory; humanoid robot; alzheimer's disease. Appl. Neuropsychol. Adult, 1–8 (2024)

31. Tahan, K., N'Kaoua, B.: Prospective memory training using the nao robot in people with dementia. In: International Conference on Human and Artificial Rationalities, pp. 281–295. Springer, Cham (2023)

32. Tierney, S.M., Bucks, R.S., Weinborn, M., Hodgson, E., Woods, S.P.: Retrieval cue and delay interval influence the relationship between prospective memory and activities of daily living in older adults. J. Clin. Exp. Neuropsychol. **38**(5), 572–584 (2016)

33. Valentí Soler, M., Agüera-Ortiz, L., Olazarán Rodríguez, J., et al.: Social robots in advanced dementia. Front. Aging Neurosci. **7**, 133 (2015). https://doi.org/10.3389/fnagi.2015.00133. PMID:26388764;PMCID:PMC4558428

34. Wada, K., Shibata, T.: Robot therapy in a care house-its sociopsychological and physiological effects on the residents. In: Proceedings 2006 IEEE International Conference on Robotics and Automation, ICRA 2006, pp. 3966–3971. IEEE (2006)

35. Wada, K., Shibata, T., Saito, T., Sakamoto, K., Tanie, K.: Psychological and social effects of one year robot assisted activity on elderly people at a health service facility for the aged. In: Proceedings of the 2005 IEEE International Conference on Robotics and Automation, pp. 2785–2790. IEEE (2005)

36. Wang, R.H., Sudhama, A., Begum, M., Huq, R., Mihailidis, A.: Robots to assist daily activities: views of older adults with Alzheimer's disease and their caregivers. Int. Psychogeriatr. **29**(1), 67–79 (2017)

37. Wilson, R.S., Beckett, L.A., Bennett, D.A., Albert, M.S., Evans, D.A.: Change in cognitive function in older persons from a community population: relation to age and Alzheimer disease. Arch. Neurol. **56**(10), 1274–1279 (1999)

38. Wilson, R.S., Leurgans, S.E., Boyle, P.A., Bennett, D.A.: Cognitive decline in prodromal Alzheimer disease and mild cognitive impairment. Arch. Neurol. **68**(3), 351–356 (2011)

39. Wu, Y.H., Wrobel, J., Cornuet, M., Kerhervé, H., Damnée, S., Rigaud, A.S.: Acceptance of an assistive robot in older adults: a mixed-method study of human–robot interaction over a 1-month period in the Living Lab setting. Clin. Interv. Aging, 801–811 (2014)

40. Zöllig, J., Martin, M., Kliegel, M.: Forming intentions successfully: differential compensational mechanisms of adolescents and old adults. Cortex J. Devoted Study Nerv. Syst. Behav. **46**(4), 575–589 (2010). https://doi.org/10.1016/j.cortex.2009.09.010

41. Tulsulkar, G., Mishra, N., Thalmann, N.M., Lim, H.E., Lee, M.P., Cheng, S.K.: Can a humanoid social robot stimulate the interactivity of cognitively impaired elderly? A thorough study based on computer vision methods. Vis. Comput. **37**(12), 3019–3038 (2021)

42. Yu, C., Sommerlad, A., Sakure, L., Livingston, G.: Socially assistive robots for people with dementia: systematic review and meta-analysis of feasibility, acceptability and the effect on cognition, neuropsychiatric symptoms and quality of life. Ageing Res. Rev. **78**, 101633 (2022)

Does Replacing the Experimenter with an Ignorant Student Robot Improve the Success of Children with ASD in the False Belief Task?

Marion Dubois-Sage[1,2], Yohann Mosset-Cancel[1], Frank Jamet[1,2,4], and Jean Baratgin[1,2,3]([✉])

[1] UFR de Psychologie, Université Paris 8, 93526 Saint-Denis, France
`jean.baratgin@paris-reasoning.eu`
[2] Laboratoire Cognitions Humaine et Artificielle (RNSR 200515259U), Saint-Denis, France
[3] Association P-A-R-I-S, 75005 Paris, France
[4] UFR d'Education, CY Cergy Paris Université, 95000 Cergy Pontoise, France

Abstract. Theory of mind is commonly assessed with the false belief task, which involves attributing a false belief to a character. Children with Autism Spectrum Disorders are said to perform less well on this test than typically developing children or children with intellectual disabilities. This might suggest a deficit in theory of mind in autism, but this idea is currently being challenged. Other skills seems to be involved in the false-belief task, such as language pragmatics, the ability to interpret language in context. Indeed, the test question is ambiguous: it can lead to different interpretations. Failure could therefore be due to a difference in interpretation between the child and the experimenter. For people with limited pragmatic language skills (such as people with autism or typically developing children under 4), the pragmatic ambiguity of the test question could be an obstacle to success, even if they are otherwise able to attribute a false belief to others. Nevertheless, changing the context could improve performance: the test question, usually asked by an adult experimenter, could become less ambiguous when asked by a NAO robot presented as an ignorant student (mentor-child paradigm). 26 children with autism (and 24 children with intellectual disabilities) passed the false-belief task with a human or with a robot, in order to assess whether they perform better in this new context. The results show no improvement in performance with this paradigm, but children with ASD may have been distracted by the robot during the test, which would have diminished their success.

Keywords: Theory of mind · Autism Spectrum Disorders · False belief task · Pragmatics · Robot

J. Baratgin et al. (Eds.): HAR 2024, LNCS 15504, pp. 250–287, 2025.
https://doi.org/10.1007/978-3-031-84595-6_17

1   Introduction

People with Autism Spectrum Disorder (ASD) experience difficulties in communication and social interaction, and present restricted and repetitive patterns of
behavior, interests or activities [2]. This neurodevelopmental disorder appears
early in development and has an impact on the child's development in various
areas, including interactions with others [10]. It has long been considered that
individuals with ASD systematically present a deficit in theory of mind, which
would be at the root of the difficulties observed in interactions with others.
However, this idea is currently being criticized. We will begin by examining the
literature suggesting a theory of mind deficit in people with ASD, before recon-
sidering this data from a new perspective, which challenges this deficit. We will
focus on pragmatic theory and propose a method based on the use of a robot
as an experimenter, which could enable individuals with ASD to better mobilize
their theory of mind in the false-belief test.

1.1   Theory of Mind Deficit in Autism

Theory of Mind and False-Belief Test. Theory of Mind (ToM) refers to the
ability to attribute mental states to others [77]. These mental states may refer
to desires, thoughts or beliefs [106]. This skill plays an essential role in social
interactions since it enables us to understand and predict the behaviors of other
individuals, and thus to respond appropriately [113]. Although a wide variety
of terms are used to designate this skill (e.g. mentalizing or perspective-taking),
it is nevertheless possible to distinguish two sub-categories of ToM according
to the type of mental state that is attributed: cognitive ToM when referring to
the attribution of an epistemic state (belief, thought) and affective ToM when
referring to the attribution of an affective state (emotion) [79]. These two types
of ToM are therefore measured by different tests (e.g. the false belief task [109]
for cognitive ToM, which involves attributing a false belief to a character, and
the Reading the Mind in the Eyes Task (RMET) [16] for affective ToM, which
involves attributing a mental state from a photo of a face). In general, individuals
with ASD showed impaired performance in both types of ToM, with failure in
false belief tasks [14] and facial emotion interpretation tasks such as the RMET
[16]. Nevertheless, tasks aimed at measuring emotion attribution to others do not
necessarily correspond to a valid measure of theory of mind, as they are likely
to measure lower-level socio-cognitive processes (for example, performance on
the RMET might depend on facial expression discrimination skills rather than
emotion attribution skills) [80]. Consequently, this type of task should be set
aside in favor of valid measures of ToM. For this reason, in the present article,
we will focus on cognitive ToM using the false-belief task, which fulfils both
criteria [80].

   The false-belief test most commonly used to assess ToM in people with ASD
is Sally and Anne's unexpected transfer test [14], which is derived from the
Maxi and chocolate test originally developed by Wimmer and Perner [109] to
assess the skills of neurotypical (NT) children. Both tests involve presenting a

scenario to the child in the form of pictures or dolls, with the addition of a verbal explanation. Sally and Anne's test, however, has been adapted for children with ASD by streamlining the narrative and material, in order to lighten the verbal and attentional demands [61]: the response modality for this test does not require language, since it can be performed by pointing. The false-belief scenario is as follows: a main character puts an object away in location A. Then, in the main character's absence, another character moves the object and puts it in location B. Finally, the main character returns to retrieve the object. Not having witnessed the transfer of the object, the main character then has an erroneous belief as to the location of the object (he thinks it is in location A when in fact it is in location B). The child is then asked three questions, in a fixed order: a test question, designed to measure the child's ability to attribute a false belief to the character (ToM), and two control questions designed to ensure that the story has been correctly understood. The test question asks the child to predict where the character will go to look for the object: "Where will Maxi look for the chocolate?".

Numerous studies have indicated poorer performance on the false belief test in children with ASD compared with neurotypical children [14,15,51], leading to a consensus in the scientific literature on the ToM deficit in individuals with ASD [39]. However, we shall see that the difficulties encountered by people with ASD in responding correctly to the false belief test can be explained in other ways, and linked to skills other than ToM.

Individuals with ASD Fail the False Belief Test. An initial study comparing the performance of children with ASD with that of NT children showed that children with ASD aged 6 to 16 failed this test massively (80% failure), unlike NT children aged 3 to 5 and children with Down's syndrome aged 6 to 17 (80% success), and this despite a comparable verbal mental age between the different groups (verbal mental age above 4 years, theoretically sufficient for success) [14]. The same result has been observed in several studies [15,23,51,92]. Children with ASD and no associated Intellectual Disability (ID)[1] even perform less well on the false-belief test than children with ID (Down's syndrome or unspecified etiology) [14,15,23], specific language impairment [52,66] or delayed language development [92]. They fail different variants of the false-belief test, such as the Smarties task [66] and an image sequencing task [15]. In other tasks that do not involve consideration of others' mental states, but only their perceptual access, children with ASD perform similarly (or even better) than NT children, for example in the False Photographs task [52,66]. These findings have led some authors to argue that people with ASD present an inability to attribute mental states to others [10,12], which would not be related to a general cognitive or linguistic deficit. Nevertheless, we shall see that this failure can be interpreted in different ways. We will take a closer look at the "mindblindness" theory in order to analyze its limitations, and we will then propose the pragmatic theory,

---

[1] These children have an average Intelligence Quotient (IQ > 70), unlike children with ID (IQ < 70) [21].

which might be more relevant in explaining the performance of individuals with ASD on the false-belief test.

Developmental Theories of ToM. We will start by looking at how ToM development is conceptualized in the context of typical development, before focusing on ToM development in children with ASD.

Implicit Tasks and Developmental Theories of ToM in NT Children. In children with typical development, the explicit false-belief test is only passed from the age of 4 onwards [109]. Nevertheless, we can note a variability in performance according to the type of task used. The authors disagree on the explanation of these results and on the age at which NT children acquire ToM.

Several false-belief tasks have been developed from the Maxi and Chocolate task [109] to assess this ability in NT children. It is then important to note a distinction between explicit and implicit false-belief tasks [5]. On the one hand, explicit tasks, which constitute the standard tests, involve directly asking the child to attribute a false belief to a character (e.g. Maxi and the chocolate [109], Sally and Anne [14], or Smarties task [66]). On the other hand, implicit tasks are a simplification of explicit tasks. These may omit the final location of the object at the end of the scenario (the second character simply removes the object) [97] or make the task more interactive by playing out the scenario with puppets [84], thus enabling the success of NT children aged 2 and a half and 3 years respectively. They can also assess false-belief comprehension through gaze behaviors or physical actions performed by the child while watching the false-belief scenario presented on video (e.g. with expectation transgression [4], interactive aid [20], gaze anticipation [100], or visual preference paradigms [93]) then leading to the success of NT children aged 15 months, 18 months, 2 years and 2 and a half years respectively. In summary, explicit false-belief tasks show late success, from age 4, while implicit tasks show early success, as early as 15 months [4].

This pattern of results can be explained from two different perspectives on typical ToM development. On the one hand, according to conceptual change theory, ToM would gradually develop from the age of 4 [105], which would explain the failure of 3-year-old NT children on the explicit false-belief task [85]. A conceptual shift occurring around age 4 would enable success on explicit false-belief tasks. Early success on implicit tasks would not reflect a real understanding of the concept of belief, but rather the child's identification of statistical regularities [85]. On the other hand, according to proponents of early competence theory (or theory of task demands), NT children under the age of 4 have already developed ToM [5]. Their failure on explicit false-belief tasks would be caused not by an inability to attribute mental states to others, but by the additional demands of the task. Indeed, success on the false-belief test would depend not only on mental state attribution skills, but also on general skills such as executive functions or language, notably language pragmatics. To pass the test, the child must correctly recall the story (working memory), inhibit his knowledge of the object's actual location (cognitive inhibition), and correctly interpret the experimenter's verbal question (language pragmatics skills) [95]. These skills (other than ToM) would therefore stand in the way of young children's success, but there are two main opposing theories on the type of skill involved. One theory assumes that it is the executive functions that young children lack [53]: NT children succeed on implicit tasks before age 4 because the cognitive cost of these tasks is reduced, and the development of executive functions around age 4 would then enable success on explicit tasks. Conversely, pragmatic theory assumes that it is the demands of the language pragmatics task that cause NT children to fail before age 4 [107]. We will see in the next section that this theory can also be applied to people with ASD. Given that implicit tasks have demonstrated success in NT children under 4, the theory of conceptual change seems contradicted: children under 4 are capable of attributing mental states. The theory of early competence (which assumes that failure is due to additional task demands rather than a lack of theory of mind) therefore seems more consistent with the findings of the literature. In this article, we will focus on the pragmatic theory to explain the failure of people with ASD to pass the false belief test.

Implicit Tasks and Developmental Theories of ToM in Children with ASD. Individuals with ASD perform poorly on explicit false-belief tasks. Nevertheless, in the same way, as for NT children, variants of the false-belief test enable better performance by children with ASD [24,69]. These modified versions generally involve simplification, with varying degrees of success: they can be non-verbal to minimize the impact of language on performance, based on spontaneous responses, or omit the final location of the object to reduce demands on working memory. Given that communication difficulties are among the central symptoms in autism, although they vary from child to child, it is not surprising that performance on the false belief test varies. When the tasks are non-verbal, it is possible to observe similar performance between children with ASD and NT [35,40]. In a study based on a gaze anticipation paradigm with children with ASD aged 2 to 7, the difference between the groups was only trendy (with 31% success in the children with ASD versus 63% in the NT children) [19]. Other studies based on the same paradigm show, on the contrary, a lower performance of children with ASD [95] and adults with ASD [90,94,96], even in a task not involving language. However, the discrepancy between these results can be explained by the temporal window of interest. Indeed, the majority of studies based on a gaze anticipation paradigm focus on a short delay after stimulus presentation (around 4 s). However, the processing of social information could be slowed down in individuals with ASD: when confronted with a social stimulus, they present a gaze pattern different from that of NT individuals over a short time interval, but the difference fades over a longer time interval [91]. In the first 4 s following presentation of the false belief scenario on video, children with ASD do not look at the location that corresponds to the character's false belief, unlike NT children, which is generally interpreted as a lack of spontaneous ToM in ASD [91,94–96]. But in the following 4 s (from 4 s to 8 s after stimulus presentation) children with ASD look at the location that corresponds to the character's erroneous belief, unlike NT children [40]. This indicates that children with ASD are able to spontaneously assign false beliefs when task demands are reduced, but that processing is slower than in NT children. These results are therefore consistent with the theory of task demands, according to which children with ASD have theory of mind, but do not express it to the same extent as NT children due to the additional demands of the task.

As with NT children, the poor performance of children with ASD on explicit false-belief tests can be interpreted in different ways. There are three main opposing conceptions of ToM in autism. Failure on the false-belief test could be seen as resulting from a lifelong ToM disability ("mindblindness" theory), or from a developmental delay in ToM that would result in failure on the false-belief test until later in life ("empathizing-systemizing theory"). A third, less widespread theory proposes that people with ASD fail due to a developmental delay in the additional demands of the task, particularly in the pragmatics of language (pragmatic theory, to be discussed in the next section).

In the "mindblindness" hypothesis of autism, ToM is conceptualized as a module, i.e. a cognitive skill specialized in the attribution of mental states and relatively independent of other skills. This hypothesis is in line with the theory of conceptual change: in typical development, this module would mature around 4–5 years of age, allowing success in false-belief tests. Conversely, in autism, a neurodevelopmental disorder would disrupt the maturation of this specific module, leading to failure on the tests. This central deficit in ToM would be at the root of the difficulties encountered by autistic people in social interaction and communication [23]. Nevertheless, a modular conception of ToM does not seem to correspond to the performance observed in the literature, which indicates a progressive elaboration of ToM, building on precursors such as joint attention. Moreover, it is not consistent with the clinical picture of ASD, since disturbances in social interaction and communication occur before the age of 4 [61].

Criticism of "Mindblindness" Theory. The failure of children with ASD on the false-belief test has led to the view that they lack access to theory of mind. This deficit in theory of mind in autism, considered reliable and robust, was envisaged in the following way: it would be specific to autism [14], would apply universally [13], would be replicable from one study to another, and would meet convergence validity and predictive validity. However, we shall see that the difficulties encountered by people with ASD are not consistent with the conception of this deficit: recent studies call into question the association between ASD and ToM deficit [39]. We will examine point by point the characteristics of ToM deficiency as described in the "mindblindness" theory [10], but we can already point out that failure on the false-belief test does not necessarily indicate a ToM deficit, since other factors are likely to have an impact on performance (theory of task demands). It is therefore important to distinguish between task performance on the one hand, and the child's competence on the other, as the former does not necessarily reflect the latter. Later on, we will look at the factors (not related to ToM) that may be at the root of failure in children with ASD.

Firstly, failure on the false belief test would not be specific to autism [39,104]. While children with ASD may fail the test, other populations also fail, such as children with specific language impairment [52,66] or deaf children who have learned to sign late [72]. Moreover, this deficit pattern of performance on the false belief test does not appear to be universal, i.e. it is not shared by all individuals with ASD [39,104]. Indeed, some children with ASD do manage to respond correctly, a result observed as early as the first study on the subject, with 20% of children with ASD succeeding [14]. It therefore seems important to ask ourselves about the differences between successful and unsuccessful individuals, in order to understand what stands in the way of success. ASD are highly heterogeneous, and symptoms can vary from one child to another, particularly in terms of cognitive and language skills. Yet these abilities could play a role in success on the false belief test. In some studies, adults with Asperger's syndrome (a diagnostic category that no longer exists since the DSM-V but designates people with ASD who have relatively preserved cognitive and language abilities) all pass the explicit false-belief test [96] or show performance comparable to that of NT adults [90]. The difference in performance on the explicit false belief test between NT and ASD children disappears when mental verbal age is taken into account in the analysis [95], and when the test is non-verbal there appears to be no divergence in results between the developmental groups [35][2]. Children with ASD who pass the explicit false-belief test have a higher mental verbal age than children with ASD who fail [23]. A minimum verbal mental age of 9 years would be required for children with ASD to pass the task with a 50% chance, while NT children succeed as early as age 4 [44]. However, at equal mental verbal ages, some children with ASD succeed and others fail, suggesting that a certain verbal age is necessary but not sufficient for success on the false belief test [23]. These results underline the importance of language in performance and indicate that the false-belief test mobilizes abilities other than that of attributing belief to others, such as the pragmatics of language (a theory that will be discussed in the next section). Furthermore, attempts to replicate the results obtained by Baron-Cohen et al. [14,15] have failed, showing similar performance in children with ASD compared with children with ID or with other disorders (unspecified neurodevelopmental disorders or psychiatric disorders) in a false belief test or an intentional image sequencing test [18,63]. This seems to call into question the reliability of the failure of people with ASD on this test. Moreover, the various tests used to measure ToM do not seem to meet the criterion of convergence validity [39]. There is no convergence between the results obtained on the false belief task and the other ToM tasks (e.g. the true belief task, or desire attribu-

---

[2] Given that children with ASD perform less well on the false-belief test than other children (NT or not) despite a similar verbal mental age [52,66], this seems initially consistent with the idea of a ToM deficit in autism, irrespective of language skills. However, it is possible that the measures of language ability used to construct comparable groups in terms of verbal mental age (e.g. the Picture Vocabulary Test) do not reflect the actual skills of children with ASD, particularly on the pragmatic component of language, which is particularly impacted in ASD [78].

tion) in NT children [76] and adults [36], and no correlation between the false belief task (cognitive ToM) and the RMET task (affective ToM) in adolescents with ASD [54]. Finally, ToM tests do not meet the criterion of predictive validity. If ToM is indeed a prerequisite for social engagement [13] then performance on ToM tasks should predict individuals' social-emotional abilities. However, ToM task performance fails to predict children's empathy or emotional understanding [67], social abilities in everyday interactions [47], attention and social cooperation [70], peer relations and pro-social behaviors [28], the presence of autistic traits [19,74], or repetitive behaviors [47]. In NT children, comprehension in a ToM task predicts social skills with peers, but in children with ASD the link between ToM and social competence is indirect and mediated by language skills [68]. In children with ASD, neither cognitive ToM nor affective ToM predicted parent-reported social functioning, although affective ToM skills predicted disorder severity [1]. This result suggests that limited ToM abilities are not sufficient to explain the difficulties encountered in social interactions by people with ASD.

In short, all these elements seem inconsistent with the "mindblindness" theory [14] which postulates that ToM deficiency would be the cause of autism symptoms. ToM deficits in people with ASD are not consistent with those described: they are not specific to autism, not universal, not replicable, and have no convergent or predictive validity. Furthermore, a ToM deficit does not explain the non-social symptoms of autism (restricted interests and stereotyped behaviors), nor does it encompass the full range of social symptoms [104].

In response to criticism, the same author modified this theory to refine it and bring it more into line with empirical data: the "empathizing-systemizing" theory [13]. Building on the previous theory, this theory postulates on the one hand that individuals with ASD show a reduced tendency to attribute mental states to others (empathizing) compared to the average, which would explain the social symptoms of autism (deficits in communication and social interaction). On the other hand, she argues that individuals with ASD show better-than-average abilities to spot the regularities underlying systems and to analyze these rules (systemizing). This latter aspect would explain the restricted interests and repetitive behaviours, as well as the aversion to change: the preference of people with ASD is for predictable systems. The non-social symptoms of autism would then result from the desire of individuals with ASD to maintain a predictable environment. This update qualifies the previous theory, particularly with regard to ToM deficits. The "empathizing-systemizing" theory maintains that individuals with ASD show delayed ToM development and poorer ToM skills than the average individual, but there is no longer any question of a universal, systematic ToM deficit in autism.

1.2   Pragmatic Theory to Explain the Failure of Children with ASD in the False-Belief Test

We have seen two theories that have been proposed to explain the failure of individuals with ASD and NT on the false belief test ("mindblindness" and "empathizing-systemizing" theory). If we consider that failure on this task reflects an absence of theory of mind, then NT children under 4 and a majority of children with ASD would lack this ability. Nevertheless, success on this task may depend on abilities other than ToM, which could impact on the performance of neurotypical children and children with ASD. Within the theories of task demands already discussed above, we will take a particular interest in pragmatic theory [107]. Indeed, the false belief task requires advanced language pragmatics skills in order to understand the expected interpretation of the test question. However, people with ASD frequently encounter difficulties in language pragmatics, which could hinder their success. After defining pragmatics and its impact in ASD, we will analyze the extent to which the explicit false-belief test question is ambiguous, and explain how this ambiguity is likely to bias the performance of people with ASD. We will then show that it is possible to reduce the ambiguity of the test question by modifying the context of the experiment with the mentor-child paradigm [7–9,33].

Defining Language Pragmatics. The pragmatics of language correspond to "the use that is made of language by integrating the context" in which it is uttered [60]. During a conversation, the addressee seeks to understand the meaning communicated by the speaker's utterance (which often goes beyond the literal meaning) and does so by means of inferences [41]. Pragmatic abilities enable the interpretation of the meaning communicated by an utterance, taking into account the context of enunciation. The context of enunciation thus plays a crucial role in the interpretation of an utterance, so it is important to analyze the context in which the false-belief test takes place.

Pragmatics seems to influence the performance of young NT children on reasoning tests involving dialogue with the experimenter, also known as conversational tests [75], and it could also impact that of children with ASD. In many classic psychological tests, the child's success depends on his or her answer to the test question. However, this question is ambiguous for the child, as it can lead to different interpretations. To give the right answer to the test question, the child must have the same interpretation of the test question as the experimenter. If the child's interpretation differs from that of the experimenter, the child will not provide the expected answer, but this does not mean that they do not know the correct answer. Thus, the child's interpretation of the test question is constrained by their level of development in pragmatics. The poor performance of young NT children on this type of task would therefore be due to a divergence of interpretation between the experimenter and the child regarding the test question [75]. This could also be the case for people with ASD, as they also display limited pragmatic abilities. First, we will look at the pragmatic difficulties involved, before explaining how the false-belief test question is ambiguous.

Pragmatic Difficulties in ASD. Individuals with ASD present difficulties in social interaction and communication. Despite some variability in language abilities in ASD, difficulties in language pragmatics appear to be frequent in ASD [89] with poorer performance compared to NT children [6,22], children with language delay of the same age [42], or children with ID [29, 38]. This result is observable even with high-functioning children with ASD [3]. The pragmatic difficulties encountered by people with ASD are notably observed during conversations, with more transgression of conversational maxims [102], but also in the comprehension of figurative language [48], metaphors [22, 43, 62] or irony [27, 43, 64]. The pragmatics score seems to be a good predictor of ASD diagnosis, and of the disorder's level of severity [30].

Nevertheless, there is wide variability in pragmatic abilities from one individual to another within the autistic population [83]. For this reason, as already emphasized for ToM deficits, pragmatic deficits in ASD should be nuanced. As a recent study points out, certain aspects of language pragmatics seem to be preserved in adolescents and adults with ASD [27]. They show performance comparable to that of NT children on a test of comprehension of indirect requests (such as "Can you move [...]?" or "Is it possible for you to move [...]?"). However, the performance of individuals with ASD follows a different pattern to that of NT children since they respond in the same way for both formulations of the request, unlike NTs who respond less to the second formulation of the question (an unconventional formulation, since it is generally used to ask informative questions rather than to ask something of another person). This suggests that NT adolescents and adults have a finer comprehension of indirect requests than ASD adolescents and adults, which could be explained by better abilities in language pragmatics. Furthermore, in the same study, individuals with ASD showed less comprehension of ironic sentences than NT individuals, while no difference was observed for comprehension of literal statements. Consequently, individuals with autism may show preserved performance in understanding indirect requests, but marked difficulties in understanding irony. This aspect of pragmatics represents a greater level of difficulty than the other aspects, which could explain this divergence of results: understanding irony would require more advanced pragmatic abilities than understanding indirect requests [27] or metaphors [43]. These results indicate that pragmatic difficulties persist into adulthood, at least in certain areas such as understanding irony. As we shall see, this skill plays a crucial role in the success of the false-belief test.

Linking ToM and Pragmatics. ToM and pragmatics seem to overlap in certain respects. The difficulties encountered in language pragmatics by individuals with ASD were initially attributed to the ToM deficit [10, 11]. Thus, some authors have linked pragmatic communication to ToM abilities [43] based on relevance theory [101], which states that in conversation, taking into account the communicative intention of the interlocutor (in other words, attributing mental states to the interlocutor) plays a crucial role in understanding the message, especially when it is ambiguous. More precisely, the interpretation of an utterance (hence the pragmatics of language) would depend on a specific form of mental state attribution that would be dedicated to intention communication [49, 59]. According to this conception, pragmatics would be a specific form of ToM, conceptualized as a submodule of ToM that would be specific to pragmatic processing [101]. People with ASD would be less competent in the pragmatics of language because their ToM deficit prevents them from attributing an intention to the interlocutor, which is necessary for understanding non-literal language [43]. ToM would therefore be necessary for understanding non-literal language.

Nevertheless, pragmatic and ToM skills do not fully overlap [17]. Performance in pragmatics seems to be better predicted by semantic skills than by ToM skills, and success on the false belief task is not sufficient to guarantee metaphor comprehension [62]. In other words, some children pass the false-belief task (ToM) but fail to understand metaphors (pragmatics). Moreover, when children with ASD have relatively preserved language skills (no marked language deficit), their performance on a metaphor comprehension task is comparable to that of NT children [62]. This suggests that language pragmatics and ToM are two distinct concepts, which is important to note for the next section.

Reducing the Pragmatic Ambiguity of the False-Belief Test to Improve Success

Pragmatic Ambiguity of the False-Belief Test Question. In the false-belief test, the test question is ambiguous. According to the principle of cooperation [41], in a conversation, the recipient of a message seeks to understand the interlocutor's expectations in order to respond to them in the best possible way. To grasp the communicative intention of the interlocutor, the recipient will make inferences, based on the utterance and the context. Various aspects define the situational context, such as the respective encyclopedic knowledge of the different interlocutors, the social relations established between them, or the spatiotemporal framework. When an utterance is ambiguous, i.e. when it can refer to several interpretations, the choice of the appropriate interpretation will be guided by relevance theory [101]. The more cognitive effort a statement requires to be processed, the less relevant it is. Conversely, the more a statement produces contextual effects (changes in the context), the more relevant it is. Thus, relevance is seen as an optimization of the ratio between cognitive cost and contextual effects. We shall see that the failure of young NT children and children with ASD to pass the false-belief test can be explained on the basis of relevance theory.

In the false-belief task, the statement of the test question "Where will Maxi look for the chocolate?" is ambiguous, as it can refer to two different interpretations, and can express two different expectations on the part of the experimenter. On the one hand, the experimenter may be seeking to test the child's knowledge of the world ("Tell me where Maxi should look to find the chocolate"), and on the other, he or she may be seeking to test the child's false-belief abilities ("Tell me where Maxi thinks he will find the chocolate"). According to relevance theory, the child will select, from among these interpretations, the one that seems most relevant to the context in which the question is uttered, i.e., the one that seems least costly from a cognitive point of view and has the most contextual effect in a given context. Various aspects of the context in which the false-belief test is administered are likely to influence the performance of young NT children and children with ASD, by accentuating the ambiguity of the test question. Firstly, the adult is seen by the child as possessing knowledge superior to his or her own, and as having more encyclopedic knowledge. In addition, the experimenter is presented as a teacher, an authority figure whom the child must obey. Finally, the time and place of enunciation are also important. The tests are generally carried out at the school or childcare facility, and take place during school time, which could reinforce the idea of a school test, requiring the display of knowledge. In short, all these factors in the situational context lead the child to assume that school behavior (the display of knowledge) is expected of him, as is usually the case in the school setting. In this context, the child interprets the adult's question as testing his knowledge of the real state of the world, rather than his understanding of the character's false belief. NT children under 4 and children with ASD would therefore have the ability to attribute false beliefs to others, but would not be able to infer the expected interpretation of the test question, due to insufficient pragmatic abilities. On the contrary, for older NT children and adults, who have further developed their pragmatic abilities through social experience, it would be less costly to understand what the test question relates to in this context and to identify the experimenter's real expectation [107]. Furthermore, children with ASD would fail the false belief test more than children with ID or language delay because these populations have better pragmatic abilities [29,38,42]. We will see that the same mechanism could operate for children with ASD, causing them to fail.

Several studies suggest a link between pragmatic language skills and success on ToM tests. One study of children with ASD compared pragmatic abilities and degree of disorder severity according to performance on a ToM task [83]. Children with low scores on ToM tasks also had fewer pragmatic abilities and a higher degree of disorder severity (including fewer adaptive behaviors). Another study has also shown that children with ASD who have the best pragmatic abilities are also those who perform best on the false belief test, pragmatic ability being the best predictor of success [37]. In a neuroimaging study, no behavioral differences are noted between adults with ASD and NT adults on the false-belief test [99]. At the cerebral level, there is no difference in activation when the false-belief scenario is presented, but discrepancies appear when the experimenter asks the test question. This suggests that the difficulty for people with ASD lies in interpreting the test question, rather than understanding the scenario itself, which is consistent with pragmatic theory. The success observed at a later age in children with ASD could therefore be explained by the fact that they show a slower development of their pragmatic abilities compared to NT children [108]. Children with ASD show delayed development of pragmatics [108], which would explain why most of them fail the false belief test if they are not at least 9 years old [44]. Moreover, certain aspects of pragmatic abilities could remain deficient in adulthood [27], which could explain why some adults with ASD still fail the false belief test.

The pragmatic difficulties encountered by individuals with ASD could also explain the difference in ToM developmental sequence observed in the literature between NT children and children with ASD. The hidden emotions test is the most complex of the ToM scale for NT children [106], but it is better performed by children with ASD than the false belief test [71–73]. It is possible that the hidden emotion test requires less language pragmatics than the false belief test, which would explain the divergent performance of children with ASD. In the hidden emotion test, the test questions are explicit: "How did Matt really feel? Does he feel happy, sad or neutral?" and "What emotion is Matt showing on his face? Does he look happy, sad or neutral?" [106]. Since these questions do not involve pragmatic ambiguity, they do not require inference of the experimenter's intention and are therefore simpler to understand for people with ASD whose pragmatic abilities are limited. This pattern of results seems consistent with the pragmatic theory according to which individuals with ASD would have access to the ToM but fail the false belief test due to difficulties in language pragmatics.

Reducing Pragmatic Ambiguity by Modifying the Test Question. The performance of NT children under 3 and children with ASD can be improved when the test question is made more explicit, which is consistent with the pragmatic theory that interpreting the test question can make the task more complex.

Modifying the test question by adding a temporal precision "Where will Sally look *first* for the marble?" improves the success of 3-year-old NT children [98,103] but does not improve the success of children with ASD aged 7 to 18 [103]. This suggests that in young NT children, limited pragmatic language skills may be an obstacle to success: they don't understand the experimenter's intention when asked the test question. When the question is made more explicit by adding the term "first", their performance improves significantly. The fact that the performance of children with ASD was not improved was initially interpreted as evidence that children with ASD fail not because of pragmatic difficulties (additional demands of the task), but because of difficulties in representing the mental states of others. Nevertheless, another study shows that among children with ASD aged 6 to 17 who fail the classic false-belief test, half can nevertheless succeed when the wording of the test question is modified in this way [37,78]. The performance of children with ASD on the false-belief test is therefore improved when the test question is made more explicit. We shall see that it is also possible to reduce the pragmatic ambiguity of the false-belief test by modifying the context of enunciation.

Reducing Pragmatic Ambiguity by Modifying the Context of Enunciation. It is also possible to reduce the pragmatic ambiguity of the test question by simply modifying the context of enunciation, while retaining the original wording of the question. As we have seen, the interpretation of the pragmatically ambiguous test question is based on the context. When the experimenter who asks the test question is not present during the object transfer (a second experimenter replaces the first after the transfer of the object's location in the scenario), 3-year-old NT children perform better on the [87] test. In this new context, the children no longer have common knowledge with the experimenter as to the actual location of the object. It would then be easier to understand that the experimenter, when asking the test question, is seeking to test their understanding of the character's false belief and not their knowledge about the world. Similarly, when the false-belief test is presented in a context of competition for a reward (marble, sticker or car toy), the performance of children with ASD increases considerably (from 13% success to 74% success) [69]. More precisely, in this study, the child is warned that he will be able to leave with the object if the agent has not found it, and he must choose which of two agents will be able to open a box before him (one agent has witnessed the object transfer, the other has not). Children overwhelmingly choose the agent who doesn't know the object's actual location, suggesting that they are able to assign an erroneous belief. Finally, the context in which the test is presented seems to play a crucial role. When the instructions and questions of the false-belief test are given by computer rather than by an experimenter physically present with the participants, children with ASD show no difference in performance compared with NT children [24]. In the condition in which the experimenter is physically present, NT children show better performance than children with ASD. The authors interpret this result as indicating that NT children benefit from the presence of the experimenter in the room, whereas this is not the case for children with ASD (they show comparable performance in both conditions).

Reducing Pragmatic Ambiguity with the Mentor-Child Paradigm. Furthermore, it is also possible to improve the performance of 3-year-old NT children on the false-belief test using the mentor-child paradigm [7–9,31]. In this paradigm, the adult asks the child to assume the role of teacher towards a humanoid robot NAO presented as a naive and ignorant student. Replacing the human experimenter with a robot offers numerous methodological advantages in terms of neutralising pragmatic and contextual effects [45,55–58]. This methodology is also interesting for children, who can attribute intentions and states of mind to the robot [32]. The conversation with the adult experimenter envisaged in the original false-belief paradigm is therefore replaced by a conversation with a robot presented as a naïve student, thus altering the experimenter's social status. To enable the establishment of this master/student relationship between child and robot, we draw on the tendency of young NT children to help others. The adult tells the child "You have to help NAO learn things because he has never been to school". NAO's ignorance is reinforced during the interaction, as he tells the child that he doesn't know anything and needs his help to learn. When the false-belief test question is asked by an ignorant person, such as the NAO robot, the new context in which the question is stated makes it easier for young NT children to select the expected interpretation (as this weakens the alternative interpretation that the robot would like to test their knowledge about the world). On the one hand, the results show that the typical development of the theory of mind takes place earlier than Wimmer and Perner [109] assumed since NT children are able to attribute false beliefs before the age of 4 [7]. On the other hand, they confirm that the pragmatic factors identified by [75] are an obstacle to young children's success in the original false belief test. The pragmatic factors involved in the original test would therefore lead to an underestimation of NT children's ability to attribute mental states. They could also hinder the success of other populations with limited pragmatic abilities, such as people with ASD.

We therefore sought to adapt the mentor-child paradigm for children with ASD [31]. The aim of the present article is to determine whether this new enunciation context improves their performance on the false-belief test. Given that the mentor-child paradigm involves the child taking on the role of robot teacher, its implementation with children with ASD relies on several prerequisites [31]. The child must perceive the robot as a social agent (1), accept to help the robot (2), understand the role of a student and the concept of ignorance (3), and understand the role of a teacher who transmits knowledge (4). Children with ASD appear to interact with robots as they do with humans [110,114], and may exhibit social behaviors similar to those of NT children, such as eye contact, vocalizations and smiles [88]. They can spontaneously express helping behaviors towards others to the same extent as NT children [34], or even more [65]. On the other hand, a study of adults with ASD and NT adults indicates that participants with ASD are more motivated to help an agent presented as having autistic features (e.g. difficulties in understanding social interactions) than an agent presented with neurotypical features (no difficulties in understanding social interactions) [50]. Consequently, it might be relevant to present the robot as endowed with autistic characteristics in the mentor-child paradigm, in order to increase children's motivation to help it. Furthermore, children with ASD are able to play the role of teacher to a robot to impart knowledge [112], and can correct its errors to point out the correct answer [46]. They therefore seem able to help an ignorant robot acquire knowledge, adopting the status of a naive robot's teacher, as is the case in the mentor-child paradigm.

1.3 Children with ASD Might Perform Better on the False Belief Test with an Ignorant Robot as Experimenter

Studies suggest that children with ASD fail the false belief test massively, with poorer performance compared to NT children and children with ID [14,15,51,103]. Nevertheless, the idea of a systematic ToM deficiency in ASD has recently been called into question [39]. Furthermore, the pragmatic ambiguity of the false-belief test seems to complicate the task [7,107], which could lead to an underestimation of their abilities. For these reasons, it is questionable whether children with ASD will perform better on the false-belief test in the mentor-child paradigm (robot experimenter) than in the control condition (human experimenter), and whether their performance will then be comparable to that of children with ID who have more pragmatic abilities [29,38]. If children with ASD perform better with the mentor-child paradigm, this would suggest that the pragmatic demands of the task are indeed impeding their success. Moreover, if their performance is then similar to that of children with ID (in the robot experimenter condition), this would confirm that the poor performance of children with ASD found in the literature is not due to a specific ToM deficit, but rather to pragmatic difficulties related to the test-taking context. For this reason, children with ID are also involved in the present experiment, in order to quantitatively compare their performance with that of children with ASD.

We expect to observe better performance in children with ASD in the robot experimenter condition compared with the human experimenter condition (H1). According to the literature, children with ASD should perform less well than children with ID in the human experimenter condition (H2a). Conversely, in the robot experimenter condition, we expect to observe similar success between children with ASD and children with ID using the mentor-child paradigm (H2b).

2 Method

2.1 Participants

The sample consisted of 50 participants, all verbal, including 26 children and adolescents aged 7–19 with a DSM-V-based diagnosis of ASD [2], without associated ID ($Mage = 13.73$; $SDage = 3.75$; 5 girls; $IQrange : 71-119$), and 24 participants aged 11–17 with a diagnosis of mild ID (except one child with moderate ID) ($Mage = 14.5$; $SDage = 2.06$; 6 girls; $IQrange : 48-69$). All children and adolescents were recruited from the Roland Chavance Center (France, Hautes-Pyrénées), and were being cared for in two structures attached to the center: a Medical and Educational Institute (Institut Médico-Educatif) and a Special Education and Home Care Service (Service d'Education Spéciale et de Soins à Domicile). After the experimental protocol had been validated by the Roland Chavance Center and the agreement of the legal guardians obtained, the experiment was proposed to the children in the form of an activity.

The children were randomly divided into two groups: the first group interacted with the robot (experimental condition), while the second group interacted with the human (control condition). Among children with ASD, 13 performed the test with the robot as experimenter, and 13 performed the test with the human as experimenter. Among children with ID, 12 performed the test with the robot, and 12 performed the test with the human (see Table 1 for sample characteristics)

Table 1. Sample characteristics. *ASD = Participants with Autism Spectrum Disorders, ID = Participants with Intellectual Disability; Human = Human experimenter, Robot = Robot experimenter*

Development type	ASD			ID		
Condition	Robot	Human	Total	Robot	Human	Total
n	13	13	26	12	12	24
Mean age (SD)	13.85 (4.26)	13.61 (3.33)	13.73 (3.75)	14.41 (2.11)	14.58 (2.11)	14.5 (2.06)
Age range (years)	7–19	8–19	7–19	11–17	11–17	11–17
Mean IQ (SD)	89.23 (12.89)	88.31 (12.90)	88.77 (12.65)	61.58 (6.60)	60.42 (8.09)	61 (7.25)
IQ range	71–108	78–119	71–119	48–69	48–69	48–69
Ratio of boys:girls	11:2	10:3	21:5	9:3	9:3	18:6

2.2 Materials

NAO Robot. We used a NAO robot created by Aldebaran Robotics (version 5 - "Evolution"). This humanoid-looking robot is 58 cm tall, has a movable head, arms and legs, and is equipped with LEDs on its eyes, head and torso. As the NAO robot is the most widely used for interaction with children with ASD [82, 86], and is well accepted by this population [81], it therefore seems appropriate for the mentor-child paradigm. The robot's programming is based on the Wizard of Oz method: the robot is semi-teleoperated from a computer using Choregraphe software, i.e. programs are prepared in advance, but are launched in real time by an experimenter. This type of programming allows for greater flexibility in interaction with the children, who may have variable reaction times. In addition, the Wizard of Oz method implies that the experimenter programming the robot is not visible to the participant, in order to give the illusion that the robot is acting autonomously. In the present study, the experimenter programming the robot is hidden behind a folding screen.

False Belief Task. We used an adapted version of the classic false-belief task [109] with color images [36] rather than Sally and Anne's test [14] which was in black and white. This task takes the form of three vignettes (9 cm × 9 cm) (see Vignettes), to which a verbal explanation is added as follows: in the first image, Maxi puts his chocolate away in the blue cupboard, and then leaves the room. In the second image, in Maxi's absence, his mom moves the chocolate into the green cupboard. In the last image, Maxi returns to fetch his chocolate. The experimenter then asks the child three questions in a fixed order. First, the false-belief test question: "Where will Maxi look for the chocolate?" followed by two control questions, designed to check that the child has understood the story: "Where is the chocolate at the end of the story?" (reality question) and "Where was the chocolate at the beginning of the story?" (memory question). To pass the test, the child must correctly answer the test question and the control questions [14]. The questions asked are strictly identical in the robot and human conditions.

Colored Balloons. Before presenting the false-belief test to the child, the experimenter asked him to indicate the color of two balloons (blue and green) to check that the child could distinguish between these colors (which are also those of the closets in the false-belief scenario). In the robot experimenter condition, this also reinforced the robot's ignorance, as it told the child that it didn't know the colors. Conversely, in the human experimenter condition, these questions reinforce the adult's status as a teacher who knows.

Video Recording. In addition to the computer used to teleoperate the robot, the equipment also involved the use of a camera mounted on a tripod to film the passage of instructions, in particular, to analyze the number of fixations on the experimenter and the time spent looking away during the presentation of instructions (distraction time).

2.3   Procedure

The experiment took place in three locations attached to the Roland Chavance Center. The test was carried out individually, in a quiet room. All three testing rooms were of a similar size ($12\,\text{m}^2$), and each of these rooms had an area hidden by a folding screen, allowing the Wizard of Oz technique to be used. The experiment required the presence of two adults. The first adult (psychologist of the center) accompanied the children to the testing room, while the second adult (experimenter and trainee psychologist) set up the testing room, programmed the robot in the robot experimenter condition, and acted as the human agent in the human experimenter condition. As a result, the children already knew the two adults involved in the experiment.

The procedure consisted of 3 stages (see Fig. 1), described below.

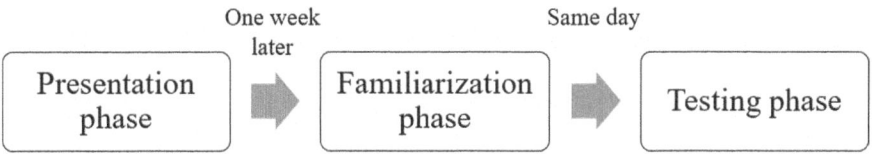

Fig. 1. Diagram of the experimental procedure.

Presentation Phase. A week before the test, the psychologist presented a photo of the NAO robot to the children concerned, telling them that they would be carrying out an activity with the robot. The other half of the children were told that they would carry out an activity with the trainee psychologist.

Familiarization Phase. On the day of the test, the psychologist accompanied the child to the experiment room. In the robot condition, the experimenter installed the robot in a sitting position on the table, facing the child's chair (at a distance of around 1 m). In the human condition, the experimenter sits on a chair behind the table. In this way, when the child enters the room, accompanied by the psychologist, the agent (robot or human) is already seated at the table (or on the table in the case of the robot). The blue and green balloons are also on the table, positioned on either side of the agent (see Image of the experimental set-up). The psychologist first asked the child to sit down on the chair provided, then introduced the agent. The procedure thus consisted of a familiarization phase, followed by a test phase.

In the robot experimenter condition, the psychologist introduces the child to the robot: "I'm going to introduce you to NAO, the robot I told you about last week. He needs your help to learn things and understand a story. NAO doesn't have any friends because he has difficulty talking to other people. He doesn't like to be touched, so be careful not to touch him.". The last two sentences were intended to suggest that the robot has autistic traits (social difficulties but also tactile sensory characteristics, i.e. atypical responses to tactile stimuli) in order to increase the child's tendency to help the robot. This also reduces the likelihood of the child trying to touch the robot (the robot being fragile). When the child entered the room, the psychologist first address the robot: "NAO, I'd like you to meet [participant's first name], who has agreed to help you learn things...", then the participant: "I'll leave you with NAO for a little while, listen carefully to what he says and try to help him. See you later.". Once the participant had settled in, the robot introduced itself to the child: "Hello, my name is NAO and I'm a robot. I need you to help me learn things because I've never been to school.". The robot then asked the child to point out two colored balloons: "I don't know colors, can you help me learn? Can you show me the blue/green balloon?". This step was designed to reinforce the robot's ignorance and ensure that the child knows the two colors (which are the closet colors in the false-belief scenario).

In the human experimenter condition, the psychologist told the child that he was going to carry out an activity with the trainee psychologist: "He's going to tell you a story and ask you some questions.". When the child returned to the room, the psychologist first addressed the adult: "Yohann, [participant's first name] has agreed to answer your questions" and then the participant: "I'll leave you with Yohann for a little while, so listen carefully to what he says. See you later.". Once the participant was settled, the adult introduced himself to the child: "Hello, I'm going to ask you some questions to check that you know the colors well.". The adult then asked the child to point out two colored balloons: "Can you show me the blue/green balloon?" This step reinforces the adult's status as a person with knowledge.

Testing Phase. The participant then performs the false belief test.

In the robot experimenter condition, the robot solicits the child's help: "I've been told a story that I don't quite understand, and I need your help.". He then tells the child the false-belief scenario: "In the first picture, Maxi puts his chocolate away in the blue cupboard and leaves the room. In the second image, when Maxi isn't there, his mom moves the chocolate and puts it in the green cupboard. In the third picture, Maxi comes back to eat.". The robot then asks the child three questions, in a fixed order: test question, reality question and memory question.

In the human experimenter condition, the adult presents the story to the child: "I'm going to tell you a story and ask you some questions.". He then tells the child the false-belief scenario and asks the questions in exactly the same way as in the robot experimenter condition.

In both conditions, the adult immediately notes the child's responses on a scoring sheet. Prompt sentences were provided if the child did not respond to the robot's first prompt: "Show me with your finger where Maxi will look to get the chocolate.". At the end of the test, the experimenter (human or robot) thanks the child and informs the psychologist, who then escorts the child back to the classroom.

Rating. We adopted the same methodology as in the literature [14] to rate the performance: the false-belief task is considered successful when the child has correctly answered all three questions (the test question and the two control questions). In addition, the participant's gaze time on the scenario vignettes and the number of visual fixations on the experimenter during the story presentation were rated manually from the video recordings. Although we made no initial hypothesis about the children's visual behavior, we analyzed it for a finer-grained understanding of performance. Specifically, we calculated the child's distraction time during the story presentation, i.e. the number of seconds during which he or she did not look at the vignettes (distraction time = total story presentation time − time during which the participant looked at the vignettes). The number of fixations corresponds to the number of times the child looks at the experimenter during the presentation of the scenario. All of these measures are quantitative.

3   Results

3.1   Generalized Linear Models

All statistical analyses were performed using R 4.3.0 software (see R code and data). The aim was (i) to test whether success at the false belief task varies according to condition (human experimenter vs. robot experimenter), (ii) according to developmental type (ASD vs. ID), and (iii) according to distraction time and number of fixations.

Success at the False Belief Task Predictors. We sought to determine the predictors of success in the false belief test using generalized linear models (logistic regression) on the binary categorical variable of success (ranging from 0 to 1). Since all the children answered the control questions correctly, the analyses are based on their success on the test question (see Fig. 2).

A null model including no predictors was calculated (model A0). At level 1, this null model was compared with several models including different predictors (experimenter, developmental type, gender, age, IQ, distraction time, number of fixations). Predictor quality was determined according to the Akaike Information Criterion (AIC): the predictor with the lowest AIC represents the predictor that best fits the data. We also took into account the Bayes Information Criterion (BIC) in our model selection, a fit index that penalizes complex models. As with the AIC, the lowest BIC represents the predictor that best fits the data. The model with the best predictor was retained for further analysis. We first compared simple models with the null model, then combined models including several predictors (see Table 2 in Appendix). We retained the model with the lowest AIC and BIC. Among the simple models, the model containing the number of fixations on the experimenter (model A7) predicted success better than all the other models, including the model containing no predictor, on both AIC and BIC [Model A7: AIC = 64.587, BIC = 68.412; Model A0: AIC = 71.315, BIC = 73.227, BFA7, A0 = 11.107]. Although we have not made any assumptions about the number of fixations, we can see that this is the factor that best predicts the data. Among the combined models, our aim was to see whether another model could generate better predictions of success than Model A7. However, no model was better than the A7 model. This model indicates that the number of fixations on the experimenter predicts success in the false belief task: the higher the number of fixations, the less successful the participant. According to the A7 model, there is no effect of experimenter or developmental type on success. Children with ASD performed as well in the robot experimenter condition as they did in the human experimenter condition. Moreover, in both conditions, children with ASD performed as well as children with ID. There is no effect of gender, age and IQ on success.

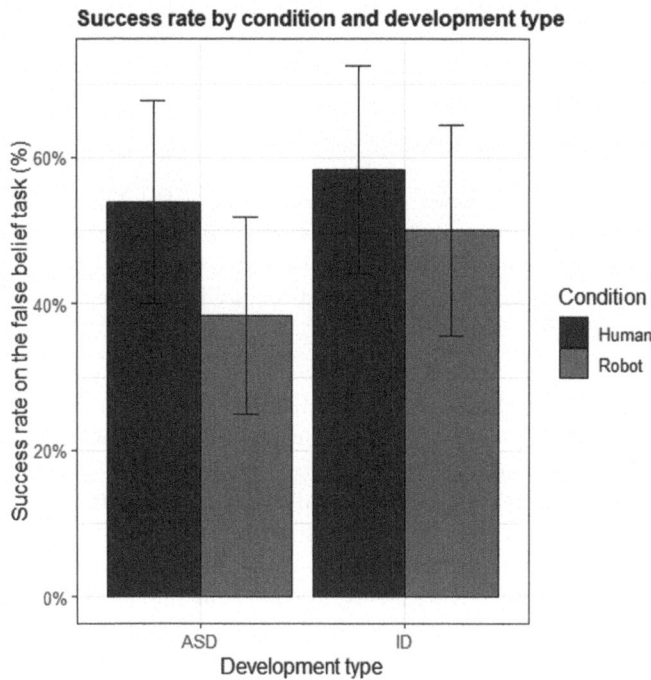

Fig. 2. Success rate on the false belief task by experimenter and type of development (bars represent standard errors). *ASD = Participants with Autism Spectrum Disorders, ID = Participants with Intellectual Disability; Human = Human experimenter, Robot = Robot experimenter.*

Number of Fixations Predictors. We sought to determine the predictors of the number of fixations using generalized linear models on the number of fixations variable (ranging from 0 to 5). We began our analysis using a Poisson regression, but then opted for a Gaussian regression as this did not change the preferred model and allowed for simpler analysis. The aim was to assess whether the number of fixations on the experimenter varied according to condition and type of development.

A null model including no predictors was calculated (model B0). At level 1, this null model was compared with several models including different predictors (experimenter, developmental type, gender, age, IQ, distraction time). Given that the number of fixations is the best predictor of success in the false-belief task, we excluded the inverse models (number of fixations according to success at the test question). We first compared simple models with the null model, then combined models including several predictors (see Table 3 in Appendix). We retained the model with the lowest AIC and BIC. Among the simple models, the model containing the distraction time (model B6) predicted the number of fixations better than all the other models, including the model containing no predictor, on both AIC and BIC [Model B6: AIC = 112.065, BIC = 117.801; Model B0: AIC = 168.141, BIC = 171.965, BFB7, B0 = 5.77429E+11]. Among the combined models, our aim was to see whether another model could generate better predictions of the number of fixations than Model B6, but no model was better than the B6 model. This model indicates that the distraction time predicts the number of fixations on the experimenter: the longer the time spent looking away, the greater the number of fixations on the experimenter. According to the B6 model, there is no effect of condition or developmental type on the number of fixations on the experimenter. The number of fixations in the robot experimenter condition was similar to the number of fixations in the human experimenter condition. Moreover, the number of fixations was comparable between children with ASD and those with ID. There is no effect of gender, age and IQ on the number of fixations.

Distraction Time Predictors. We sought to determine the predictors of the distraction time using generalized linear models on the distraction time variable (ranging from 0 to 18 s). The aim was to assess whether the distraction time varied according to condition and type of development.

A null model including no predictors was calculated (model C0). At level 1, this null model was compared with several models including different predictors (experimenter, developmental type, gender, age, IQ). Given that the distraction time is the best predictor of the number of fixations on the experimenter, we excluded the inverse models (distraction time according to number of fixations and success at the test question). We first compared simple models with the null model, then combined models including several predictors (see Table 4 in Appendix). We retained the model with the lowest AIC and BIC. Among the simple models, the model containing the experimenter (model C1) predicted the distraction time better than all the other models, including the model containing no predictor, on both AIC and BIC [Model C1: AIC = 260.446, BIC = 266.182; Model C0: AIC = 270.379, BIC = 274.203, BFC1, C0 = 55.168].

Among the combined models, our aim was to see whether another model could generate better predictions of the distraction time than Model C1. With regard to BIC, no model was better than the C1 model. However, the AIC gave the C12 model (containing the interaction between condition and type of development in addition to gender) as better [Model C12: AIC = 252.943, BIC = 270.152; Model C1: AIC = 260.446, BIC = 266.182; BFC12, C1 = 7.58]. According to the C12 model, there is an interaction effect of condition, developmental type and gender on the distraction time. In the robot experimenter condition, children with ASD show more distraction time than children with ID (and especially for women with ASD). In the human experimenter condition, the distraction time was comparable between children with ASD and those with ID (see Fig. 3). This result should be treated with caution, however, as the C12 model is not the best in terms of BIC. The difference between girls and boys must also be treated sparingly given the small number of girls in the sample (5 girls compared with 21 boys in children with ASD and 6 girls compared with 18 boys in children with ID). Like the C1 model, this model also indicates that the condition significantly predicts distraction time: distraction time is higher in the robot experimenter condition than in the human experimenter condition.

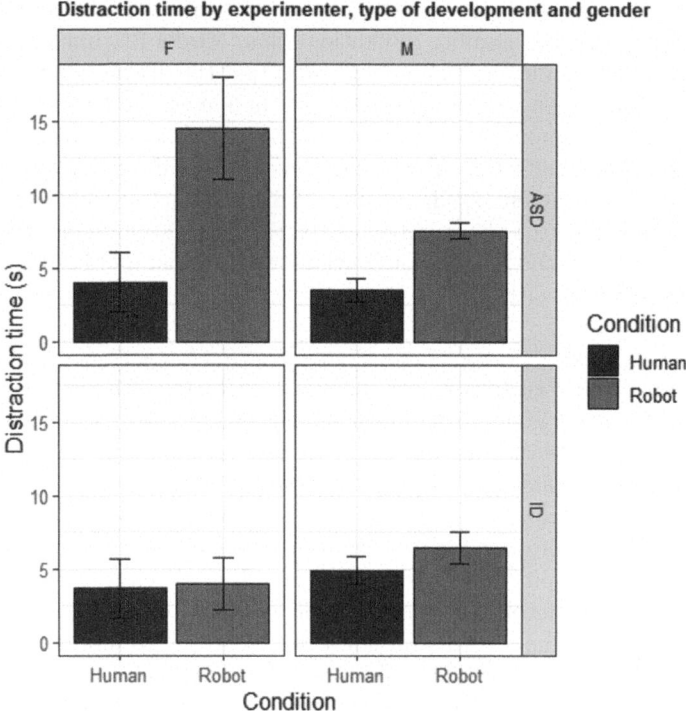

Fig. 3. Distraction time by experimenter, type of development and gender (bars represent standard errors). *ASD = Participants with Autism Spectrum Disorders, ID = Participants with Intellectual Disability; Human = Human experimenter, Robot = Robot experimenter; F = Female participants; M = Male participants.*

4   Discussion

Statistical analyses of generalized linear models indicate that the number of fixations factor best predicts success at the test question. The tendency of participants to give the correct answer to the test question depends mainly on the number of times they looked at the experimenter during the presentation of the false-belief scenario. The more participants looked at the experimenter in terms of number of fixations, the less successful they were on the test, regardless of condition (robot experimenter vs. human) or type of development (ASD vs. ID). There was therefore no direct effect of experimenter or type of development on success, which contradicts our initial hypothesis. Nevertheless, we shall see that the condition and type of development could have an indirect impact on success. These results will be discussed below.

We had assumed that children with ASD would perform better on the false-belief test in the robot experimenter condition than in the human experimenter condition (H1). Our results contradict this hypothesis: the robot experimenter condition leads to a statistically comparable performance to the human experimenter condition (respectively 38.46% vs. 53.85% success rate). In other words, the mentor-child paradigm did not improve the success of children with ASD. At the descriptive level, it even seems to have reduced performance. There was no effect of gender, age or IQ on success on the test question: boys gave as many correct answers as girls, younger participants performed as well as older participants, and participants with higher IQs performed as well as participants with lower IQs. These results can be interpreted in two ways. On the one hand, it could contradict the pragmatic theory, according to which reducing the ambiguity of the test question could improve the performance of children with ASD. The lack of performance improvement with the mentor-child paradigm would indicate that children fail the false-belief test because of a difficulty in ToM rather than a difficulty in pragmatics, thus supporting the "mindblindness" or "empathizing-systemizing" theory. On the other hand, it is possible that the mentor-child paradigm is unsuited to individuals with ASD, being insufficient to reduce pragmatic difficulty. For example, participants with ASD might have difficulty understanding the concept of an ignorant student, and would then interpret the robot experimenter's test question in the same way as when asked by the human experimenter, i.e. as being asked by an interlocutor with more knowledge than themselves. However, we shall see that another interpretation of these results can be proposed, which could perhaps lead to an improvement of the mentor-child paradigm.

We also expected to observe poorer performance by children with ASD compared with children with ID in the human experimenter condition (H2a), in line with the literature [14,15]. In the robot experimenter condition, by contrast, we expected to observe similar success between children with ASD and children with ID using the mentor-child paradigm (H2b). However, our results contradict the first hypothesis (H2a) and show no statistical difference according to the participants' developmental type: in the human experimenter condition, children with ASD pass the test question to the same extent as children with ID (respectively

53.85% vs. 58.33% success rate, i.e. 7 children with ASD out of 13 giving the right answer vs. 7 children with ID out of 12 giving the right answer)[3]. The same pattern of results was observed in the robot experimenter condition, with no statistical difference between children with ASD (38.46% success rate, 5 out of 13 children giving the right answer) and children with ID (50% success rate, 6 out of 12 children giving the right answer). In the present study, the success rate of children with ASD was comparable to that reported by Peterson et al. [72] (47% success rate), but higher than that obtained by Baron-Cohen et al. [14] (20% success rate) or by Leslie et al. [52] (28% success rate). Yet the mean IQ of participants with ASD in the present study ($MIQ = 88.77$; range: 71–119) is comparable to that in the study by Baron-Cohen et al. [14] ($MIQ = 82$; range: 70–108). We therefore fail to replicate these results indicating poor performance by children with ASD, as already pointed out by other authors [39]. Furthermore, all the children in our sample answered both control questions correctly, whatever the experimental condition, suggesting that our test may be simpler than the classic test by Baron-Cohen et al. [14]. This can be explained from a methodological point of view: some studies presented the false-belief scenario in the form of scenes played out by dolls [14,52], which requires the child to retain in memory all the steps in the story and the different locations of the object in order to answer the questions. Conversely, in our version of the test, the scenario is presented to the children in the form of vignettes that remain visible when the questions are asked, thus minimizing the amount of information to be retained, which would facilitate success. Our results are closer to those of another study that tests false belief understanding using an unexpected content test (such as the Smarties task), which also requires less retention of details in memory, resulting in greater success [72]. This result therefore seems consistent with processing demand theory: the false-belief test would involve cognitive resources other than ToM, such as executive functions [53] or pragmatics [107], likely to lower the performance of individuals with ASD. Performance heterogeneity could also reflect the significant inter-individual variability observed within ASD [89].

Analyses of the number of fixations and distraction time also provide some interesting insights. Although we did not hypothesize the effect of fixation number and distraction time on child performance, we did investigate whether these variables could predict success on the false-belief test. We shall see that, although experimental condition and developmental type have no direct effect on success, they could have an indirect impact. Indeed, the higher the number of fixations on the experimenter (i.e. the number of times the child looked at the experimenter during the presentation of the false-belief scenario), the more the participant failed to give the correct answer to the test question. The number of fixations seems itself to be predicted by distraction time (i.e. the time spent looking away from the scenario vignettes during storytelling): the more time the child spends looking away, the higher the number of fixations on the experimenter. Interestingly, analysis of the predictors of distraction time suggests an effect of the experimental condition: in the robot experimenter condition, distraction time is

---

[3] It is important to note that the average chronological age of the two groups is similar.

greater than in the human experimenter condition. More precisely, there seems to be an interaction effect between condition and type of development: children with ASD show more distraction time with the robot than with the human, and the magnitude of the difference in distraction time with the robot and with the human is greater in children with ASD than in children with ID. This effect would be even stronger in girls with ASD, which could be explained by their tendency to exhibit more social behaviors compared to boys with ASD [26,111]. However, the latter result should be interpreted with caution, given the small number of girls with ASD included in the sample (n = 5). Neither participant age nor IQ had any effect on distraction time. This result suggests that the child's distraction is linked to the experimenter. Indeed, if distraction time alone had been the best predictor of success, this might have indicated that participants failed more because they looked at something other than the scenario vignettes (e.g. the testing room), without the experimental condition having any impact. However, it was the number of fixations on the experimenter that best predicted success, indicating that the distraction was induced by the experimenter himself. So it is not the presence of the robot as such that impacts the performance of children with ASD, but the fact that they would be distracted by the robot experimenter, and look at it more during the scenario presentation. We will summarize the predictors of success in the false-belief test by adding these analyses. Distraction time appears to be predicted by experimental condition (robot vs. human experimenter) and by developmental type (as well as by gender to a lesser extent). Children with ASD, and particularly girls, are more distracted in the robot experimenter condition than in the human experimenter condition, unlike children with ID. This distraction time predicts the number of fixations on the experimenter: as distraction time increases, so does the number of fixations on the experimenter, suggesting that children with ASD looked more at the experimenter in the robot condition than in the human condition. Yet, as we have seen, the higher the number of fixations on the experimenter, the lower the child's performance on the test. This result can then be interpreted as follows: as children with ASD were more distracted in the robot condition, they showed more fixations on the experimenter, which reduced their performance on the test question in this experimental condition. Thus, the lack of benefit of the mentor-child paradigm on the performance of children with ASD could be explained by an increased interest in the robot compared with the human, an effect that has already been highlighted in the literature [33,114]. This strong interest in robots would explain why children with ASD look more at the robot experimenter than at the human experimenter, unlike children with ID. Thus, the robot could potentially help reduce the pragmatic ambiguity of the task, but this effect would be counteracted by the interest of children with ASD in the robot, which leads them to turn away from the task, thus hindering their performance.

To confirm this interpretation, it would be appropriate to repeat this experiment, establishing a robot habituation paradigm [25] beforehand. In other words, the children should be offered several interactions with the robot before the test phase, to ensure that they are accustomed to the robot's presence and do not react to the novelty. We presented an image of the robot before the experiment, but this was not enough to familiarize the children with the robot. The habituation paradigm would thus control the baseline level of the child's visual response to the robot at the start of the experiment, to ensure that the child has become accustomed to its presence. It might also be advisable to replace the human experimenter with an adult unknown to the children, since in the present study the adult was familiar to them, which may have increased the difference in distraction time between human and robot experimenters (since the children were used to the adult's presence but not to the robot's). However, the results of the present article remain uncertain given the small size of our sample.

Finally, the success rate of children with ASD confirms that they are capable of attributing false beliefs to others, and to the same extent as children with ID. This result contradicts the "mindblindness" and "empathizing-systemizing" theories, which argue that ToM would be more impacted in people with ASD than in other populations (such as people with ID). The task demands theory seems more consistent with the results, but it is not yet possible to confirm the pragmatic theory. Further studies are needed to draw a conclusion and determine the role played by pragmatics in the performance of children with ASD on the false-belief test.

5   Conclusion

The mentor-child paradigm did not lead to better performance for children with ASD on the false-belief test. Children showed statistically similar performance with the robot experimenter and the human experimenter, which seems to contradict the pragmatic theory: reducing the pragmatic ambiguity of the false-belief test with the mentor-child paradigm does not lead to improved performance. Nevertheless, this population's marked interest in robots may have biased the results. Indeed, children with ASD were more distracted in the robot experimenter condition, and therefore looked more at the experimenter and less at the scenario vignettes in this condition, which may have led to poorer performance on the test question. Thus, the implementation of the mentor-child paradigm should be improved in the future, notably by including a robot habituation phase for children with ASD to limit their distraction time during the presentation of the false-belief scenario. Although the results do not support the pragmatic theory that people with ASD fail the false belief test because of the ambiguity of the test question, they do show that a certain proportion of children with ASD can pass the test, and to the same extent as children of the same age with ID. This seems to contradict the "mindblindness" theory of autism. Further studies are needed to confirm this result and clarify the impact of pragmatics on the performance of children with ASD on the false belief test.

Acknowledgements. We would like to thank the Roland Chavanne Center for their welcome and interest in this research project, as well as all the children who took part in the study and their families. We would also like to thank Baptiste Jacquet and Maxime Bourlier for their help with the statistical analysis.

Appendix

Table 2. Models predicting success at the false belief task. *Best model(s) are shown in bold; BF x0: Bayes Factor which indicates how the given model (AX) compares to the null model (A0). The higher the BF, the better the model.*

	AIC	BIC	BF x0
Level 1			
Simples model vs. Random			
Model A0 (Null)	71.315	73.227	1
Model A1 (Experimenter)	72.593	76.417	0.203
Model A2 (Development)	72.994	76.818	0.166
Model A3 (Gender)	73.198	77.022	0.15
Model A4 (Age)	73.260	77.084	0.145
Model A5 (IQ)	73.132	76.956	0.155
Model A6 (Distraction time)	66.310	70.134	4.695
Model A7 (Number of fixations)	**64.587**	**68.412**	**11.107**
Combined models vs. Random			
Model A8 (Experimenter+Development)	74.267	80.003	0.034
Model A9 (Experimenter*Development)	76.204	83.852	0.005
Model A10 (Experimenter*Gender)	72.424	80.072	0.033
Model A11 (Experimenter*Age)	76.410	84.058	0.004
Model A12 (Experimenter*IQ)	76.192	83.840	0.005
Model A13 (Number of fixations+Experimenter)	66.484	72.220	1.654
Model A14 (Number of fixations*Experimenter)	68.182	75.830	0.272
Model A15 (Number of fixations+Development)	66.587	72.323	1.571
Model A16 (Number of fixations*Development)	68.583	76.231	0.223
Model A17 (Number of fixations+Gender)	66.587	72.323	1.571
Model A18 (Number of fixations*Gender)	67.478	75.126	0.387
Model A19 (Number of fixations+Age)	66.540	72.276	1.609
Model A20 (Number of fixations*Age)	65.769	73.417	0.909
Model A21 (Number of fixations+IQ)	66.573	72.310	1.582
Model A22 (Number of fixations*IQ)	68.506	76.154	0.231
Model A23 (Number of fixations+Distraction time)	66.475	72.211	1.662
Model A24 (Number of fixations*Distraction time)	65.268	72.916	1.168
Model A25 (Distraction time+Experimenter)	68.134	73.870	0.725
Model A26 (Distraction time*Experimenter)	69.379	77.027	0.15
Improving model A7			
Model A27 (Number of fixations+Distraction time+Development)	68.474	76.122	0.235
Model A28 (Number of fixations*Distraction time+Development)	67.250	76.810	0.167
Model A29 (Number of fixations*Distraction time*Development)	73.042	88.338	0.001
Model A30 (Number of fixations+Experimenter+Development)	68.484	76.132	0.234
Model A31 (Number of fixations*Experimenter+Development)	70.154	79.714	0.039
Model A32 (Number of fixations*Experimenter*Development)	74.383	89.679	0
Model A33 (Number of fixations+Distraction time+Experimenter)	68.295	75.943	0.257
Model A34 (Number of fixations*Distraction time+Experimenter)	66.910	76.470	0.198
Model A35 (Number of fixations*Distraction time*Experimenter)	71.934	87.230	0.001
Model A36 (Number of fixations*Distraction time+Development+Experimenter)	68.878	80.350	0.028
Model A37 (Number of fixations*Distraction time+Development*Experimenter)	69.904	83.288	0.007
Model A38 (Number of fixations*Distraction time+Development*Experimenter)	84.884	115.477	0

Table 3. Models predicting the number of fixations on the experimenter during the false-belief task. *Best model(s) are shown in bold; BF x0: Bayes Factor which indicates how the given model (BX) compares to the null model (B0). The higher the BF, the better the model.*

	AIC	BIC	BF x0
Level 1			
Simples model vs. Random			
Model B0 (Null)	168.141	171.965	1
Model B1 (Experimenter)	161.411	167.147	11.125
Model B2 (Development)	168.480	174.216	0.324
Model B3 (Gender)	169.687	175.423	0.177
Model B4 (Age)	170.126	175.862	0.143
Model B5 (IQ)	168.572	174.308	0.31
Model B6 (Distraction time)	**112.065**	**117.801**	**5.77429E+11**
Combined models vs. Random			
Model B7 (Experimenter+Development)	161.427	169.075	4.243
Model B8 (Experimenter*Development)	157.988	167.548	9.103
Model B9 (Experimenter*Gender)	164.616	174.176	0.331
Model B10 (Experimenter*Age)	165.363	174.923	0.228
Model B11 (Experimenter*IQ)	162.361	171.922	1.022
Improving model B6			
Model B12 (Distraction time+Experimenter)	113.997	121.645	84500948779
Model B13 (Distraction time*Experimenter)	115.992	125.552	11981268694
Model B14 (Distraction time+Development)	113.202	120.850	1.25749E+11
Model B15 (Distraction time*Development)	113.942	123.502	33394953149
Model B16 (Distraction time+Experimenter+Development)	115.094	124.654	18768806717
Model B17 (Distraction time+Experimenter*Development)	116.176	127.648	4199702996
Model B18 (Distraction time*Experimenter*Development)	112.932	130.140	1208500857
Model B19 (Distraction time*Experimenter+Development)	116.970	128.442	2824347210

Table 4. Models predicting distraction time during the false belief task. *Best model(s) are shown in bold; BF x0: Bayes Factor which indicates how the given model (CX) compares to the null model (C0). The higher the BF, the better the model.*

	AIC	BIC	BF x0
Level 1			
Simples model vs. Random			
Model C0 (Null)	270.379	274.203	1
Model C1 (Experimenter)	**260.446**	**266.182**	**55.168**
Model C2 (Development)	271.520	277.256	0.217
Model C3 (Gender)	272.356	278.092	0.143
Model C4 (Age)	271.689	277.425	0.2
Model C5 (IQ)	271.743	277.479	0.194
Combined models vs. Random			
Model C6 (Experimenter+Developpement)	261.353	269.001	13.475
Model C7 (Experimenter*Developpement)	258.393	267.953	22.753
Model C8 (Experimenter*Gender)	263.835	273.395	1.498
Model C9 (Experimenter*Age)	263.500	273.060	1.771
Model C10 (Experimenter*IQ)	262.902	272.462	2.388
Improving model C7			
Model C11 (Experimenter*Development+Gender)	260.087	271.559	3.75
Model C12 (Experimenter*Development*Gender)	**252.943**	**270.152**	**7.58**
Model C13 (Experimenter*Development+Age)	259.464	270.936	5.12
Model C14 (Experimenter*Development*Age)	261.433	278.641	0.109
Model C15 (Experimenter*Development+IQ)	260.387	271.859	3.229
Model C16 (Experimenter*Development*IQ)	263.348	280.556	0.042

References

1. Altschuler, M., et al.: Measuring individual differences in cognitive, affective, and spontaneous theory of mind among school-aged children with autism spectrum disorder. J. Autism Dev. Disord. **48**(11), 3945–3957 (2018). https://doi.org/10.1007/s10803-018-3663-1
2. American Psychiatric Association: Diagnostic and Statistical Manual of Mental Disorders. American Psychiatric Association, 5th edn. (2013). https://doi.org/10.1176/appi.books.9780890425596
3. Angeleri, R., Gabbatore, I., Bosco, F., Sacco, K., Colle, L.: Pragmatic abilities in children and adolescents with autism spectrum disorder: a study with the ABaCo battery. Minerva Psichiatr. **57**, 93–103 (2016)
4. Baillargeon, R., Scott, R.M., Bian, L.: Psychological reasoning in infancy. Ann. Rev. Psychol. **67**, 159–186 (2016). https://doi.org/10.1146/annurev-psych-010213-115033

5. Baillargeon, R., Scott, R.M., He, Z.: False-belief understanding in infants. Trends Cogn. Sci. **14**(3), 110–118 (2010). https://doi.org/10.1016/j.tics.2009.12.006

6. Baixauli-Fortea, I., Miranda Casas, A., Berenguer-Forner, C., Colomer-Diago, C., Roselló-Miranda, B.: Pragmatic competence of children with autism spectrum disorder. Impact of theory of mind, verbal working memory, ADHD symptoms, and structural language. Appl. Neuropsychol. Child **8**(2), 101–112 (2017). https://doi.org/10.1080/21622965.2017.1392861

7. Baratgin, J., Dubois-Sage, M., Jacquet, B., Stilgenbauer, J.L., Jamet, F.: Pragmatics in the false-belief task: let the robot ask the question! Front. Psychol. **11** (2020). https://doi.org/10.3389/fpsyg.2020.593807

8. Baratgin, J., Jacquet, B., Dubois-Sage, M., Jamet, F.: "Mentor-child and Naive-pupil-robot" paradigm to study children's cognitive and social development. In: Workshop: Interdisciplinary Research Methods for Child-Robot Relationship Formation, HRI-2021 (2021)

9. Baratgin, J., Jamet, F.: Le paradigme de "l'enfant mentor d'un robot ignorant et naïf" comme révélateur de competences cognitives et sociales précoces chez le jeune enfant. In: WACAI 2021. Centre National de la Recherche Scientifique [CNRS], Saint Pierre d'Oléron, France (2021). https://hal.archives-ouvertes.fr/hal-03377546

10. Baron-Cohen, S.: Social and pragmatic deficits in autism: cognitive or affective? J. Autism Dev. Disord. **18**(3), 379–402 (1988). https://doi.org/10.1007/BF02212194

11. Baron-Cohen, S.: Autism: a specific cognitive disorder of & 'mind-blindness'. Int. Rev. Psychiatry **2**(1), 81–90 (1990). https://doi.org/10.3109/09540269009028274

12. Baron-Cohen, S.: Mindblindness: An Essay on Autism and Theory of Mind. The MIT Press (1995). https://doi.org/10.7551/mitpress/4635.001.0001

13. Baron-Cohen, S.: Autism: the empathizing-systemizing (E-S) theory. Ann. N. Y. Acad. Sci. **1156**(1), 68–80 (2009). https://doi.org/10.1111/j.1749-6632.2009.04467.x

14. Baron-Cohen, S., Leslie, A.M., Frith, U.: Does the autistic child have a "theory of mind"? Cognition **21**(1), 37–46 (1985). https://doi.org/10.1016/0010-0277(85)90022-8

15. Baron-Cohen, S., Leslie, A.M., Frith, U.: Mechanical, behavioural and Intentional understanding of picture stories in autistic children. Br. J. Dev. Psychol. **4**(2), 113–125 (1986). https://doi.org/10.1111/j.2044-835X.1986.tb01003.x

16. Baron-Cohen, S., Wheelwright, S., Hill, J., Raste, Y., Plumb, I.: The "reading the mind in the eyes" test revised version: a study with normal adults, and adults with Asperger syndrome or high-functioning autism. J. Child Psychol. Psychiatry Allied Disciplines **42**(2), 241–251 (2001). https://doi.org/10.1017/S0021963001006643

17. Bosco, F.M., Tirassa, M., Gabbatore, I.: Why pragmatics and theory of mind do not (completely) overlap. Front. Psychol. **9** (2018). https://www.frontiersin.org/articles/10.3389/fpsyg.2018.01453

18. Buitelaar, J.K., Wees, M.V.D., Swaab-Barneveld, H., Gaag, R.J.V.D.: Theory of mind and emotion-recognition functioning in autistic spectrum disorders and in psychiatric control and normal children. Dev. Psychopathol. **11**(1), 39–58 (1999). https://doi.org/10.1017/s0954579499001947

19. Burnside, K., Wright, K., Poulin-Dubois, D.: Social motivation and implicit theory of mind in children with autism spectrum disorder. Autism Res.: Official J. Int. Soc. Autism Res. **10**(11), 1834–1844 (2017). https://doi.org/10.1002/aur.1836

20. Buttelmann, D., Carpenter, M., Tomasello, M.: Eighteen-month-old infants show false belief understanding in an active helping paradigm. Cognition **112**(2), 337–342 (2009). https://doi.org/10.1016/j.cognition.2009.05.006

21. Cai, J., Hu, X., Guo, K., Yang, P., Situ, M., Huang, Y.: Increased left inferior temporal gyrus was found in both low function autism and high function autism. Front. Psychiatry **9** (2018). https://www.frontiersin.org/articles/10.3389/fpsyt.2018.00542

22. Cardillo, R., Mammarella, I.C., Demurie, E., Giofrè, D., Roeyers, H.: Pragmatic language in children and adolescents with autism spectrum disorder: do theory of mind and executive functions have a mediating role? Autism Res. **14**(5), 932–945 (2021). https://doi.org/10.1002/aur.2423

23. Charman, T., Baron-Cohen, S.: Understanding drawings and beliefs: a further test of the metarepresentation theory of autism: a research note. J. Child Psychol. Psychiatry **33**(6), 1105–1112 (1992). https://doi.org/10.1111/j.1469-7610.1992.tb00929.x

24. Chevallier, C., Parish-Morris, J., Tonge, N., Le, L., Miller, J., Schultz, R.T.: Susceptibility to the audience effect explains performance gap between children with and without autism in a theory of mind task. J. Exp. Psychol. Gen. **143**(3), 972–979 (2014). https://doi.org/10.1037/a0035483

25. Cohen, L.B.: Habituation of infant visual attention. In: Habituation, pp. 207–238. Routledge (1976)

26. Del Bianco, T., et al.: EU-AIMS LEAP Group: unique dynamic profiles of social attention in autistic females. J. Child Psychol. Psychiatry **63**(12), 1602–1614 (2022). https://doi.org/10.1111/jcpp.13630

27. Deliens, G., Papastamou, F., Ruytenbeek, N., Geelhand, P., Kissine, M.: Selective pragmatic impairment in autism spectrum disorder: indirect requests versus irony. J. Autism Dev. Disorders **48**(9), 2938–2952 (2018). https://doi.org/10.1007/s10803-018-3561-6

28. Devine, R.T., Hughes, C.: Silent films and strange stories: theory of mind, gender, and social experiences in middle childhood. Child Dev. **84**(3), 989–1003 (2013). https://www.jstor.org/stable/23469324

29. Diken, Ö.: Pragmatic language skills of children with developmental disabilities: a descriptive and relational study in turkey. Eurasian J. Educ. Res. **55**, 109–122 (2014). https://doi.org/10.14689/ejer.2014.55.7

30. Dolata, J.K., Suarez, S., Calamé, B., Fombonne, E.: Pragmatic language markers of autism diagnosis and severity. Res. Autism Spectr. Disord. **94**, 101970 (2022). https://doi.org/10.1016/j.rasd.2022.101970

31. Dubois-Sage, M., Jacquet, B., Jamet, F., Baratgin, J.: The mentor-child paradigm for individuals with autism spectrum disorders. In: CONCATENATE Social Robots Personalisation - International Conference on Human Robot Interaction (HRI) 2023. Association for Computing Machinery (2023). https://doi.org/10.48550/arXiv.2312.08161

32. Dubois-Sage, M., Jacquet, B., Jamet, F., Baratgin, J.: We do not anthropomorphize a robot based only on its cover: context matters too! Appl. Sci. **13**(15) (2023). https://doi.org/10.3390/app13158743

33. Dubois-Sage, M., Jacquet, B., Jamet, F., Baratgin, J.: People with autism spectrum disorder could interact more easily with a robot than with a human: reasons and limits. Behav. Sci. (Basel) **14**(2), 131 (2024). https://doi.org/10.3390/bs14020131

34. Dunfield, K.A., Best, L.J., Kelley, E.A., Kuhlmeier, V.A.: Motivating moral behavior: helping, sharing, and comforting in young children with autism spectrum disorder. Front. Psychol. **10**, 25 (2019). https://doi.org/10.3389/fpsyg.2019.00025
35. Durrleman, S., Franck, J.: Exploring links between language and cognition in autism spectrum disorders: complement sentences, false belief, and executive functioning. J. Commun. Disord. **54**, 15–31 (2015). https://doi.org/10.1016/j.jcomdis.2014.12.001
36. Duval, C., Piolino, P., Bejanin, A., Eustache, F., Desgranges, B.: Age effects on different components of theory of mind. Conscious. Cogn. **20**(3), 627–642 (2011). https://doi.org/10.1016/j.concog.2010.10.025
37. Eisenmajer, R., Prior, M.: Cognitive linguistic correlates of 'theory of mind' ability in autistic children. Br. J. Dev. Psychol. **9**(2), 351–364 (1991). https://doi.org/10.1111/j.2044-835X.1991.tb00882.x
38. Escudero, S.C., Sepúlveda, E.M.: Pragmatic competence in people with dual diagnosis: down syndrome and autism spectrum disorder. BMC Psychol. **12**(1), 74 (2024). https://doi.org/10.1186/s40359-023-01508-5
39. Gernsbacher, M.A., Yergeau, M.: Empirical failures of the claim that autistic people lack a theory of mind. Arch. Sci. Psychol. **7**(1), 102–118 (2019). https://doi.org/10.1037/arc0000067
40. Glenwright, M., Scott, R.M., Bilevicius, E., Pronovost, M., Hanlon-Dearman, A.: Children with autism spectrum disorder can attribute false beliefs in a spontaneous-response preferential-looking task. Front. Commun. **6**, 669985 (2021). https://doi.org/10.3389/fcomm.2021.669985
41. Grice, H.P.: Logic and conversation. In: Cole, P., Morgan, J.L. (eds.) Speech Acts, Syntax and Semantics, vol. 3, pp. 43–58. Academic Press, New York (1975)
42. Hage, S.V.R., Sawasaki, L.Y., Hyter, Y., Fernandes, F.D.M.: Social communication and pragmatic skills of children with autism spectrum disorder and developmental language disorder. Codas **34**(2), e20210075 (2021). https://doi.org/10.1590/2317-1782/20212021075
43. Happé, F.G.E.: Communicative competence and theory of mind in autism: a test of relevance theory. Cognition **48**(2), 101–119 (1993). https://doi.org/10.1016/0010-0277(93)90026-R
44. Happé, F.G.E.: The role of age and verbal ability in the theory of mind task performance of subjects with autism. Child Dev. **66**(3), 843–855 (1995). https://doi.org/10.2307/1131954
45. Jamet, F., Masson, O., Jacquet, B., Stilgenbauer, J.L., Baratgin, J.: Learning by teaching with humanoid robot: a new powerful experimental tool to improve children's learning ability. J. Robot. **2018** (2018). https://doi.org/10.1155/2018/4578762
46. Jimenez, F., Yoshikawa, T., Furuhashi, T., Kanoh, M., Nakamura, T.: Feasibility of collaborative learning and work between robots and children with autism spectrum disorders. In: Otake, M., Kurahashi, S., Ota, Y., Satoh, K., Bekki, D. (eds.) New Frontiers in Artificial Intelligence, pp. 454–461. Springer, Cham (2017). https://doi.org/10.1007/978-3-319-50953-2_32
47. Joseph, R.M., Tager-Flusberg, H.: The relationship of theory of mind and executive functions to symptom type and severity in children with autism. Dev. Psychopathol. **16**(1), 137–155 (2004). https://doi.org/10.1017/S095457940404444X
48. Kalandadze, T., Norbury, C., Nærland, T., Næss, K.A.B.: Figurative language comprehension in individuals with autism spectrum disorder: a meta-analytic review. Autism **22**(2), 99–117 (2018). https://doi.org/10.1177/1362361316668652

49. Kissine, M.: Pragmatics, cognitive flexibility and autism spectrum disorders. Mind Lang. **27**(1), 1–28 (2012). https://doi.org/10.1111/j.1468-0017.2011.01433.x

50. Komeda, H., Kosaka, H., Fujioka, T., Jung, M., Okazawa, H.: Do individuals with autism spectrum disorders help other people with autism spectrum disorders? An investigation of empathy and helping motivation in adults with autism spectrum disorder. Front. Psych. **10**, 376 (2019). https://doi.org/10.3389/fpsyt.2019.00376

51. Leekam, S.R., Perner, J.: Does the autistic child have a metarepresentational deficit? Cognition **40**(3), 203–218 (1991). https://doi.org/10.1016/0010-0277(91)90025-y

52. Leslie, A.M., Frith, U.: Autistic children's understanding of seeing, knowing and believing. Br. J. Dev. Psychol. **6**(4), 315–324 (1988). https://doi.org/10.1111/j.2044-835X.1988.tb01104.x

53. Leslie, A.M., German, T.P., Polizzi, P.: Belief-desire reasoning as a process of selection. Cogn. Psychol. **50**(1), 45–85 (2005). https://doi.org/10.1016/j.cogpsych.2004.06.002

54. Lukito, S., et al.: Specificity of executive function and theory of mind performance in relation to attention-deficit/hyperactivity symptoms in autism spectrum disorders. Mol. Autism **8**, 60 (2017). https://doi.org/10.1186/s13229-017-0177-1

55. Masson, O., Baratgin, J., Jamet, F.: Nao robot and the "endowment effect". In: 2015 IEEE International Workshop on Advanced Robotics and its Social Impacts (ARSO), Lyon, France, pp. 1–6 (2015). https://doi.org/10.1109/ARSO.2015.7428203

56. Masson, O., Baratgin, J., Jamet, F.: Nao robot as experimenter: social cues emitter and neutralizer to bring new results in experimental psychology. In: Proceedings of the International Conference on Information and Digital Technologies, IDT 2017, pp. 256–264 (2017). https://doi.org/10.1109/DT.2017.8024306

57. Masson, O., Baratgin, J., Jamet, F.: Nao robot, transmitter of social cues: what impacts? In: Benferhat, S., Tabia, K., Ali, M. (eds.) Advances in Artificial Intelligence: From Theory to Practice, pp. 559–568. Springer, Cham (2017). https://doi.org/10.1007/978-3-319-60042-0_62

58. Masson, O., Baratgin, J., Jamet, F., Ruggieri, F., Filatova, D.: Use a robot to serve experimental psychology: some examples of methods with children and adults. In: International Conference on Information and Digital Technologies (IDT-2016), Rzeszow, Poland, pp. 190–197 (2016). https://doi.org/10.1109/DT.2016.7557172

59. Mazzarella, D., Noveck, I.: Pragmatics and mind reading: the puzzle of autism (Response to Kissine). Language **97**(3), e198–e210 (2021). https://doi.org/10.1353/lan.2021.0037

60. Moeschler, J.: La pragmatique après Grice: contexte et pertinence. L'information grammaticale **66**(1), 25–31 (1995). https://doi.org/10.3406/igram.1995.3044. Company: Persée - Portail des revues scientifiques en SHS Distributor: Persée - Portail des revues scientifiques en SHS Institution: Persée - Portail des revues scientifiques en SHS Label: Persée - Portail des revues scientifiques en SHS Publisher: Peeters

61. Nader-Grosbois, N.: La théorie de l'esprit. Entre cognition, émotion et adaptation sociale, Questions de personne, vol. 1re éd. De Boeck Supérieur, Louvain-la-Neuve (2011). https://www.cairn.info/la-theorie-de-l-esprit--9782804163235.htm

62. Norbury, C.F.: The relationship between theory of mind and metaphor: evidence from children with language impairment and autistic spectrum disorder. Br. J. Dev. Psychol. **23**(3), 383–399 (2005). https://doi.org/10.1348/026151005X26732

63. Oswald, D.P., Ollendick, T.H.: Role taking and social competence in autism and mental retardation. J. Autism Dev. Disord. **19**(1), 119–127 (1989). https://doi.org/10.1007/BF02212723

64. Panzeri, F., Mazzaggio, G., Giustolisi, B., Silleresi, S., Surian, L.: The atypical pattern of irony comprehension in autistic children. Appl. Psycholinguist. **43**(4), 757–784 (2022). https://doi.org/10.1017/S0142716422000091

65. Paulus, M., Rosal-Grifoll, B.: Helping and sharing in preschool children with autism. Exp. Brain Res. **235**(7), 2081–2088 (2017). https://doi.org/10.1007/s00221-017-4947-y

66. Perner, J., Frith, U., Leslie, A.M., Leekam, S.R.: Exploration of the autistic child's theory of mind: knowledge, belief, and communication. Child Dev. **60**(3), 689–700 (1989). https://doi.org/10.2307/1130734

67. Peterson, C.: Theory of mind understanding and empathic behavior in children with autism spectrum disorders. Int. J. Dev. Neurosci. **39**, 16–21 (2014). https://doi.org/10.1016/j.ijdevneu.2014.05.002

68. Peterson, C., Slaughter, V., Moore, C., Wellman, H.M.: Peer social skills and theory of mind in children with autism, deafness, or typical development. Dev. Psychol. **52**(1), 46–57 (2016). https://doi.org/10.1037/a0039833

69. Peterson, C.C., Slaughter, V., Peterson, J., Premack, D.: Children with autism can track others' beliefs in a competitive game. Dev. Sci. **16**(3), 443–450 (2013)

70. Peterson, C.C., Slaughter, V., Wellman, H.M.: Nimble negotiators: How theory of mind (ToM) interconnects with persuasion skills in children with and without ToM delay. Dev. Psychol. **54**(3), 494–509 (2018). https://doi.org/10.1037/dev0000451

71. Peterson, C.C., Wellman, H.M.: Longitudinal theory of mind (ToM) development from preschool to adolescence with and without ToM delay. Child Dev. **90**(6), 1917–1934 (2019). https://doi.org/10.1111/cdev.13064

72. Peterson, C.C., Wellman, H.M., Liu, D.: Steps in theory-of-mind development for children with deafness or autism. Child Dev. **76**(2), 502–517 (2005). https://doi.org/10.1111/j.1467-8624.2005.00859.x

73. Peterson, C.C., Wellman, H.M., Slaughter, V.: The mind behind the message: advancing theory-of-mind scales for typically developing children, and those with deafness, autism, or Asperger syndrome. Child Dev. **83**(2), 469–485 (2012). https://doi.org/10.1111/j.1467-8624.2011.01728.x

74. Pieslinger, J.F., Wiskerke, J., Igelström, K.: Contributions of face processing, social anhedonia and mentalizing to the expression of social autistic-like traits. Front. Behav. Neurosci. **16** (2022). https://doi.org/10.3389/fnbeh.2022.1046097

75. Politzer, G.: The class inclusion question: a case study in applying pragmatics to the experimental study of cognition. Springerplus **5**(1), 1133 (2016). https://doi.org/10.1186/s40064-016-2467-z

76. Poulin-Dubois, D., Yott, J.: Probing the depth of infants' theory of mind: disunity in performance across paradigms. Dev. Sci. **21**(4), e12600 (2018). https://doi.org/10.1111/desc.12600

77. Premack, D., Woodruff, G.: Does the chimpanzee have a theory of mind? Behav. Brain Sci. **1**(4), 515–526 (1978). https://doi.org/10.1017/S0140525X00076512

78. Prior, M., Dahlstrom, B., Squires, T.L.: Autistic children's knowledge of thinking and feeling states in other people. J. Child Psychol. Psychiatry **31**(4), 587–601 (1990). https://doi.org/10.1111/j.1469-7610.1990.tb00799.x

79. Quesque, F., et al.: Defining key concepts for mental state attribution. Commun. Psychol. **2** (2024). https://doi.org/10.1038/s44271-024-00077-6

80. Quesque, F., Rossetti, Y.: What do theory-of-mind tasks actually measure? Theory and practice. Perspect. Psychol. Sci. **15**(2), 384–396 (2020). https://doi.org/10.1177/1745691619896607
81. Rakhymbayeva, N., Amirova, A., Sandygulova, A.: A long-term engagement with a social robot for autism therapy. Front. Robot. AI **8**, 669972 (2021). https://doi.org/10.3389/frobt.2021.669972
82. Raptopoulou, A., Komnidis, A., Bamidis, P.D., Astaras, A.: Human-robot interaction for social skill development in children with ASD: a literature review. Healthcare Technol. Lett. **8**(4), 90–96 (2021). https://doi.org/10.1049/htl2.12013
83. Rosello, B., Berenguer, C., Baixauli, I., García, R., Miranda, A.: Theory of mind profiles in children with autism spectrum disorder: adaptive/social skills and pragmatic competence. Front. Psychol. **11** (2020). https://www.frontiersin.org/articles/10.3389/fpsyg.2020.567401
84. Rubio-Fernández, P., Geurts, B.: How to pass the false-belief task before your fourth birthday. Psychol. Sci. **24**(1), 27–33 (2013). https://doi.org/10.1177/0956797612447819
85. Ruffman, T.: To belief or not belief: children's theory of mind. Dev. Rev. **34**(3), 265–293 (2014). https://doi.org/10.1016/j.dr.2014.04.001
86. Saleh, M.A., Hanapiah, F.A., Hashim, H.: Robot applications for autism: a comprehensive review. Disab. Rehabil.: Assistive Technol. **16**(6), 580–602 (2021). https://doi.org/10.1080/17483107.2019.1685016
87. Salter, G., Breheny, R.: Removing shared information improves 3- and 4-year-olds' performance on a change-of-location explicit false belief task. J. Exp. Child Psychol. **187**, 104665 (2019). https://doi.org/10.1016/j.jecp.2019.104665
88. Scassellati, B.: How social robots will help us to diagnose, treat, and understand autism. In: Thrun, S., Brooks, R., Durrant-Whyte, H. (eds.) Robotics Research. Springer Tracts in Advanced Robotics, vol. 28, pp. 552–563. Springer, Heidelberg (2007). https://doi.org/10.1007/978-3-540-48113-3_47
89. Schadenberg, B.R., Reidsma, D., Heylen, D.K.J., Evers, V.: Differences in spontaneous interactions of autistic children in an interaction with an adult and humanoid robot. Front. Robot. AI **7** (2020). https://doi.org/10.3389/frobt.2020.00028
90. Schneider, D., Slaughter, V.P., Bayliss, A.P., Dux, P.E.: A temporally sustained implicit theory of mind deficit in autism spectrum disorders. Cognition **129**(2), 410–417 (2013). https://doi.org/10.1016/j.cognition.2013.08.004
91. Schuwerk, T., Vuori, M., Sodian, B.: Implicit and explicit Theory of Mind reasoning in autism spectrum disorders: the impact of experience. Autism **19**(4), 459–468 (2015). https://doi.org/10.1177/1362361314526004
92. Schwartz Offek, E., Segal, O.: Comparing theory of mind development in children with autism spectrum disorder, developmental language disorder, and typical development. NDT **18**, 2349–2359 (2022). https://doi.org/10.2147/NDT.S331988
93. Scott, R.M., He, Z., Baillargeon, R., Cummins, D.: False-belief understanding in 2.5-year-olds: evidence from two novel verbal spontaneous-response tasks. Dev. Sci. **15**(2), 181–193 (2012). https://doi.org/10.1111/j.1467-7687.2011.01103.x
94. Senju, A.: Spontaneous theory of mind and its absence in autism spectrum disorders. Neuroscientist **18**(2), 108–113 (2012). https://doi.org/10.1177/1073858410397208
95. Senju, A., et al.: Absence of spontaneous action anticipation by false belief attribution in children with autism spectrum disorder. Dev. Psychopathol. **22**(2), 353–360 (2010). https://doi.org/10.1017/S0954579410000106

96. Senju, A., Southgate, V., White, S., Frith, U.: Mindblind eyes: an absence of spontaneous theory of mind in Asperger syndrome. Science **325**(5942), 883–885 (2009). https://doi.org/10.1126/science.1176170

97. Setoh, P., Scott, R.M., Baillargeon, R.: Two-and-a-half-year-olds succeed at a traditional false-belief task with reduced processing demands. Proc. Natl. Acad. Sci. **113**(47), 13360–13365 (2016). https://doi.org/10.1073/pnas.1609203113

98. Siegal, M., Beattie, K.: Where to look first for children's knowledge of false beliefs. Cognition **38**(1), 1–12 (1991). https://doi.org/10.1016/0010-0277(91)90020-5

99. Sommer, M., et al.: False belief reasoning in adults with and without autistic spectrum disorder: similarities and differences. Front. Psychol. **9**, 183 (2018). https://doi.org/10.3389/fpsyg.2018.00183

100. Southgate, V., Senju, A., Csibra, G.: Action anticipation through attribution of false belief by 2-year-olds. Psychol. Sci. **18**(7), 587–592 (2007). https://doi.org/10.1111/j.1467-9280.2007.01944.x

101. Sperber, D., Wilson, D.: Relevance: Communication and Cognition, 2nd edn. Relevance: Communication and cognition. Blackwell Publishing, Malden (1995)

102. Surian, L.: Are children with autism deaf to Gricean maxims? Cogn. Neuropsychiatry **1**(1), 55–72 (1996). https://doi.org/10.1080/135468096396703

103. Surian, L., Leslie, A.: Competence and performance in false belief understanding: a comparison of autistic and normal 3-year-old children. Br. J. Dev. Psychol. **17**, 141–155 (1999). https://doi.org/10.1348/026151099165203

104. Tager-Flusberg, H.: Evaluating the theory-of-mind hypothesis of autism. Curr. Dir. Psychol. Sci. **16**(6), 311–315 (2007). https://doi.org/10.1111/j.1467-8721.2007.00527.x

105. Wellman, H.M.: Making Minds: How Theory of Mind Develops. Oxford University Press, New York (2014)

106. Wellman, H.M., Liu, D.: Scaling of theory-of-mind tasks. Child Dev. **75**(2), 523–541 (2004). https://doi.org/10.1111/j.1467-8624.2004.00691.x

107. Westra, E., Carruthers, P.: Pragmatic development explains the theory-of-mind scale. Cognition **158**, 165–176 (2017). https://doi.org/10.1016/j.cognition.2016.10.021

108. Whyte, E.M., Nelson, K.E.: Trajectories of pragmatic and nonliteral language development in children with autism spectrum disorders. J. Commun. Disord. **54**, 2–14 (2015). https://doi.org/10.1016/j.jcomdis.2015.01.001

109. Wimmer, H., Perner, J.: Beliefs about beliefs: representation and constraining function of wrong beliefs in young children's understanding of deception. Cognition **13**(1), 103–128 (1983). https://doi.org/10.1016/0010-0277(83)90004-5

110. Wood, L.J., Dautenhahn, K., Rainer, A., Robins, B., Lehmann, H., Syrdal, D.S.: Robot-mediated interviews - how effective is a humanoid robot as a tool for interviewing young children? PLoS ONE **8**(3), e59448 (2013). https://doi.org/10.1371/journal.pone.0059448

111. Wood-Downie, H., Wong, B., Kovshoff, H., Cortese, S., Hadwin, J.A.: Research review: A systematic review and meta-analysis of sex/gender differences in social interaction and communication in autistic and nonautistic children and adolescents. J. Child Psychol. Psychiatry **62**(8), 922–936 (2021). https://doi.org/10.1111/jcpp.13337

112. Zaraki, A., et al.: A novel reinforcement-based paradigm for children to teach the humanoid kaspar robot. Int. J. Soc. Robot. **12**(3), 709–720 (2020). https://doi.org/10.1007/s12369-019-00607-x

113. Zhang, Y., et al.: Theory of robot mind: false belief attribution to social robots in children with and without autism. Front. Psychol. **10**, 1732 (2019). https://doi.org/10.3389/fpsyg.2019.01732

114. Šimleša, S., Stošić, J., Bilić, I., Cepanec, M.: Imitation, focus of attention and social behaviours of children with autism spectrum disorder in interaction with robots. Interact. Stud. Soc. Behav. Commun. Biol. Artif. Syst. **23**(1), 1–20 (2022). https://doi.org/10.1075/is.21037.sim

Education

Enhancing Pharmacology Education: Investigating the Efficacy of Serious Game-Based Learning Through Virtual Simulation

Florian Laronze[✉], Solène Delsuc, and Bernard N'Kaoua

University of Bordeaux, Bordeaux, France
florian.laronze@hotmail.fr

Abstract. In 2010, the Member States of the E.U. established a directive concerning the protection of animals, stipulating the limitations of the use of animals for scientific or education purposes only in cases where all measures aimed at avoiding such use have failed. One of the solutions, within the educational framework, where the observed results during experimentation are expected, is the use of simulation with serious games (SG). The use of SG has demonstrated their relevance in general college education, even though not all studies agree on the game elements of SG that promote academic outcomes. However, in pharmacology education, few studies have looked at the effect of serious games on these various academic outcomes, and in particular at the game elements that can improve them. In this context, the aim of this work was to have a better understanding of the role of a serious game in the education of pharmacology students by studying the effect of environment (virtual/non-virtual) and the ability to interact with it (active/passive) on cognitive (acquisition of knowledge, cognitive load), behavioral (engagement) and affective (motivation, emotions) outcomes. In total, 46 first year undergraduate students participated in a pharmacology lesson through SG (involving a virtual laboratory) or a course slideshow, both in active or passive modality. Results have shown that the virtual environment in active mode can lead to better behavioral (engagement) and affective (enjoyment, intrinsic motivation, etc.) outcomes, but to similar cognitive outcomes, in comparison with slideshows. These results and proposals for future work are discussed.

Keywords: Serious game · Pharmacology · Virtual environment · Active learning · Gamified learning · Academic outcomes

1 Introduction

In France, in college education, since the law of February 2013, the use of vertebrate animals has been prohibited (Journal officiel de la République française, 2013) and the Code Civil and its Article 515–14 now acknowledges that animals are "sentient beings endowed with sensitivity" (Journal officiel de la République française, 2015). Additionally, the Member States of the European Union have established a directive concerning the protection of animals, stipulating the limitation of the use of animals for scientific or educational purposes only in cases where all measures aimed at avoiding such use have failed. However, for certain study areas such as pharmacology, students need to be able to carry out practical work involving animals in order to understand and acquire certain theoretical and practical knowledge (Dewhurst & Ward, 2014). In this context, serious games give students access to virtual environments, such as virtual laboratories, that simulate practical and theoretical work involving animals.

A serious game can be defined as "a game in which education (in its various forms) is the primary objective, rather than entertainment" (Michael & Chen, 2006, p. 17). Nowadays, serious games are mostly digital and are being developed in various disciplines (e.g. science, mathematics, medicine, economics, geography, languages, etc.), at various levels of study (primary, secondary, university), in different genres (e.g. action, adventure, combat, roleplay, puzzle, platform, etc.) and through different media (computers, virtual reality, telephones, etc.; Connolly et al., 2012; Cheng et al., 2015).

When investigating the general academic outcomes of serious games, scholars typically distinguish between behavioral outcomes (e.g.: engagement, participation, social collaboration, etc.), cognitive outcomes (e.g.: knowledge acquisition, critical thinking, creative thinking, cognitive load) and affective outcomes (e.g.: motivation, satisfaction, emotions, immersion, attitudes towards technology, etc.; (Bloom, 1956; Vlachopoulos & Makri, 2017; Krath et al., 2021). Most studies have shown that students learning with a serious game tend to have better behavioral outcomes and affective outcomes compared to control groups with classic education (Vlachopoulos & Makri, 2017). The results of the effects of serious games on cognitive outcomes, and in particular knowledge acquisition, are much more controversial (Vlachopoulos & Makri, 2017). While some studies show an improvement in knowledge acquisition with serious games (Connolly et al., 2012; Vlachopoulos & Makri, 2017), others show that this is only the case through the indirect effects of behavioral and affective factors (Bai et al., 2020; Qian & Clark, 2016), and still others show that even high levels of satisfaction or motivation are not necessarily predictive of better learning (Iten & Petko, 2016; Zhonggen, 2019). In addition, certain factors such as study area, game type or level of study can also have an impact on these results (Lamb et al., 2018).

Despite the use of educational serious games within the education context there have been few studies in the pharmacology area, and published research is still sparse (Chang et al., 2015; Dabbous et al., 2023). Lancaster et al., (2014) have shown in online serious gaming simulation a pre-post significant increase of students' (n = 79) knowledge of patient-controlled analgesia (abilities to recognize signs and symptoms of opioid overdose). Students were also satisfied and confident engaging in this activity. Cheesman et al. (2014) investigated the effect of a virtual computer program on improving knowledge acquisition and practical laboratory skills and the transfer of these abilities under real-world conditions in a second-year pharmacology class (n = 233). Results reported an increase in students' confidence in successfully completing a live practical experiment after the virtual training and a significant decrease in the mean completion time. However, no effect of the virtual computer program was found on knowledge acquisition. Abdel Haleem et al. (2023) studied medical students' perceptions (n = 60) of the ability of a virtual simulation system to improve their learning in pharmacology. Only half of the students replied that this system could be useful to them to facilitate their practice in real conditions and to acquire theoretical knowledge. Thus, the results of serious games on the different academic outcomes are very diverse and need to be explored in greater depth.

While research into the effectiveness of serious games has long studied them as a single system, recent theoretical models stress the crucial importance of identifying the specific elements of this effectiveness (Landers, 2014; Krath et al., 2021). For example, Landers (2014) points out in his "Theory of gamified learning" that researchers should take a specific interest in each game element that may be present in a serious game in order to assess its effectiveness on behavioral, affective and cognitive outcomes. Nine categories of game elements are thus identified: action language (e.g.: active/passive), assessment, conflict/challenge, control, environment (e.g.: virtual/non-virtual), game fiction, human interaction, immersion, and rules/goals (Bedwell and al., 2012; Landers, 2014). In the case of serious games simulating a virtual environment (e.g.: virtual laboratories), the two central game element to be studied are: 1) the effect of the virtual environment in itself, compared with a more traditional learning material (e.g. slideshow with videos and text, text reading, videos), on cognitive, behavioral and affective outcomes: 2) the effect of being active, i.e. interacting with the various elements in the virtual environment (action language element), compared with passive condition, on cognitive, behavioral and affective outcomes.

Indeed, numerous studies have shown that, compared with traditional learning environments, virtual environments enable serious games to have better effects on the acquisition of theoretical and practical knowledge (Papastergiou, 2009; Suh et al, 2010; Wrzesien, 2010), intrinsic motivation (when involvement in a task is due to the individual's attraction to the task itself, Ryan & Deci, 2000;), satisfaction, enjoyment and interest (Moreno et al., 2001; Wrzesien et al.; 2010) and engagement in the activity (Annetta et al., 2009; Barab et al., 2012). Some studies also stress the importance of taking into account the effect of the virtual environment on cognitive load (cognitive processes that distract from the active processing of learning content; Mayer, 2014), which can sometimes lead to visual fatigue, nausea, headaches, etc. (Saredakis et al., 2020). Systematic reviews and meta-analyses have also shown that being active in serious games has better effects on cognitive, motivational and behavioral outcomes (Wouters et al., 2013; Vlachopoulos & Makri, 2017). For example, Ritterfeld et al. (2009) compared the effects of the environment (serious game vs text) and interactivity (active vs passive (replay)) on the knowledge acquisition and motivation of undergraduate students ($n = 100$) on the subject of the human digestive system. Their results showed superior effects of the serious game and active modalities (even better combined) on knowledge acquisition and motivation.

However, while current work on serious games in pharmacology is already methodologically limited due to the recurrent absence of control groups to evaluate their effectiveness (Lancaster et al., 2014; Cheesman et al., 2014; Abdel Haleem et al., 2023), to our knowledge no work has studied the specific effects of the environment and interactivity on cognitive, behavioral and affective outcomes.

In this context, the aim of this work is to have a better understanding of the role of a serious game in the education of pharmacology students by studying the effect of environment (virtual/non-virtual) and the ability to interact with it (active/passive) on cognitive outcomes (acquisition of knowledge, cognitive load), behavioral outcomes (engagement) and affective outcomes (motivation, emotions).

Our hypotheses are as follows:

H1: Virtual environment and active interaction will lead to higher acquisition of knowledge and cognitive load than non-virtual environment and passive interaction.

H2: Virtual environment and active interaction will lead to higher engagement than non-virtual environment and passive interaction.

H3: Virtual environment and active interaction will lead to higher motivation, higher positive emotions and fewer negative emotions than non-virtual environment and passive interaction.

2　Method

2.1　Participants

Participants were recruited via email sent to the secretariats of programs related to pharmacology (Pharmacy and first year of medicine: PASS) or presenting an option "License Access Health (LAS)" in all fields of study in the Universities of Bordeaux, Besançon, Aix-Marseille, Poitiers, Nantes, Strasbourg, and Amiens (France).

In total, 46 first year undergraduate students participated in our study. These 46 students had an average age of 19.6 years, among them 28 were pursuing a degree in PASS, 16 in Pharmacy, 2 in other areas. Among them, 19 of the participants were male, 25 were female, and 2 chose not to respond to the question. All participants were randomly allocated to each group (virtual active = 13; virtual passive = 10; non-virtual active = 13; non-virtual passive = 10).

2.2　Course Content and Variables Modalities

The course content tested in this work is a first-year university theoretical and practical lesson focusing on the alteration of GABAA receptors and includes the implementation of 3 experiments: an observation and counting experiment of mouse behaviors placed in a maze, a timing experiment of mouse righting reflex, and a timing experiment of mouse awakening. GABA is the acronym for γ-Aminobutyric acid. GABA is the primary inhibitory neurotransmitter in the central nervous system and is a key coordinator of brain activity. GABA's inhibitory effects are mediated by two types of receptors including GABAA receptor and GABAB. GABAA receptors play a role in mediating the effects of benzodiazepines (anxiolytics). An alteration in these receptors impacts the effect of these drugs, decreasing their efficacy and modifying tolerance. In mice that have been subjected to restraint stress, this stress results in an increase in the self-directed behaviors characteristic of anxiety. The aim of this lesson is to enable students to observe this behavioral difference in mice with this impairment by comparing them with healthy mice.

Methodologically, our work includes 2 independent variables, with 2 modalities corresponding to the two game elements studied: type of environment (virtual/non-virtual) and interaction mode (active/passive).

The virtual modality designates the serious-game condition which is a dynamic, realistic condition in which the participant is immersed in a virtual laboratory in which he can move around with his computer mouse to carry out different activities, move the camera angle, etc…

The non-virtual modality designates the PowerPoint condition, a static condition in which only PPT slides are shown, but which presents content (texts,images, video) similar to the virtual condition.

Interaction mode (active/passive) refers to the ability to interact with the environment.

Our work also includes 3 main categories of academic outcomes, each with one or more dependent variables on which the effects of the 2 independent variables are studied: cognitive outcomes (knowledge acquisition, cognitive load), affective outcomes (intrinsic motivationn, enjoyment, hope, pride, hopelessness, and boredom) and behavioral outcomes (engagement).

Thus, in order to assess the effect of the independent variables on the various dependent variables, the course content has been duplicated identically in four different presentation modalities which varied only on the two parameters (virtual/non-virtual and active/passive) of interest: a serious game (virtual active), a video of a serious game (virtual passive), an interactive course slideshow (non-virtual active), and a video of a course slideshow (non-virtual passive). These four presentation modalities correspond to the four groups into which our participants are divided.

Serious Game (virtual active)
For the virtual active modality, participants were required to complete a serious game including a virtual lab, NeuroLabo, designed and developed by the company Practeex. Within this simulation, participants had to: view a slideshow with course content, answer a quiz, move sticky notes to order the steps of the experimental protocol, select personal protective equipment, assemble the groups of mice they wanted to observe, and conduct three interactive experiments (which included for each experiment choosing the syringe, the dose of product to inject, and observing behaviors). At each stage, multiple choices were presented to participants, and they had to find the correct answer (progression in the simulation was contingent upon correct answers). The virtual active modality lasts approximately 40 min.

The serious game used in this work is a low-immersive virtual environment because its present high level of interactivity, i.e. the ability of the environment to change based on the user's actions in real time (Steuer 1992; Ristor et al., 2023). However, it does not involve high vividness characteristics, as head-mounted displays or multi-sensory stimulation (e.g.: smell, touch, etc.), not allowing it to be considered as a high-level immersive virtual environment (Fig. 1).

Serious Game Video (virtual passive).
The content of this virtual passive mode is a replay video of the virtual active modality created by recording the screen of a fictitious session. In order to maintain the quiz effect and the trial-and-error learning, intentional errors were kept in the various steps. The video lasts approximately 40 min.

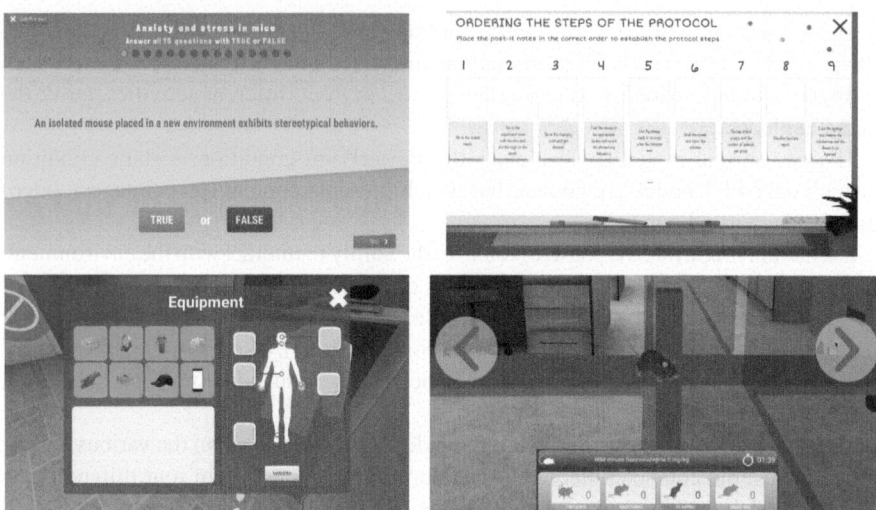

Fig. 1. Screenshots of the NeuroLabo serious game. Top left: In the virtual lab, participants had to answer a quiz related to the virtual lab theme, here on animal experimentation and animal behaviors (In this example, the quiz question asks the participant if a mouse placed in a new environment exhibits stereotyped behaviors (the correct answer was "false")). Top right: In the virtual lab office, participants had to move sticky notes representing the different steps of the experimental protocol to put them in order. Bottom left: In the virtual lab locker room, participants had to choose the appropriate personal protective equipment from a list to enter the lab. Bottom right: Observation experiment of animal behaviors in a maze; the modeled mouse is placed in a transparent maze to observe its movements and behaviors. Thanks to the buttons on the screen, participants could count the number of times each behavior occurred (green: freezing, orange: grooming, red: rearing up, blue: sniffing).

Active Course Slideshow (non-virtual active)

For the non-virtual active modality participants were required to follow a slideshow and to answer questions to progress through it. The content of the slideshow mirrored identically that of the simulation: slideshow, quiz, selection of protocol steps, choice of personal protective equipment, and videos of mouse groups. The session lasts approximately 40 min.

Passive Course Slideshow (non-virtual passive)

The content of this non-virtual passive modality is a replay video of the non-virtual active modality created by recording the screen of a fictitious session. In order to maintain the quiz effect and the trial-and-error learning, intentional errors were kept in the various steps. The video lasts approximately 40 min (Fig. 2).

2.3 Procedure

The participants who agreed to take part in this study were randomly allocated to our 4 groups and had to complete the following 3 Phases consecutively from their computer.

During Phase 1, participants were asked to complete the knowledge questionnaire.

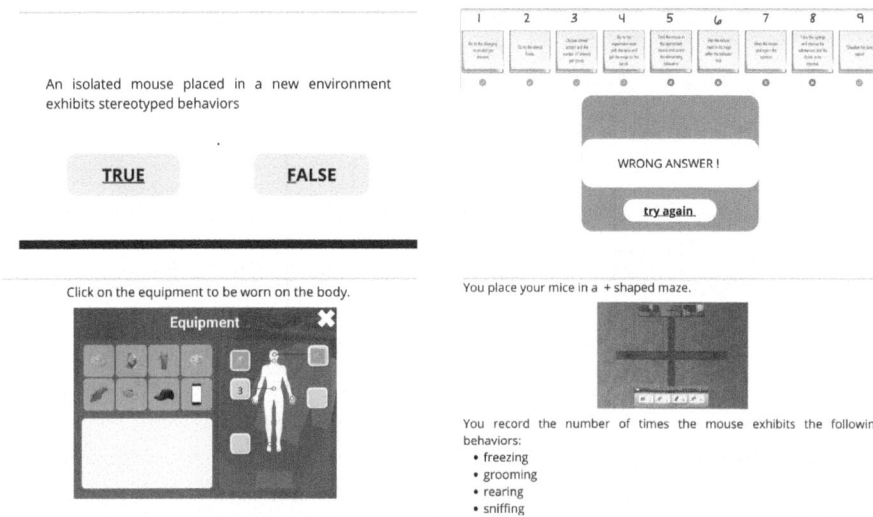

Fig. 2. Screenshots of the slideshow. Top left: Slideshow version of the quiz related to the theme of the virtual lab, here on animal experimentation and animal behaviors, present in the virtual lab (In this example, the quiz question asks the participant if a mouse placed in a new environment exhibits stereotyped behaviors (the correct answer was "false")). Top right: Slideshow version of the sticky notes representing the different steps of the experimental protocol that participants had to move to put in order in the virtual lab office. Bottom left: Slideshow version of the selection step, choosing the appropriate personal protective equipment from a list to enter the lab in the virtual lab locker room. Bottom right: Slideshow version of the animal behavior observation experiment in a maze; the modeled mouse is placed in a transparent maze to observe its movements and behaviors.

In Phase 2, participants were assigned to one of four groups: serious game (virtual active), serious game video (virtual passive), interactive course slideshow (non-virtual active), passive slideshow (non-virtual passive), and were required to perform the associated activity.

Finally, during Phase 3, participants were asked to respond to the same knowledge questionnaire as in Phase 1. Additionally, they completed scales measuring cognitive load, engagement, and evaluation of emotions and motivation scale. In order to mitigate recency effects, primacy effects, and fatigue induced by the duration of the experience, counterbalancing was applied in Phase 3 for administering the various scales.

The protocol procedure is shown in Fig. 3.

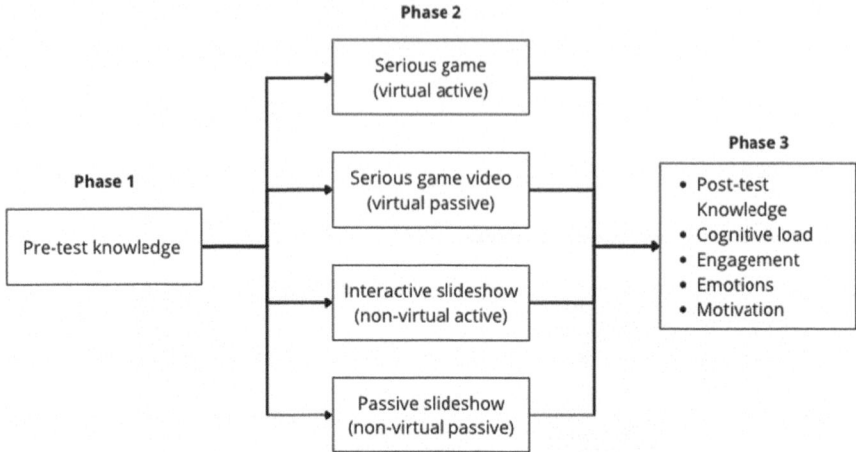

Fig. 3. Flowchart of the different stages of the procedure

2.4 Scales

Cognitive Outcomes
To assess knowledge acquisition, a questionnaire consisting of 10 questions related to the course content was designed. The questions were multiple-choice questions covering various aspects of the course: the content of the slides, establishment of the experimental protocol, implementation of the protocol (creation of animal groups, choice of materials), as well as identification of mouse behaviors. For analysis, participants' responses were converted using the following scoring: all correct answers = 1 point, one error (omission or incorrect answer) = 0.5 point, two or more errors = 0 points. Subsequently, the difference between pre- and post-scores was calculated to establish a score corresponding to knowledge acquisition.

Cognitive load was assessed using the Cognitive Load Index (CLI; Leppink, 2013) in its adapted and translated version into French by Fontaine et al. (2019). The CLI is a self-reported measurement scale used to assess three types of cognitive load in learning contexts. These three types of cognitive load are: intrinsic, extraneous, and germane. In this study, we were particularly interested in intrinsic cognitive load which corresponds to the task complexity in itself (Fontaine et al., 2019). The scale consists of 10 items on a 7-point Likert scale (e.g.: "The topics covered during this activity were complex.").

As for all the other scales (with the exception of knowledge acquisition), cognitive load was assessed only at post-test.

Behavioral Outcomes
To measure participant engagement, we used the User Engagement Scale - Short Form (O'Brien & Toms, 2010) in its adapted and translated version into French by Fontaine et al. (2019). This scale consists of 12 items presented on a 5-point Likert scale and

is divided into four subscales to measure attributes of engagement in digital contexts: Focused Attention, Perceived Usability, Aesthetic Appeal, and Reward. In this study, we were particularly interested in focused attention and reward. The focused attention corresponds to "the feeling of being absorbed in the interaction and losing track of time" (e.g.: "This experience absorbed me so much that I lost track of time."). The reward component corresponds to "the overall success of the interaction and user's willingness to recommend an app to others or engage it in future" (e.g.: "Using NeuroLabo was worth it.").

Affective Outcomes

Participant motivation was assessed using the SIMS, a Situational Motivation Scale (Guay et al., 2000) in its French version. The scale consists of 16 items presented on a 7-point Likert scale. The SIMS evaluates intrinsic motivation, identified regulation, external regulation, and amotivation. In this study, we were particularly interested in intrinsic motivation, which assesses the motivation to carry out an activity for the pleasure that the activity itself produces (e.g.: "I participate in this activity because it is truly enjoyable.").

To assess emotions, we utilized the AEQ-S (Academic Emotions Questionnaire - Short version), specifically focusing on the scale related to emotions experienced in the classroom (Pekrun et al., 2009; Bieleke, 2021). The AEQ-S is a questionnaire presented on a 5-point Likert scale covering emotions such as enjoyment, hope, pride, anger, anxiety, shame, hopelessness, and boredom. This scale thus provides a much more accurate measure of what students feel during a learning experience, which is less the case with ordinary unidimensional satisfaction scales (Pekrun et al., 2009). An example of an item is "This activity bored me.".

3   Results

Mean and standard deviation of the different academic outcomes according to the type of environment (virtual/non-virtual) and mode of interaction (active/passive) are shown in Table 1.

Table 1. Mean and standard deviation of the different academic outcomes according to the type of environment (virtual/non-virtual) and mode of interaction (active/passive)

	Virtual active	Virtual passive	Non-virtual active	Non-virtual passive
Knowledges acquisition (pre-post difference)	3.04 (SD = 2.99)	2.15 (SD = 2.19)	3.73 (SD = 2.06)	2.40 (SD = 2.88)
Intrinsic cognitive load	2.23 (SD = 1.24)	2.57 (SD = 0.66)	2.64 (SD = 0.96)	2.33 (SD = 1.1)
Intrinsic motivation	5.94 (SD = 1.08)	4.42 (SD = 1.28)	4.50 (SD = 1.45)	3.40 SD = 1.78)
Enjoyment	4.08 (SD = 1.12)	3.40 (SD = 0.56)	3.62 (SD = 1.04)	2.70 (SD = 1.57)
Hope	5.69 (SD = 2.90)	4.30 (SD = 2.50)	4.69 (SD = 3.04)	2.60 (SD = 0.84)
Hopelessness	1.69 (SD = 0.75)	1.50 (SD = 0.52)	1.85 (SD = 0.68)	2.50 (SD = 0.85)
Boredom	1.85 (SD = 0.98)	2.90 (SD = 1.10)	2.48 (SD = 1.27)	3.70 (SD = 1.42)
Pride	3.38 (SD = 0.96)	2.70 (SD = 0.67)	3.15 (SD = 0.89)	2.30 (SD = 1.06)

3.1 Cognitive Outcomes

Regarding knowledge acquisition, we began by ensuring that participants in different groups started with similar levels of knowledge. To do this, we conducted a two-way ANOVA only on knowledge acquisition score with two between-participants factors: environment type (virtual/non-virtual) and interaction mode (active/passive). The ANOVA revealed no significant effect of environment type ($F(1, 42) = 0.238$, $p = 0.6$), no significant effect of interaction mode ($F(1,42) = 0.357$, $p = 0.5$), and no significant interaction effect between the two factors ($F(1,42) = 3.81$, $p = 0.9$). Thus, the results indicate that there were no significantly different knowledge levels between groups during pre-tests.

Then, we carried out two-factor ANOVAs to study the effects on knowledge acquisition and cognitive load scores according to the environment type and interaction mode factors.

For knowledge acquisition, the ANOVA revealed no significant effect of environment type ($F(1, 42) = 0.3819$, $p = 0.53$), a significant effect of interaction mode ($F(1,42) = 2.11$, $p = 0.04$), and no significant interaction effect between the two factors ($F(1,42) = 0.08$, $p = 0.77$). Thus, results indicated that knowledge acquisition between pre and post-tests was influenced only by the interaction mode factor and that being active led to a better acquisition of knowledge than being passive.

For cognitive load, the two-way ANOVA indicated that there was no effect of environment type (F(1,42) = 0.6, p = 0.44), no effect of interaction mode (F(1,42) = 0.28, p = 0.59) and no interaction effect between the two factors (F(1,42) = 0.31, p = 0.57). Thus, results indicate that there were no differences in cognitive load between the different activities induced by the environment type or the interaction mode.

3.2 Behavioral Outcomes

For the focus engagement, the two-way ANOVA indicated that there were no significant effect of environment type (F(1,42) = 0.93, p = 0.33), a significant effect of interaction mode (F(1,42) = 11.7, p = 0.001), and no interaction effect (F(1,42) = 0.14, p = 0.7). The active group (M = 2.96, SD = 1.08) thus had a higher level of focus engagement than the passive group (M = 1.98, SD = 0.74).

Regarding the reward component, the two-way ANOVA indicated that there were a significant effect of environment type (F(1,42) = 5.8, p = 0.02), a significant effect of interaction mode (F(1,42) = 10.89, p = 0.002), and no interaction effect (F(1,42) = 0.05, p = 0.82). The virtual modality (M = 3.34, SD = 0.63) and active modality (M = 3.4, SD = 0.78) groups had respectively higher levels of reward engagement than the non-virtual group (M = 2.81, SD = 0.97) and the passive group (M = 2.66, SD = 0.77).

3.3 Affective Outcomes

Regarding the intrinsic motivation, the two-way ANOVA indicated a significant effect of environment type (F(1,42) = 8.78, p = 0.005), a significant effect of interaction mode (F(1,42) = 9.88, p = 0.003), and no interaction effect between the two factors (F(1,42) = 0.251, p = 0.6). There was therefore a significant higher intrinsic motivation score for the virtual (M = 5.28, SD = 1.37) compare to non-virtual (M = 4.02, SD = 1.66) groups. This was also the case for the interaction mode factor, with an active group having significantly higher intrinsic motivation (M = 5.22, SD = 1.45) than the passive group (M = 3.91, SD = 1.60).

Regarding emotions, the two-way ANOVA indicated that there was a significant effect of environment type for hopelessness (F(1,42) = 7.36, p = 0.01) and boredom (F(1,42) = 3.96, p = 0.05), and a strong trend for enjoyment (F(1,42) = 3.05, p = 0.08) and hope (F(1,42) = 3.15, p = 0.08). Two-way ANOVA also indicated that there was a significant effect of interaction mode on enjoyment (F(1,42) = 5.73, p = 0.021), hope (F(1,42) = 5.25, p = 0.02), pride (F(1,42) = 8.03, p = 0.007) and boredom (F(1,42) = 10.4, p = 0.002).

No interaction effects were found for the different emotions.

The significant environmental effects were due to the fact that virtual environment brings less hopelessness (Virtual: M = 1.61, SD = 0.65; Non-virtual: M = 2.13, SD = 0.8), less boredom (Virtual: M = 2. 23, SD = 1.15; Non-virtual: M = 3, SD = 1.45), more enjoyment (Virtual: M = 3.78, SD = 0.95; Non-virtual: M = 3.22, SD = 1.35), and more hope (Virtual: M = 5.09, SD = 2.76; Non-virtual: M = 3.78, SD = 2.54), compared to non-virtual environment. Moreover, the significant effect of interactive mode is due to the fact that the active modality brings more enjoyment (Active: M = 3.85, SD = 1.08; Passive: M = 3.05, SD = 1.19), hope (Active: M = 5.19, SD = 2.95, SD = 1.08;

Passive: M = 3. 45, SD = 2.01), pride (Active: M = 3.27, SD = 0.9; Passive: M = 2.5, SD = 0.88) and less boredom (Active: M = 2.15, SD = 1.16; Passive: M = 3.30, SD = 1.30) than the passive modality.

4 Discussion

With the introduction of regulations limiting the use of sentient animals for student education (Journal officiel de la République française, 2015), serious games, and in particular virtual laboratory simulations, appear to be a solution for enabling students to acquire theoretical and practical knowledge in pharmacology. While many studies have shown that serious games have shown positive effects on affective and behavioral outcomes, and more varied results on cognitive outcomes (Vlachopoulos & Makri, 2017), very few studies have looked at their effects on pharmacology education. In particular, no work has been done on the study of two very important game elements in serious games: 1) the effect of the virtual environment in itself, compared with a more traditional learning material (e.g. slideshow with videos and text, text reading, videos) 2) the effect of being active, i.e. interacting with the various elements in the virtual environment (action language element), compared with passive condition.

In this context, the aim of this work was to have a better understanding of the role of a serious game in the education of pharmacology students by studying the effect of environment (virtual/non-virtual) and the ability to interact with it (active/passive) on cognitive outcomes (acquisition of knowledge, cognitive load), behavioral outcomes (engagement) and affective outcomes (motivation, emotions).

Firstly, for the affective and behavioral outcomes, our results showed that the virtual environment and active interaction led to higher scores on the different scales.

For intrinsic motivation, our results showed higher scores in virtual and active modalities. In addition, in these modalities (virtual and active), the participants also experienced more positive emotions, such as enjoyment, pride in having understood the content and hope that the content had been well understood. In these modalities, participants also felt fewer negative emotions, such as less boredom and hopelessness at not having understood the content of the lessons. These results are in line with the studies which have shown that serious games can improve various affective outcomes (motivation, satisfaction, etc.; Moreno et al., 2001; Wrzesien et al.; 2010 Annetta et al., 2009; Barab et al., 2012) but extends them to the field of pharmacology teaching which, until then, showed disparate results (Lancaster et al., 2014; Cheesman et al., 2014; Abdel Haleem et al., 2023).

Our results also showed that the virtual environment and active modalities showed greater engagement for students who had taken the pharmacology course. Once again, these results are consistent with the literature that has shown that serious games allow students to be more engaged in their learning (Vlachopoulos & Makri, 2017). Our results also show that students found the virtual environment and active conditions to be more engaging to the extent that students would be more likely to recommend these modalities to other students and that they would be more likely to engage in similar activities again in the future (O'Brien, 2012).

Serious games are known to make activities more engaging, motivating, fun, etc. (Lancaster et al., 2014). Our results confirm this by showing effects particularly on behavioral (engagement) and affective (intrinsic motivation, enjoyment, boredom, etc.) outcomes in pharmacology education.

Secondly, for the cognitive outcomes, our results showed no difference between the environment type or interaction mode modalities on cognitive load. This means that the virtual environment was not more demanding cognitively than the slideshow, nor was being active more demanding than being passive. Thus, the fact that the active virtual environment use in our study is no more distracting and cognitively costly than the passive slideshow is a relevant point. Indeed, the cognitive load theory postulate that adding too much material and features in virtual environment, with the aim of making them highly immersive, can create cognitive processing that is not relevant to the learning objective (Sweller, 1994; Sweller et al., 2011; Morélot et al., 2021). As the serious game used in this work is a low-immersive virtual environment, i.e. it does not involve head-mounted displays or multi-sensory stimulation (e.g.: smell, touch, etc.), it is consistent with the fact that the level of cognitive load is similar to the non-immersive slideshow.

Our results also showed that, in our four condition (virtual active, virtual passive, non-virtual active, non-virtual passive) there is an increase of knowledge acquisition. However, with regard to the effect of the factors handled, there is a higher knowledge acquisition in the active interaction mode compared with the passive, but no difference of the type of environment. This result joins literature showing more robust effect of active/passive compared with virtual/non-virtual on knowledge acquisition. Indeed, numerous studies have already shown an active/passive effect on traditional education (e.g.: active note-taking versus passive lecture; Chi et al., 2014), as well as in work on serious games (Ritterfeld et al., 2009; Vlachopoulos & Makri, 2017). However, the literature shows greater variability in the results on the effects of the virtual environment compared to control devices on knowledge acquisition (e.g.: slideshow, text document, etc.; Iten & Petko, 2016; Zhonggen, 2019).

One explanation for this lack of difference between virtual and non-virtual environments could potentially be linked to our low-immersive serious game which, although relevant for a low cognitive load, may not be sufficiently immersive to generate better knowledge acquisition. Indeed, some studies have shown that the higher acquisition of knowledge induced by a virtual environment was linked to a high level of immersion (Elangovan & Ismail, 2014; Checa & Bustillo, 2019). However, this explanation needs to be nuanced, as other authors have shown greater effects for high-immersive environments on procedural learning but not for conceptual knowledge (Morélot et al., 2021), or even higher knowledge acquisition in low-immersive environment (Parong & Mayer, 2018). Future work will therefore have to control the effect of immersion and its impact on cognitive load and knowledge acquisition in pharmacology serious games.

Another possibility to the similar level of knowledge acquisition between virtual and non-virtual environment is that the effect of the virtual environment is not direct on the acquisition of knowledge, but comes through an indirect mechanism. Indeed, according to "Theory of gamified learning" (Landers, 2014), game elements have a direct effect on behavioural and affective outcomes, which in turn have an effect on cognitive outcomes. In our work, it is therefore possible that the effect of the active modality on the acquisition of knowledge stems from the direct effect on emotions, motivation and commitment, which themselves have an effect on the acquisition of knowledge, as some studies have already shown in the past (Bai et al., 2020; Qian & Clark, 2016). This hypothesis can thus be tested in the future by other work using mediation models (Hayes et al., 2018).

5 Limitations and Implications

Our work has limitations and implications for future work.

First of all, it is important to note that this work focused on a specific field of study (pharmacology), a specific course content (alteration of GABAA receptors); a specific serious game (NeuroLabo) and specific game elements (type of environment and interaction mode). It is therefore important not to over-extrapolate the results obtained. However, a significant advantage of having studied specific game elements has been to identify their specific contributions, so that they can be tested and reused by university teachers with the aim of gamifying their courses (Landers, 2014).

Secondly, although this work has focused on academic outcomes that are essential and widespread in the literature on gamified learning and virtual reality (Huang et al., 2020; Ritzhaupt et al., 2021; Ristor et al., 2023), other factors may be studied in the future. This is the case, for example, with the sense of presence, the subjective experience of feeling present in the virtual environment (Nowak and Biocca 2003), which can also have indirect effects on the acquisition of knowledge in virtual environments (Ristor et al., 2023).

Finally, future work could also examine the active/passive effect in greater depth by looking at the role of collaboration between students. Indeed, certain models (e.g.: ICAP, Chi & Wylie., 2014) show that the acquisition of knowledge is better in learning activities when students are actively involved, compared with passive learning, but that collaborative learning, i.e. with interaction between students, makes it possible to reinforce this learning even more. Future work could therefore test this hypothesis in collaborative pharmacology serious games by comparing pairs of students with single students.

6 Conclusion

In conclusion, this work has shown that for university pharmacology teaching, serious games involving active virtual environments can lead to better behavioral and affective outcomes, and similar cognitive outcomes, in comparison with lecture slideshows. By comparing, in a strictly controlled manner, virtual and non-virtual environments, as well as active and passive modalities, our results show that serious games can therefore be relevant for pharmacology college education. Thus, the teaching of certain theoretical and practical concepts, which in the past involved the lives of many sentient animals, can now be replaced, whenever possible, by the use of digital devices as serious games.

References

Abdel Haleem, S.E.A., Ahmed, A.A., El Bingawi, H., Elswhimy, A.: Medical students' perception of virtual simulation-based learning in pharmacology. Cureus 15(1), e33261 (n.d.). https://doi.org/10.7759/cureus.33261

Annetta, L.A., Minogue, J., Holmes, S.Y., Cheng, M.-T.: Investigating the impact of video games on high school students' engagement and learning about genetics. Comput. Educ. **53**(1), 74–85 (2009). https://doi.org/10.1016/j.compedu.2008.12.020

Bai, S., Hew, K.F., Huang, B.: Does gamification improve student learning outcome? Evidence from a meta-analysis and synthesis of qualitative data in educational contexts. Educ. Res. Rev. **30**, 100322 (2020). https://doi.org/10.1016/j.edurev.2020.100322

Barab, S., Pettyjohn, P., Gresalfi, M., Volk, C., Solomou, M.: Game-based curriculum and transformational play: designing to meaningfully positioning person, content, and context. Comput. Educ. **58**(1), 518–533 (2012). https://doi.org/10.1016/j.compedu.2011.08.001

Bedwell, W.L., Pavlas, D., Heyne, K., Lazzara, E.H., Salas, E.: Toward a taxonomy linking game attributes to learning: an empirical study. Simul. Gaming **43**(6), 729–760 (2012). https://doi.org/10.1177/1046878112439444

Bieleke, M., Gogol, K., Goetz, T., Daniels, L., Pekrun, R.: The AEQ-S: a short version of the achievement emotions questionnaire. Contemp. Educ. Psychol. **65**, 101940 (2021). https://doi.org/10.1016/j.cedpsych.2020.101940

Bloom, B.S.: Taxonomy of Educational Objectives, Handbook: The Cognitive Domain. (David McKay) (1956)

Chang, H.Y., Poh, D.Y.H., Wong, L.L., Yap, J.Y.G., Yap, K.Y.-L.: Student preferences on gaming aspects for a serious game in pharmacy practice education: a cross-sectional study. JMIR Med. Educ. **1**(1), e2 (2015). https://doi.org/10.2196/mededu.3754

Checa, D., Bustillo, A.: A review of immersive virtual reality serious games to enhance learning and training. Multimed. Tools Appl. **79**(9), 5501–5527 (2020). https://doi.org/10.1007/s11042-019-08348-9

Cheesman, M.J., Chen, S., Manchadi, M.-L., Jacob, T., Minchin, R.F., Tregloan, P.A.: Implementation of a Virtual Laboratory Practical Class (VLPC) module in pharmacology education. Pharmacogn. Commun. **4**(1), 2 (2014). https://doi.org/10.5530/pc.2014.1.2

Cheng, M.-T., Chen, J.-H., Chu, S.-J., Chen, S.-Y.: The use of serious games in science education: a review of selected empirical research from 2002 to 2013. J. Comput. Educ. **2**(3), 353–375 (2015). https://doi.org/10.1007/s40692-015-0039-9

Chi, M.T.H., Wylie, R.: The ICAP framework: linking cognitive engagement to active learning outcomes. Educ. Psychol. **49**(4), 219–243 (2014). https://doi.org/10.1080/00461520.2014.965823

Connolly, T.M., Boyle, E.A., MacArthur, E., Hainey, T., Boyle, J.M.: A systematic literature review of empirical evidence on computer games and serious games. Comput. Educ. **59**(2), 661–686 (2012). https://doi.org/10.1016/j.compedu.2012.03.004

Dabbous, M., et al.: Instructional educational games in pharmacy experiential education: a quasi-experimental assessment of learning outcomes, students' engagement and motivation. BMC Med. Educ. **23**(1), 753 (2023). https://doi.org/10.1186/s12909-023-04742-y

Décret N° 2013-118 Du 1er Février 2013 Relatif à La Protection Des Animaux Utilisés à Des Fins Scientifiques

Dewhurst, D., Ward, R.: The virtual pharmacology lab—a repository of free educational resources to support animal-free pharmacology teaching. Altern. Lab. Anim. ATLA **42**(1), P4-8 (2014). https://doi.org/10.1177/026119291404200115

Elangovan, T., Ismail, Z.: The effects of 3D computer simulation on biology students' achievement and memory retention. Asia-Pac. Forum Sci. Learn. Teach. **15**(2) (2014)

Fontaine, G., et al.: Traduction, adaptation et évaluation psychométrique préliminaire d'une mesure d'engagement et d'une mesure de charge cognitive en contexte d'apprentissage numérique. Pédagogie Médicale **20**(2), Article 2 (2019). https://doi.org/10.1051/pmed/2020009

Guay, F., Vallerand, R., Blanchard, C.: On the assessment of situational intrinsic and extrinsic motivation: the situational motivation scale (SIMS). Motiv. Emot. **24**, 175–213 (2000). https://doi.org/10.1023/A:1005614228250

Iten, N., Petko, D.: Learning with serious games: is fun playing the game a predictor of learning success? Br. J. Edu. Technol. **47**(1), 151–163 (2016). https://doi.org/10.1111/bjet.12226

Journal officiel de la République française. (n.d.). Article 515–14—Code civil—Légifrance. Accessed 26 May 2024. https://www.legifrance.gouv.fr/codes/article_lc/LEGIARTI0000302 50342

Krath, J., Schürmann, L., von Korflesch, H.F.O.: Revealing the theoretical basis of gamification: a systematic review and analysis of theory in research on gamification, serious games and game-based learning. Comput. Hum. Behav. **125**, 106963 (2021). https://doi.org/10.1016/j.chb.2021. 106963

Lamb, R.L., Annetta, L., Firestone, J., Etopio, E.: A meta-analysis with examination of moderators of student cognition, affect, and learning outcomes while using serious educational games, serious games, and simulations. Comput. Hum. Behav. **80**, 158–167 (2018). https://doi.org/10. 1016/j.chb.2017.10.040

Lancaster, R.J.: Serious game simulation as a teaching strategy in pharmacology. Clin. Simul. Nurs. **10**(3), e129–e137 (2014). https://doi.org/10.1016/j.ecns.2013.10.005

Leppink, J., Paas, F., Van der Vleuten, C.P.M., Van Gog, T., Van Merriënboer, J.J.G.: Development of an instrument for measuring different types of cognitive load. Behav. Res. Methods **45**(4), 1058–1072 (2013). https://doi.org/10.3758/s13428-013-0334-1

Mayer, R.E.: The Cambridge Handbook of Multimedia Learning. Cambridge University Press, Cambridge (2014)

Michael, D., Sande, C.: Serious games: games that educate, train and inform. Thomson Course Technology (2006)

Morélot, S., Garrigou, A., Dedieu, J., N'Kaoua, B.: Virtual reality for fire safety training: Influence of immersion and sense of presence on conceptual and procedural acquisition. Comput. Educ. **166**, 104145 (2021). https://doi.org/10.1016/j.compedu.2021.104145

Moreno, R., Mayer, R.E., Spires, H.A., Lester, J.C.: The case for social agency in computer-based teaching: do students learn more deeply when they interact with animated pedagogical agents? Cogn. Instr. **19**(2), 177–213 (2001). https://doi.org/10.1207/S1532690XCI1902_02

O'Brien, H.L., Toms, E.G.: The development and evaluation of a survey to measure user engagement. J. Am. Soc. Inform. Sci. Technol. **61**(1), 50–69 (2010). https://doi.org/10.1002/asi. 21229

Papastergiou, M.: Digital game-based learning in high school computer science education: impact on educational effectiveness and student motivation. Comput. Educ. **52**(1), 1–12 (2009). https:// doi.org/10.1016/j.compedu.2008.06.004

Parong, J., Mayer, R.E.: Learning science in immersive virtual reality. J. Educ. Psychol. **110**(6), 785–797 (2018). https://doi.org/10.1037/edu0000241

Pekrun, R., Elliot, A.J., Maier, M.A.: Achievement goals and achievement emotions: testing a model of their joint relations with academic performance. J. Educ. Psychol. **101**, 115–135 (2009). https://doi.org/10.1037/a0013383

Qian, M., Clark, K.R.: Game-based learning and 21st century skills: a review of recent research. Comput. Hum. Behav. **63**, 50–58 (2016). https://doi.org/10.1016/j.chb.2016.05.023

Ritterfeld, U., Shen, C., Wang, H., Nocera, L., Wong, W.L.: Multimodality and interactivity: connecting properties of serious games with educational outcomes. Cycberpsychol. Behav. **12**(6), 691–697 (2009). https://doi.org/10.1089/cpb.2009.0099

Ryan, R.M., Deci, E.L.: Self-determination theory and the facilitation of intrinsic motivation, social development, and well-being. Am. Psychol. **55**, 68–78 (2000). https://doi.org/10.1037/ 0003-066X.55.1.68

Saredakis, D., Szpak, A., Birckhead, B., Keage, H.A. D., Rizzo, A., Loetscher, T.: Factors associated with virtual reality sickness in head-mounted displays: a systematic review and meta-analysis. Front. Hum. Neurosci. **14** (2020). https://doi.org/10.3389/fnhum.2020.00096

Suh, S., Kim, S.W., Kim, N.J.: Effectiveness of MMORPG-based instruction in elementary English education in Korea. J. Comput. Assisted Learn. **26**(5), 370–378 (2010). https://doi.org/10.1111/j.1365-2729.2010.00353.x

Sweller, J.: Cognitive load theory, learning difficulty, and instructional design. Learn. Instr. **4**(4), 295–312 (1994). https://doi.org/10.1016/0959-4752(94)90003-5

Sweller, J., Ayres, P., Kalyuga, S.: Measuring Cognitive Load. In: Sweller, J., Ayres, P., Kalyuga, S. (eds.), Cognitive Load Theory, pp. 71–85. Springer, Cham (2011). https://doi.org/10.1007/978-1-4419-8126-4_6

Vlachopoulos, D., Makri, A.: The effect of games and simulations on higher education: a systematic literature review. Int. J. Educ. Technol. High. Educ. **14**(1), 22 (2017). https://doi.org/10.1186/s41239-017-0062-1

Wouters, P., van Nimwegen, C., van Oostendorp, H., van der Spek, E.D.: A meta-analysis of the cognitive and motivational effects of serious games. J. Educ. Psychol. **105**(2), 249–265 (2013). https://doi.org/10.1037/a0031311

Wrzesien, M., Alcañiz Raya, M.: Learning in serious virtual worlds: evaluation of learning effectiveness and appeal to students in the E-Junior project. Comput. Educ. **55**(1), 178–187 (2010). https://doi.org/10.1016/j.compedu.2010.01.003

Zhonggen, Y.: A meta-analysis of use of serious games in education over a decade. Int. J. Comput. Games Technol. **2019**, e4797032 (2019). https://doi.org/10.1155/2019/4797032

Video-Based Analysis of the Mechanical Design Process in a Student Dyadic Activity

Sylvain Luc Agbanglanon[1]([✉]) [iD] and Vassilis Komis[2] [iD]

[1] Cheikh Anta Diop University of Dakar, Dakar, Senegal
luc.agbanglanon@ucad.edu.sn
[2] University of Patras, Patras, Greece

Abstract. The aim of this study is to rationalize the dyadic collaborative design activity of students. It explores the characteristics of the process underway in a mechanical design activity, from the point of view of the nature of actions initiated and external representations mobilized. This work is based on video recordings of the collaborative activity of six dyads of students engaged in designing solutions to improve an existing mechanical system. The workspace of each dyad is filmed by a camera and synchronized with the capture of the modelling actions on the dyads' computer screens. The coding of the video recordings is, on the one hand, based on the characteristics of the external representations used. On the other hand, it is inspired a priori by the micro-strategies defined by Purcell et al. [1] and adopted by Gero and McNeil [2]. The identified categories of actions are enriched a posteriori by anchoring them to the data. By implementing the cSPADE algorithm to search for patterns, our study was able to identify the most frequent sequences of action categories and the most frequent sequences of mobilized external representations. In addition, a CBA classification algorithm makes it possible to detect which external representations are most frequently mobilized to support which categories of actions. These results therefore help to initiate a characterization of the design activity under study.

Keywords: Video-based analysis · Pattern mining · cSPADE algorithm · Design process · Dyadic activity

1 Introduction

Design activity can be understood as being in essence an activity of solving ill-defined and ill-structured problems which do not lead to a single solution. In the design of mechanical systems, or of any other nature, the engineer or designer aims to: define objects and systems from the point of view of their constitution and operation in order to satisfy specific objectives [3], or to define specifications enabling the construction of objects to meet specific requirements. This activity is considered not as that of an individual, but rather a collective activity [4, 5] which has a socio-cognitive dimension in addition to the technical and instrumental ones [4]. In a collaborative modelling learning context, the categories of actions identified fall under analysis, synthesis, control-execution, cognitive support, technical support and social interactions [6]. Design learning inherits the

J. Baratgin et al. (Eds.): HAR 2024, LNCS 15504, pp. 308–322, 2025.
https://doi.org/10.1007/978-3-031-84595-6_19

practices in progress in the professional environment, because in technology education in general or specifically in engineering education, "…we like to teach about technology as we can see it being practised" [7].

This justifies the growing interest in collaborative learning in mechanical design. Moreover, the complexity of collaborative design activity has led to the need to understand the inter-individual interactional dynamics, which are expressed through the actions of design teams-members. Understanding the interrelationships that occur within design teams, whether in a learning situation or in a professional situation, requires recourse to a characterization of the nature of the actions that take place there and of the dynamics of the interactions within the teams. In relation to collaborative design activities in general, the taxonomy developed by Ostergaard & Summers [8] and revisited by Righter et al. [9], highlights a number of factors related to the functioning of design teams. The diversity of these factors establishes the complexity of the interactions in design teams and of the design activity itself.

Research has looked at the place of external representations, to show their importance [10], and the contribution of digital drawing or design assistance tools to the activity of the draughtsman or design subject in both architecture and mechanical engineering [11–16]. In design, external representations can be considered as aids to reflexion [10, 17], and the advantages of external representations of a graphic nature over those of a textual nature have been noted [18]. Thus, the usefulness of sketching in the emergence of original solutions and the contribution of modelling in the production of solutions with a more complete functional character have been noted [19]. The question of the variation in the way external representations are used according to the type of design activity and the phase of the design process leads to the observation that the sketch is one of the essential means of representation throughout the design process. It supports the design and communication activity in particular and plays a major role in the framing phase of the problem to be solved [20].

The complex nature of the design activity leads us to seek to shed more light on it, despite the existence of previous works which have studied the question. Similarly, the interactions between members of design teams, on the one hand, and between designers and external representations of the problems they are called upon to solve, on the other, are fairly complex. This complexity calls for a more thorough rationalization of the mechanical design activity in order to make it more intelligible, with the aim of better informing design learning. One of the issues that needs to be addressed is the influence of designers' actions on the design process as a whole. This is in order to understand how design scholars can relate design practices in particular situations to design processes [21].

We adress this topic through the following question: How, in mechanical design, is the dyadic activity of students structured from the point of view of the actions initiated and the external representations mobilized?

In this paper, we aim to describe the way in which the design process studied is structured from the point of view of the actions initiated and the external representations mobilized during these actions. In the remainder of the article, we present a review of the literature in order to establish the theoretical foundations on which the methodological construction discussed below is based. This is followed by the results and their discussion in the light of previous work.

2 Theoretical Background: Design Process Made of Design Actions Supported by External Representations

In this section, we aim to establish the background that will enable us to clarify the methodological construction that we are going to implement in order to collect our data. This construction is based on the clarification of the design process and the interrelation between design actions and the mobilization of external representations in support of these actions.

2.1 The Design Process Stucture

The design process has been the subject of recent research, which studied the design process through protocol, narrative, and ethnographic approaches. These studies focused on micro- and macro-level and variance theory versus process theory. Protocol-based approaches focus more on the micro level, whereas narrative and ethnographic approaches target the macro level [21, 22].

Design is a complex process that consists of a variety of tasks and teamwork elements. Design teams are able to perform a wide range of actions and goals across team members, resulting in substantial heterogeneity in composition and progression over time. Recent studies show that design teams dynamically combine taskwork and teamworks elements and a broad range of goal/action combinations with substantial variation across the team members [22].

2.2 Mobilization of External Representations During Actions: An Integrated Prism for Reading the Activity

The activity which is studied in this work, involves students and is rather routine according to the meaning used by Darses et al. [23], Brown & Chandrasekaran [24] and by Ullman [25]. It is a collaborative activity which is rather co-located, synchronous, collective, direct, and models the nature of the interactions [26, 27]. The design activity can be read according to the areas of the problem invested or according to the design strategies deployed [2]. The problem areas refer either to the characteristics of the design object, or to the degree of abstraction with which the problem is addressed. The designers' strategies come from the typology of the actions they initiate. The distributed nature of this co-design activity calls for taking into account the place of external representations with which interactions contribute to the construction of internal representations. The design activity brings human agents into contact with their environment, in a continuous process of mutual adjustment between internal representations and external representations, made of a permanent interaction between external cognitive aid and internal individual cognition [28]. This suggests a relationship between the actions initiated and the external representations which, un this context, constitute cognitive aids with which designers interact when they initiate design actions during their activity.

2.3 Typology of External Representations Mobilized

The characterization of the external representations present in the environment is based on their attributes [29]. These characteristics come from the sensory channel requested, from the modality of representation referring to the homogeneity or heterogeneity of its nature, from the level of abstraction, from the specificity of the representation, from the type of external representation, from its integrated or non-integrated character, its dynamism and its dimensionality. As a first characteristic of external representations, we consider the sensory channel mobilized by external representations in the activity we are studying; they are essentially graphic in nature and use the visual channel.

Table 1. Typology of external representations

	Type	Dimension	Dynamism	Abstract
On digital media	Realistic 3D drawing	3D	Yes	---
	Animated 3D drawing	3D	Yes	---
	Animated 3D diagram	3D	Yes	--+
On paper	3D drawing	3D	No	--+
	Plan drawing	2D	No	-++
	Sketch	2D	No	-++
	Text	2D	No	+++

From the point of view of dimensionality, the representations on digital support are all spatial, whereas those on paper are mainly unidimensional apart from the realistic drawing on paper. Similarly, the dynamic nature or not of the external representations introduces a clear distinction between those on digital media which are dynamic and those on paper which are essentially static. Note that the external representations on digital support are all homogeneous, while those on paper are heterogeneous with the exception of text and realistic drawing on paper. Likewise, the diagrams and the text are essentially general in character while the drawings are specific. The level of abstraction of representations on paper is higher compared to those on digital. Relative to the degree of integration, only the plan drawing on paper has this character.

The essential assumptions on which our work is based is that the design activity is sequential and that it makes use of the external representations available in the designers' environment. We therefore postulate, with reference to previous work, that the design process follows a sequencing which starts with the analysis of the problem, the development of solutions and the evaluation of solutions.

3 Methods

3.1 Data Collection

The participants in our study are students with a 2-year higher education qualification in mechanical engineering. This study is essentially based on the co-localized collaborative design activity of six dyads of students, formed according to affinities and composed of 11 boys and a girl. These students are committed to improving an existing mechanical system. Their activity is based on plans and kinematic diagrams of a mechanical system which is provided on paper and/or digital media. The filmed activity allows us to collect the types of external representations mobilized during the various design actions initiated. Within each team, an installed software allows video recording of the computer screen, synchronized with that of the workspace. The shooting of this space is done by means of a USB camera, which overhangs it according to the spatial arrangement illustrated in Fig. 1. The sound is recorded, by the built-in microphone of the team's computer, in the goal of capturing verbal protocols during the design process [30, 31].

Fig. 1. System put in place to collect data

The material arrangement adopted aims to film the use of the various tools and work supports [32]. Video capture from the computer screen allows us to record the team's work, performed with the modelling tools, as well as viewing the animated kinematic diagram or animated 3D drawing videos that are provided.

3.2 Data Analysis

3.2.1 Processing of Primary Data: Coding of Video

The video data, lasting one hour for each team, meaning six hours in total, was coded with Nvivo using a codebook shown in Table 2, focusing on the external representations

Table 2. Codebook of actions.

Category	Action	Code
Evaluation of solution	Solution approval request (*requires approval or validation of the proposal from the peer*)	sar
	Questioning a solution (*about a proposed solution, points out a problem in the layout, in the shape of an element, in the assembly technology between two elements, a precaution to be taken in relation to the existing elements provided, rejects or refutes a proposed solution*)	qs
	Validation or consolidation of proposed solution (*providing reasons for a solution, issuing a favorable judgement on the solution*)	vcs
Expression of difficulties	Acknowledgement of lack of understanding of elements of the problem	lup
	Report of a lack, absence or difficulty with a tool	rdt
	Expression or observation of the absence of not having a solution at the moment	eas
Graphic elaboration of solution	Proposition of solution representation (*proposes or says he will represent his solution*)	psr
	Request for solution representation (*explicitly or implicitly asks a friend to represent a solution*)	rsr
	Representation of solution (*presentation of a trace of the solution in a visual form that remains: sketch, written text, drawing, diagram..., this representation can be silent or accompanied by a speech*)	rs
	Suggests a solution representation (*suggests a type, a form of representation or dimensions to a partner who is sketching or drawing a solution*)	ssr
Indecision	Suspended, undetermined or inintelligible statement	sus
	Silent phase or thinking	spr
Interaction regulation	Expression of agreement, understanding, or attention to the elements put forward (*accepts what is said or the gestural or graphic representation made, says he/she agrees, expresses or says he/she understands*)	eau
	Request for understanding expression (*ask his/her teammate if he/she has understood*)	rue

(*continued*)

Table 2. (*continued*)

Category	Action	Code
	Regulation of speaking turn	rtp
Problem analysis	Explanation or precision of the problem (*precision or reformulation of the elements of the problem by integrating its own formulation: terms, turns of phrase, gestures*)	epp
	Proposition of problem explanation (*proposes to explain or says he will explain elements of the problem*)	ppe
	Information search or consultation of information sources on the problem (*search, collection, reading or recording of information, textual or graphic in paper or digital format, or questioning of the elements of the problem*)	isc
	Rejection of problem explanation	reje
	Request for explanation-precision of the problem	reqe
Strategy focus	Explanation of analogy or strategy deployed (*tells how it does it, expresses an analogy with an existing system*)	ead
	Questions about the steps to take	qst
	Suggestion related to the organization or conduct of work (*suggests a form of organization of the dyad in terms of working conditions, focus or division of work*)	so
Tools management	Providing or suggesting tools or work aids (*placing tools or materials on the workspace that were not in the foreground, opening a window or running a video on the computer, offering to provide a work tool*)	aost
	Request for documents or working tools	rdwt
	Request for expression of satisfaction in relation to tools	rest
	Removing documents or working tools (*removing tools or equipment from the workspace, relegating them to the background, closing a window or stopping a video on the computer*)	rewt
Verbal elaboration of solution	Explanation or precision of solutions (*ephemeral verbal or pictorial expression indicating or specifying characteristics of the solution such as shapes, dimensions, positions...*)	eps

(*continued*)

Table 2. (*continued*)

Category	Action	Code
	Proposition of solution explanation (proposes to explain or says he will explain his solution)	pse
	Request for explanation or search for precision of solutions or strategy (*explicitly or implicitly requires from his/her teammate the verbal or gestural expression of his/her solution, asks his/her teammate to propose a solution, asks to reformulate or searches, in a synthesis phase, for elements of precision of the solution such as dimensions or strategy, asks a question about elements of the solution*)	rcs

mobilized by each student, in support of the various actions. Then a validation followed, leading to a Cohen Kappa coefficient of 0.72, obtained with two external coders, on an extract of 5 min. The execution of coding requests made it possible to highlight all of the actions, which resort to the different types of external representations, initiated by each of the students, with the moment when they start and end.

Then, an Excel database of 4007 lines was built, corresponding to the 4007 actions coded, with in column the types of external representations mobilized and their characteristics presented in Table 1. Likewise, the identifiers of the authors of these actions, as well as their averages of spatial visualization scores are put in column.

3.2.2 Processing of Secondary Data

In the R software, we imported the Excel database for processing. For each type of external representation mobilized by a student, in support to a design action, the beginning time and the end time are collected from Nvivo software. The database allso contains the identifier of each student.

Using R software, we graphically represented the timeline of the different categories of actions according to the external representations they mobilize and the students who initiate them. This representation allows us to visually show how the design activity evolves over time in terms of categories of design actions initiated and external representations mobilized.

In addition, a search for categories of actions and external representations patterns were carried out using the *arulesSequences* package in R software. This package implements a cSPADE algorithm based on the Sequential Pattern Discovery using Equivalent Class (SPADE) [33]. By also implementing *arulesCBA* package, a classification were carried out with the aim of classifying the action sequences in association to external representions which are mobilized in support to these actions. The *arulesCBA* package [34] implements the CBA algorithm.

4   Results

The results of the graphic description of the categories of actions initiated and the external representations that support them (see Fig. 2) reveal a sequential contribution of the different external representations. Thus, the text on paper is mainly mobilized at the beginning of the activity to clarify the problem. This is understandable insofar as the instructions for the work and the information needed to understand the problem are given in the text. The same applies to the various graphical representations of the system under study, whether on digital or paper media, in plane projections or spatial views. On the other hand, sketches on paper come a little later to support production of solutions and reflection on them. External representations, based on digital graphic modelling tools, take up the last two-thirds of the activity's time and support actions aimed at producing and evaluating solutions.

The search for action categories patterns highlights the following results: the main pattern of categories of actions is a sequence of 2 categories of actions (Table 2). That is problem analysis followed by verbal elaboration of solution, with a support value of 0.75. This support value represents the number of sequences covered by this pattern, meaning that 75% of the total sequences are covered by this pattern. Other patterns are identified. Among the patterns of 3 categories of actions we can find problem analysis followed by verbal elaboration of solution and evaluation of solution, with a support value of 0.42. The sequence composed of verbal elaboration of solution, graphic elaboration of solution, evaluation of solution and indecision is one of the 4 categories of actions characterising the design activity. Its support value is 0.25.

The same as the search for action categories patterns, those of external representations patterns reveals (Table 4) predominance of the following sequences: < {Text on paper},{Plan on paper} >, with 0.58 as support value, < {Model on computer},{Plan on , paper} >, with 0.75 as support value, < {Text on paper},{Model on computer},{Plan on paper} >, with 0.42 as support value, and < {Text on paper},{Model on computer},{No External representation},{Plan on paper} >, with 0.25 as support value.

The search for characteristic sequences yielded the following results from the classification implemented using the *arulesCBA* package in R software [34]. This classification (see Table 3) reveals associations that characterize the process. These associations are between the following actions and external representations:

Uncertain actions supported by No external representations.
Regulation of the interaction without recourse to any external representation.
Tools management without any external representation.
Strategy focus without any external representation.
Graphic production of solutions usi ng digital graphic modelling tools.
Undefined action without recourse to any external representation (Table 5).

5   Discussion

The main objective of this study was to characterize and analyse the process underway during a collaborative design activity. To do this, we coded and analysed the actions of the pairs of participants engaged in this activity, using the tools and libraries available

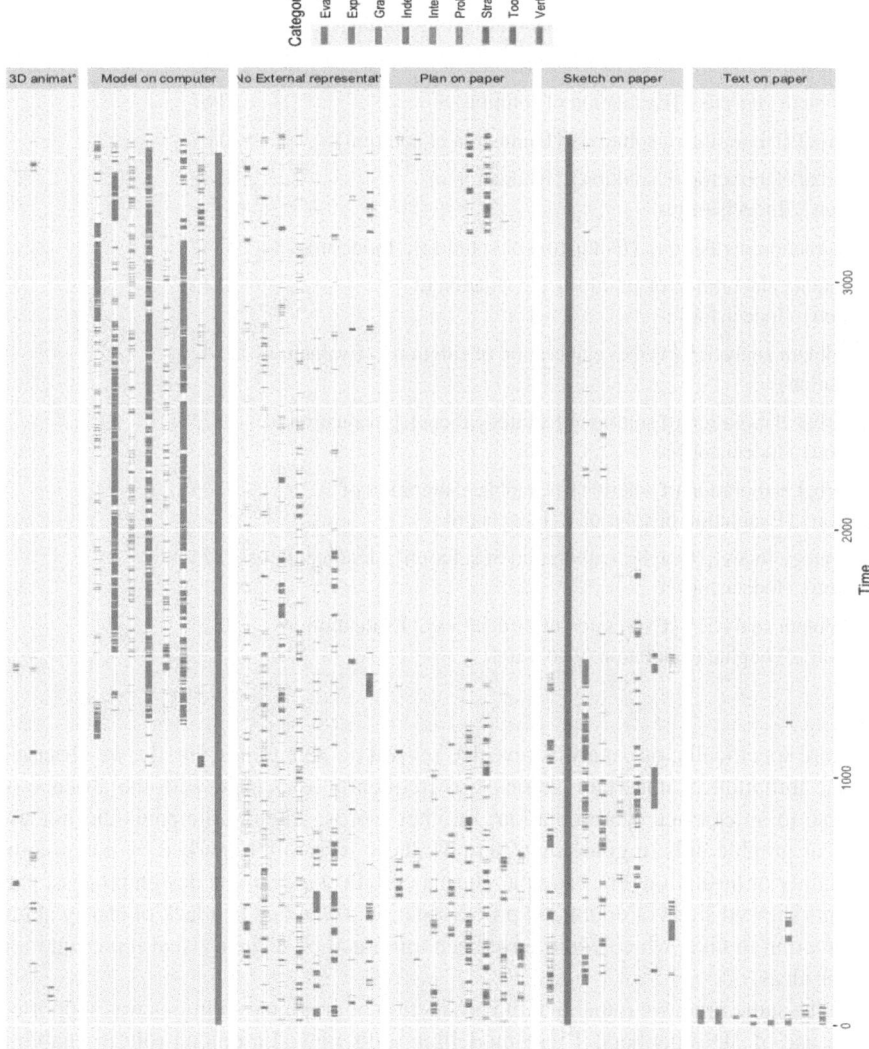

Fig. 2. Timeline of initiated actions according to external representation mobilized

in the R programming language. This methodological approach enabled us to highlight the association between the actions undertaken by the participants and the external representations they used.

Table 3. Frequent patterns of categories of actions

Sequence	Support	Pattern
<{Problem analysis},{Verbal elaboration of solution}>	0.75	2
<{Interaction regulation},{Verbal elaboration of solution}>	0.67	2
<{Evaluation of solution},{Indecision}>	0.67	2
<{Problem analysis},{Evaluation of solution}>	0.67	2
<{Verbal elaboration of solution},{Evaluation of solution}>	0.67	2
<{Verbal elaboration of solution},{Evaluation of solution},{Indecision}>	0.42	3
<{Interaction regulation},{Evaluation of solution},{Indecision}>	0.42	3
<{Graphic elaboration of solution},{Evaluation of solution},{Indecision}>	0.42	3
<{Problem analysis},{Verbal elaboration of solution},{Evaluation of solution}>	0.42	3
<{Verbal elaboration of solution},{Strategy focus},{Evaluation of solution},{Indecision}>	0.25	4
<{Verbal elaboration of solution},{Graphic elaboration of solution},{Evaluation of solution},{Indecision}>	0.25	4
<{Strategy focus},{Graphic elaboration of solution},{Evaluation of solution},{Indecision}>	0.25	4
<{Problem analysis},{Expression of difficulties},{Evaluation of solution},{Graphic elaboration of solution}>	0.25	4

More specifically, our analysis revealed several notable trends. Firstly, we observed that the initiation of uncertain actions was frequently associated with the absence of recourse to an external representation. In other words, when participants did not use visual or graphic aids to guide their actions, these were often marked by a degree of uncertainty or hesitation. For example, during certain stages of the design process, participants showed signs of doubt and procrastination, often due to a lack of clarity about the direction to take, which could have been mitigated by a clear visual representation of their ideas.

At the same time, we found that the production of graphical solutions was supported by the use of 3D modellers. These modelling tools played a crucial role in enabling participants to visualize and manipulate their ideas in a tangible way, which facilitated the generation of relevant and innovative solutions. For example, in a complex product design task, participants used 3D modelling software to create visual prototypes, which not only accelerated the development process but also improved the accuracy and quality of the proposed solutions.

These results highlight the crucial importance of external representations in the collaborative design process. Indeed, without these supports, the actions undertaken by the participants tend to be more uncertain and less effective. External representations,

Table 4. Frequent patterns of external representations

Sequence	Support	Pattern
<{Model on computer},{Plan on paper}>	0.75	2
<{Model on computer},{No External representation}>	0.75	2
<{Text on paper},{No External representation}>	0.75	2
<{Model on computer},{Sketch on paper}>	0.67	2
<{Text on paper},{Sketch on paper}>	0.67	2
<{No External representation},{Sketch on paper}>	0.58	2
<{Plan on paper},{Sketch on paper}>	0.58	2
<{Text on paper},{Plan on paper}>	0.58	2
<{Text on paper},{No External representation},{Sketch on paper}>	0.42	3
<{Text on paper},{Model on computer},{Plan on paper}>	0.42	3
<{Text on paper},{Model on computer},{No External representation}>	0.42	3
<{Model on computer},{Plan on paper},{No External representation}>	0.42	3
<{Text on paper},{Model on computer},{No External representation},{Plan on paper}>	0.25	4

Table 5. Results from the classification.

	lhs (Action Category)	Rhs (External representations)	support	confidence	coverage	lift	count	size	Covered Transactions	Total Errors
1	Indecision	=> No External representation	0.063	0.68	0.093	1.83	254	2	373	2512
2	Interaction regulation	=> No External representation	0.13	0.63	0.21	1.69	524	2	832	2366
3	Tools management	=> No External representation	0.013	0.60	0.022	1.60	53	2	89	2322
4	Strategy focus	=> No External representation	0.023	0.59	0.040	1.57	94	2	160	2246
5	Graphic elaboration of solution	=> 3D Model on computer	0.11	0.57	0.19	2.26	422	2	745	2164
6	{}	=> No External representation	0.37	0.37	1.00	1.00	4007	1	1808	2164

whether sketches, diagrams or 3D models, serve as visual and cognitive reference points that help participants to structure their ideas, clarify their thoughts and communicate more effectively with their peers. Furthermore, the fact that the production of solutions is also closely linked to the use of external representations underlines their central role in the design process. This suggests that integrating these tools from the earliest design phases could lead to more robust and innovative results.

These findings corroborate the results of previous studies [10–20], which have also highlighted the importance of external representations in various collaborative work and design contexts. For example, studies in the field of architecture have shown that sketches and mock-ups are essential for the development and communication of conceptual ideas. Similarly, in engineering, the use of CAD (computer-aided design) software has been shown to improve the accuracy of designs and reduce errors.

In sum, our study reinforces the idea that to optimize collaborative design processes, it is essential to integrate and value the use of external representations throughout these processes. This means not only providing participants with the necessary tools, but also training them to use them effectively. For example, training workshops on the use of 3D modelling software could be organized for design teams. In addition, it would be beneficial to promote a culture of visualization within teams, where the use of sketches, diagrams and prototypes is encouraged and valued.

It is also important to note that the use of external representations is not limited to digital tools. Traditional methods, such as hand-drawing or creating physical models, have also proved effective in the design process. For example, a quick sketch on paper can often capture an idea more intuitively and quickly than a complex digital model.

Finally, our study paves the way for further research into the role of external representations in different design contexts. For example, it would be interesting to examine how these representations influence group dynamics and decision-making in multidisciplinary design teams. In addition, the impact of different types of representations (2D vs. 3D, digital vs. physical) on the quality and innovation of proposed solutions also deserves to be explored.

In conclusion, external representations play an indispensable role in the collaborative design process. Their use not only improves the clarity and efficiency of the actions undertaken, but also encourages the generation of innovative, high-quality solutions. By integrating these tools into design practices, we can not only improve the results of projects, but also enrich the collaborative experience of participants.

6   Conclusion

The aim of this study was to characterize the ongoing process in students design activity by coding the actions of dyads in a collaborative design activity and processing the secondary data using the algorithms and packages available in R. The approach used highlighted the main patterns of action categories and those of external representation mobilized. Il also shows some association of action categories and external representations. The revealed association between actions and external representations reinforces the importance of external representation in design activity. The video based methodology we have implemented, which leans on the analysis of video taped design activity, makes it possible to characterize de dyadic design activity. This characterization is facilitated by the processing tools using algorithms and packages available in R Software.

References

1. Purcell, A.T., Gero, J.S., Edwards, H.M., Mc Neill, T.: The data in design protocols: the issue of data coding, data analysis in the development of models of the design process. In: Analysing Design Activity, pp. 169–187. Delft University of Technology (1994)
2. Gero, J.S., Mc Neill, T.: An approach to the analysis of design protocols. Des. Stud. **19**, 21–61 (1998). https://doi.org/10.1016/S0142-694X(97)00015-X
3. Simon, H.A.: The structure of ill structured problems. Artif. Intell. **4**, 181–201 (1973). https://doi.org/10.1016/0004-3702(73)90011-8
4. Darses, F.: Resolution collective des problemes de conception. Le Travail Humain. **72**, 43–59 (2009). https://doi.org/10.3917/th.721.0043
5. Darses, F., Falzon, P.: La conception collective: une approche de l'ergonomie cognitive. In: de Terssac, G., Friedberg, E. (eds.) Coopération et Conception. , Toulouse (1996)
6. Siampou, F., Komis, V., Tselios, N.: Online versus face-to-face collaboration in the context of a computer-supported modeling task. Comput. Hum. Behav. **37**, 369–376 (2014). https://doi.org/10.1016/j.chb.2014.04.032
7. De Vries, M.J.: Teaching about Technology: An Introduction to the Philosophy of Technology for Non-philosophers. Springer, Cham (2016). https://doi.org/10.1007/978-3-319-32945-1
8. Ostergaard, K.J., Summers, J.D.: Development of a systematic classification and taxonomy of collaborative design activities. J. Eng. Des. **20**, 57–81 (2009). https://doi.org/10.1080/09544820701499654
9. Righter, J., Blanton, A., Stidham, H., Chickarello, D., Summers, J.D.: A case study of the effects of design project length on team collaboration and leadership in senior mechanical engineering projects. In: Volume 3: 19th International Conference on Advanced Vehicle Technologies; 14th International Conference on Design Education; 10th Frontiers in Biomedical Devices, p. V003T04A019. American Society of Mechanical Engineers, Cleveland, Ohio, USA (2017). https://doi.org/10.1115/DETC2017-68197
10. Ullman, D.G., Wood, S., Craig, D.: The importance of drawing in the mechanical design process. Comput. Graph. **14**, 263–274 (1990). https://doi.org/10.1016/0097-8493(90)90037-X
11. Kurtuluş, A.: Effect of computer-aided perspective drawings on spatial orientation and perspective drawing achievement. Turkish J. Educ. Technol. **10** (2011)
12. Laisney, P., Brandt-Pomares, P.: Role of graphics tools in the learning design process. Int. J. Technol. Des. Educ. **25**, 109–119 (2015). https://doi.org/10.1007/s10798-014-9267-y
13. Martin, P., Velay, J.-L.: Do computers improve the drawing of a geometrical figure for 10 year-old children? Int. J. Technol. Des. Educ. **22**, 13–23 (2012). https://doi.org/10.1007/s10798-010-9140-6
14. Musta'amal, A.H., Norman, D.E., Hodgson, T.: Gathering empirical evidence concerning links between computer aided design (CAD) and creativity. Des. Technol. Educ. Int. J. **14**, 53–66 (2009)
15. Robertson, B.F., Radcliffe, D.F.: Impact of CAD tools on creative problem solving in engineering design. Comput. Aided Des. **41**, 136–146 (2009). https://doi.org/10.1016/j.cad.2008.06.007
16. Tovey, M.: Drawing and CAD in industrial design. Des. Stud. **10**, 24–39 (1989). https://doi.org/10.1016/0142-694X(89)90022-7
17. Cross, N., Christiaans, H., Dorst, K.: Design expertise amongst student designers. J. Art Des. Educ. **13**, 39–56 (1994). https://doi.org/10.1111/j.1476-8070.1994.tb00356.x
18. McKoy, F.L., Vargas-Hernández, N., Summers, J.D., Shah, J.J.: Influence of design representation on effectiveness of idea generation. In: Volume 4: 13th International Conference on Design Theory and Methodology, pp. 39–48. American Society of Mechanical Engineers, Pittsburgh, Pennsylvania, USA (2001). https://doi.org/10.1115/DETC2001/DTM-21685

19. Acuna, A., Sosa, R.: The complementary role of representations in design creativity: sketches and models. In: Taura, T. and Nagai, Y. (eds.) Design Creativity 2010, pp. 265–270. Springer, London (2011). https://doi.org/10.1007/978-0-85729-224-7_34
20. Cardella, M.E., Atman, C.J., Adams, R.S.: Mapping between design activities and external representations for engineering student designers. Des. Stud. **27**, 5–24 (2006). https://doi.org/10.1016/j.destud.2005.05.001
21. Wegener, F.E., Cash, P.: The future of design process research? Exploring process theory and methodology: design research society 2020. In: Proceedings of DRS 2020 Synergy, vol. 5: Processes, pp. 1977–1993 (2020). https://doi.org/10.21606/drs.2020.132
22. Cash, P., Škec, S., Štorga, M.: The dynamics of design: exploring heterogeneity in meso-scale team processes. Des. Stud. **64**, 124–153 (2019). https://doi.org/10.1016/j.destud.2019.08.001
23. Darses, F., Détienne, F., Visser, W.: 33. Les activités de conception et leur assistance: In: Ergonomie, pp. 545–563. Presses Universitaires de France (2004). https://doi.org/10.3917/puf.falzo.2004.01.0545
24. Brown, D.C., Chandrasekaran, B., Chandrasekaran, B.: Design problem solving: knowledge structures and control strategies. Pitman [u.a.], London (1989)
25. Ullman, D.G.: The Mechanical Design Process. McGraw-Hill Higher Education, Boston (2010)
26. Dominguez, G.A.R.: Caractérisation de l'activité de conception collaborative à distance: études et effets de synchronisation cognitive | Theses.fr (2005). https://theses.fr/2005INPG0137?domaine=theses
27. Schmidt, K.: Analysis of cooperative work: a conceptual framework. Risø National Laboratory, Roskilde, Denmark (1990)
28. Conein, B.: Cognition distribuée, groupe social et technologie cognitive. Réseaux **124**, 53–79 (2004). https://doi.org/10.3917/res.124.0053
29. Ainsworth, S.: DeFT: a conceptual framework for considering learning with multiple representations. Learn. Instr. **16**, 183–198 (2006). https://doi.org/10.1016/j.learninstruc.2006.03.001
30. Ericsson, K.A., Simon, H.A.: Protocol Analysis: Verbal Reports as Data. The MIT Press, Cambridge (1993). https://doi.org/10.7551/mitpress/5657.001.0001
31. van Someren, M.W., Barnard, Y.F., Sandberg, J.A.: The Think aloud Method: A Practical Guide to Modelling Cognitive Processes. Academic Press, London (1994)
32. Veillard, L.: Les méthodologies de constitution et d'analyse des enregistrements vidéo. In: Veillard, L. and Tiberghien, A. (eds.) ViSA Instrumentation de la recherche en éducation. Éditions de la Maison des sciences de l'homme (2013). https://doi.org/10.4000/books.editionsmsh.1990
33. Zaki, M.J.: SPADE: an efficient algorithm for mining frequent sequences. Mach. Learn. **42**, 31–60 (2001). https://doi.org/10.1023/A:1007652502315
34. Hahsler, M., Johnson, I., Kliegr, T., Kuchař, J.: Associative classification in R: arc, arulesCBA, and rCBA. R J. **11**, 254 (2019). https://doi.org/10.32614/RJ-2019-048

Critical Thinking and Psychorhetoric for Promoting Environmental Awareness

Laura Macchi[1] ⓘ, Laura Caravona[1](✉) ⓘ, and Elisa Palazzi[2] ⓘ

[1] University of Milan-Bicocca, Milan, Italy
{laura.macchi,laura.caravona}@unimib.it
[2] University of Turin, Turin, Italy
elisa.palazzi@unito.it

Abstract. The objective of our intervention, as part of the EduS4EL Project, is that of improving Environmental Literacy among secondary school students through innovative strategies and methodologies. The novelty of the project is to promote the overcoming of difficulties related to the consideration of climate change through the approach of cognitive and communication psychology (increase awareness of cognitive and communication biases, develop an adequate perception of risk, and of critical thinking with respect to the polarized arguments present in the public debate). In particular, the development of critical thinking and psychorhetorical analysis of fallacious arguments appears to be crucial to promote a better understanding of climate change and other environmental issues, as these disciplines aim at render people more adept at uncovering fallacies and rhetorical tricks that can be implicit in communication, help formulating correct opinions, based on reliable data and sources, and developing deliberative skills. Our intervention, composed of three main sections (education on the key topics of climate change, Critical Thinking and Psychorhetorical intervention, Debate), was carried out in an Italian high school with students of 11th and 12th grade, which performance was compared throughout a questionnaire, with a control group. Critical thinking was measured through the correct evaluation of syllogisms and selection tasks while psychorhetorical skills through the rejection of the implicit denialist message contained in newspaper articles. Overall, our results evidence a better performance by participants who attended the intervention than that of those who did not, suggesting that our intervention has been successful.

Keywords: critical thinking · psychorhetorical intervention · climate change · education · environmental awareness

1 Introduction

"An environmentally literate person [...] makes informed decisions concerning the environment; is willing to act on these decisions to improve the well-being of other individuals, societies, and the global environment; and participates in civic life" [1].

Then, she/he should know about and understand different environmental concepts, problems and issues; develop cognitive skills and abilities to understand information

J. Baratgin et al. (Eds.): HAR 2024, LNCS 15504, pp. 323–344, 2025.
https://doi.org/10.1007/978-3-031-84595-6_20

about the environment, and adopt behavioural strategies to make better decisions relating to the environment.

Despite years of effort to raise awareness about environmental issues and despite numerous initiatives to promote behaviour change, the environmental problems are still far from being solved. Educating people about these issues and encouraging them to adopt more pro-environmental behaviours is challenging. One of the reasons for this is the complexity of the topic. Environmental issues, such as climate change, are complex, with a lot of interdependent phenomena, causal chains and climate feedback loops. This makes it hard to understand these issues without having the necessary technical knowledge. Another reason is the way the human mind works, e.g., how people perceive and interpret the information they receive. Due to the existing cognitive limitations and reasoning biases, it may be difficult for individuals to properly understand environmental issues and, more importantly, to translate this understanding into action.

That is why, in order to understand what kind of interventions to undertake to promote pro-environmental behaviours in individuals, it is, first and foremost, necessary to identify the barriers to these behaviours. In particular, it is crucial to understand what makes it difficult to effectively communicate environmental issues to lay audiences, and to identify which cognitive barriers and biases come into play, making it hard for individuals to transform the acquired information into pro-environmental behaviours.

Environmental issues are undoubtedly complex, with climate change being a leading example. It is not surprising, then, that the information about these issues is difficult to communicate to a lay audience. At the same time, effective communication is a prerequisite for a successful promotion of pro-environmental and ecological actions.

To illustrate some concrete difficulties in environmental communication, we will take the example of communicating about climate change, a complex problem that needs to be understood by the general population, probably lacking the knowledge needed.

The first element that makes the climate crisis difficult to communicate is that it is a complex system, meaning it results from the interaction of two or more interdependent parts, where the whole is held together by these parts. All the subsystems that make up the climate system (the atmosphere, hydrosphere, cryosphere, biosphere, lithosphere) interact with each other in a *non-linear* manner. This non-linearity is a key characteristic of complex systems, including the climate system. Non-experts are not familiar with technical and scientific terms, such as *non-linearity, critical thresholds, or tipping points*. These concepts are inherently difficult to understand and manage, often leading people to disengage from the topic [2].

Another aspect to consider is that climate science is a science of uncertainty. The scientific method inherently includes a component of uncertainty. Communicating climate science to a non-expert audience is challenging because it often leads to the spontaneous inference that "if there is uncertainty, it means that something is not reliable". The lack of complete knowledge about a subject can be misinterpreted as a lack of certain and well-founded knowledge, leading to the erroneous conclusion that nothing is known, thereby fostering climate denialism. Climate deniers often focus on a single difficult-to-understand process, undermining the robustness that climate scientists have achieved in representing the complexity of the climate system and making future projections. This robustness is crucial and it cannot be cancelled out by a few points of uncertainty that will always exist due to the nature of the science.

Furthermore, climate science can sometimes be counterintuitive, i.e., it goes against our common feelings and perceptions. For example, a cold wave under certain circumstances is often taken as evidence that global warming is not occurring and that we need not worry about it. Our perceptions can lead us away from the idea of global warming, while science tells us that in a warming world, extreme cold events can still occur, they can even be more intense, though less frequent.

Moreover, as we are not used to seeing ourselves as part of the ecosystem, we often do not perceive the climate crisis as a threat. We basically imagine that we can do anything without it affecting us. The serious risks to our health linked to climate change are not at all conceived, seen, or felt, yet they are recognized by the scientific community, even by the World Health Organization (WHO), as one of the main causes of health problems in the near future. This contradicts the concept of "*one health*", which means that if one component of the ecosystem gets sick, we get sick too.

2 Cognitive Biases and Barriers to Behaviour Change

In the previous section, we discussed how communication challenges arise from the fact that the environmental problems are complex. In this section, we will explore how difficulties in improving environmental literacy are linked to the cognitive barriers in the human mind.

Our perception of new information, ideas or opinions about the environment can be strongly influenced by a wide variety of external and internal factors. For example, our emotions come into play when we are presented with news, or in reaction to the person who gives the information. Also, the way in which the information is presented has an impact on how we perceive it.

We often tend not to empathize with the climate crisis, perceiving it as something unrelated to our current situation, as a phenomenon that does not have an immediate impact on our lives. However, it is quite the opposite, and therefore it does require immediate action.

These examples highlight the interplay between our minds and the understanding of information and communication on environmental topics. Below, we explain some of the most relevant cognitive biases in more detail.

2.1 Too Little Worry About Climate Change

As mentioned before, one of the major difficulties in communicating about the climate crisis, is that it is often perceived as far away from our day-to-day life. Even though some consequences, such as extreme weather events, are already felt today (albeit more in certain parts of the world than in others), there are other more immediate threats like rising inflation or a potential economic recession leading to job loss, that capture people's attention. Based on research in psychology, it seems that people have only a limited capacity for worry ("a finite pool of worry"), in other words, as their worry about one risk increases, the worry about another risk diminishes [3]. Therefore, if people need to constantly worry about the immediate, pressing issues, they may not have the capacity to worry also about the future threat posed by environmental problems.

Another issue contributing to a potentially low level of worry about the environmental issues is the *optimistic bias*, whereby people tend to underestimate the risk for themselves, believing that negative events are less likely to occur to them compared to others [4]. Despite the environmental problems being global, people may think that the consequences will be primarily felt by people in other parts of the world, not by themselves. Alternatively, when it comes to global warming, people might believe they will not be affected because they expect that a technical solution will be found sooner or later [5].

Also, considering the numbers related to climate change, they may appear small (for instance, a 1.5 °C or 2 °C increase in temperature might not appear significant) and therefore could be disregarded as unimportant.

As most consequences of climate change will appear in the future, another important bias is the *relevance bias* [6], that involves people placing greater importance on immediate outcomes and gratifications rather than those that will occur later. Hence, the environmental policies which are needed may not receive the necessary attention as their effects emerge only after a long period of time. Similarly, the risks associated with climate change are perceived as happening in a distant future (*discounting effect*, e.g., 2050, which is often used as a reference year when talking about climate change issues, may seem far away), which may lead to underestimating them and promoting inaction. Finally, some actions may be influenced by what is referred to as *normalcy bias*, a cognitive bias that leads people to disbelieve or minimize threat warnings [7]. Consequently, people may underestimate the likelihood of a disaster, its timing, and its potential adverse effects.

Taken together, all these cognitive biases may lead people to underestimate the importance of the climate problems, the seriousness of the future consequences and induce them not to take the necessary actions.

2.2 Inertia, Confirmation Bias and Cognitive Dissonance

These are just a few examples in which our actions are not made consciously but guided by inertia, defined as the tendency to do nothing or to remain unchanged. Another of them is the so-called *Semmelweis reflex* [8] which is a tendency to reject new evidence or new knowledge because it contradicts established norms, beliefs, or paradigms, which leads to a cognitive confirmation of our status quo. This phenomenon is closely linked to *confirmation bias* [9] which is the tendency to recall information in a way that confirms or supports one's prior beliefs or values. It also reflects an emotions-related cognitive tool called *status quo bias* [10] which is a preference not to undertake any action to change the current situation. New information about how to correctly sort waste, for example, can be difficult to implement because it would imply a change in our long-established routine.

Whenever we encounter information that clashes with our knowledge of reality and with our current routines, our minds tend to adopt a defensive mechanism known as *cognitive dissonance* [11] which consists in a discomfort triggered by the person's belief clashing with new information perceived. According to this theory, when two actions or ideas are not psychologically consistent with each other, people do all in their power to change one of them until they align.

A person that is experiencing *cognitive dissonance* tends to make changes to justify the stressful behaviour, either by adding new parts to the cognition causing psychological dissonance through rationalization or by avoiding circumstances and contradictory information. A person, for example, might continue buying clothes from fast fashion brands, despite knowing the negative effects they have on the environment. This bias also often results in *confirmation bias* [9] with the aim of confirming previous beliefs. For example, even if we identify as environmentalists, we might still end up using a lot of plastic garbage bags when shopping. That feeling of mental discomfort about using plastic bags is a case of cognitive dissonance: we know that we could easily change our behaviour, but we often rather come up with excuses to avoid feeling guilty, thus fulfilling our confirmation bias. Another example is that we know that using fossil fuels contributes to global warming, but we still end up driving, flying, eating meat, etc.

Another critical bias to consider is linked to the concept of learned helplessness, which consists of a real or perceived lack of control over the outcome of a situation [12]. This is especially relevant to the topic of climate change, as the severity of the phenomenon and the constant exposure to predominantly negative news about it can lead people to feel a general sense of helplessness, believing their actions do not make a difference. This feeling can often translate into inaction also linked to the apparent lack of significant progress, as noticeable positive effects are usually only evident in the long term[1].

2.3 Other Cognitive Barriers

Stoknes [13] identifies *cognitive dissonance* among the main cognitive barriers to addressing climate change. Another related point discussed by the author concerns *identity,* which consists in filtering information according to the personal and cultural identity and to look for information that confirms existing ideas, beliefs and assumptions. In addition to these barriers, the author also mentions *distance* as a tendency to believe that the most severe impacts of climate change, such as melting ice, floods, fires, etc., are still far away in time and space. In fact, these phenomena often seem to affect a small part of the planet or are perceived as on the other side of the world while they are becoming less remote and increasingly frequent in reality.

The author also mentions another barrier to climate change which he defines as *condemnation,* and involves the use of catastrophic messages that portray climate change as an inevitable and irremediable disaster. These types of messages generate feelings of helplessness, and in the absence of immediate practical solutions we tend to avoid the issue (the loss aversion would lead us to avoid the problem so as not to feel helpless). Thus, catastrophic messages can backfire on us. The *negation*, on the other hand, is a defence mechanism, consisting in the denial of the reality of climate change and its causes. Another bias that comes into play is the *willpower bias* according to which we

---

[1] The feeling of helplessness has not been addressed in depth in the present work, because the issue has been investigated whitin a qualitative questionnaire that we did not report for brevity reasons.

tend to shift the responsibility onto large macro-systems (governments, companies, etc.), under the belief that climate change is a problem that has not to be resolved only by individuals, but by the institutions, the governments, etc.

It would therefore seem that communication strategies that focus on catastrophism and alarmism, or that refer to "morality" (emphasizing, for example, that global warming is a moral issue and therefore action or inaction may harm other people or even infringe on their rights) are not effective. *"We need to find a different tone to break out of the apathy of this eternal present of ours"* [5], and public communication should consider the existence and persistence of the biases [14]. Instead of talking about environmental problems in generic terms, it would be more useful and impactful to promote "personalized alerts" in which specific groups are identified, possibly perceived as close, thus bringing the phenomenon closer both geographically and in terms of possible effects on behavioural change. We should therefore question what is missing from the current representation of global warming and rethink our communication strategies accordingly.

3   How Can Behavioural Science Help?

Traditionally, pro-environmental behaviours have been encouraged through education (information provision in particular) and financial or non-financial incentives [15]. In fact, a recent review found that education and awareness-raising methods, relying on e.g. advertising campaigns, posters or handouts, were the most common method used to change people's behaviours in the environmental domain [16]. At the same time, however, they were the least successful among the methods reviewed in producing the behaviour change. Similarly, incentives seem to be among the relatively less successful methods in producing behaviour change in the environmental domain [16]. Looking at specific examples of pro-environmental behaviours, studies using education methods and financial incentives produced mixed or no effects on water use or transportation choices, but showed promising results for reducing waste production, particularly junk mail and plastic waste [15].

Considering the concept of environmental literacy as defined above, being informed about the importance and types of pro-environmental behaviours is just one aspect of being environmentally literate and may not be enough to lead to better decisions about the environmental issues. Moreover, using incentives (whether financial or non-financial) seems somewhat against the concept of environmental literacy. The idea behind such incentives is that people lack internal motivation to engage in pro-environmental behaviours and hence providing external motivation can help. However, an environmentally literate person should not lack such internal motivation, quite the opposite. Thanks to the right information, skills, competences and affective dispositions, they should be able and willing to behave in a pro-environmental way, without needing to receive any external incentives.

Thus, new approaches are needed that emphasize other aspects of environmental literacy, not just mere knowledge or awareness. These approaches should help improve competencies and skills needed to transform the knowledge into action, and should help making pro-environmental decisions and realizing pro-environmental behaviours. To this aim, understanding the psychology of human decision-making and behaviour is crucial as it allows to identify potential problems in pro-environmental decision-making and behaviour and the strategies that help overcome these problems. We present such approaches in the sections below.

3.1 Knowledge, Critical Thinking and Psychorhetoric

Our approach builds on the psychology of human decision-making and behaviour, and its aim is to increase people's reasoning and decision-making competencies. This can be achieved by improving individuals' knowledge or skills, expanding the set of decision-making tools at their disposal or changing the environment in which they make decisions [17]. The idea is that a 'boost' to specific competencies, either by strengthening the existing ones or by creating new ones, will enable specific behaviours [18]. These competences are usually those that are traditionally left out from school curricula, such as risk assessment, financial or medical decision-making, or as in our case critical thinking. Sometimes, they are specific to a certain domain (e.g., medical decision-making, climate change), while other times they are generalizable across domains. Examples of the latter category include risk literacy boosts, aimed at the ability to understand statistical information, uncertainty management boosts, which improve the ability to make assessments, predictions or decisions under conditions of uncertainty, or motivational boosts, which enhance self-control [18]. All these acquired or enhanced skills can then be used in a variety of domains, such as finances, health or environment. Below, we describe more in detail some of the generalizable competences that could be particularly useful in the environmental domain.

Critical thinking can be defined as *"the competence to engage productively in the generation, evaluation, and improvement of ideas that can result in original and effective solutions, advances in knowledge, and impactful expressions of imagination"* [19]. It is about being able to reason and argue, distinguish between facts, opinions and about the structure of reasoning. Critical thinking is conceived as a discipline that combines knowledge from different fields such as logic, scientific method and cognitive psychology, with the aim of making people better at evaluating arguments (and thus accepting or refuting them), constructively participating in debates and discussions, and uncovering fallacies and rhetorical tricks in public discourse. Actually, the aim of our intervention is to provide an important endowment to develop people's reasoning and argumentation. The intervention on Critical Thinking and psychorhetoric will make students more careful in approving and rejecting arguments, more capable of constructively intervening in discussions and debates, and more adept at uncovering fallacies and rhetorical tricks in public discussions on environmental issues. The training on Critical Thinking will equip the students with essential tools to think, formulate correct opinions (based on reliable data and sources), accept the plurality of viewpoints, and develop deliberative skills. To this aim, we have added to these skills, typically pertaining to cognitive psychology, a new section on how the formulation of the text (with regard to what it implicitly communicates, for example) affects reasoning (the relationship between language and thought) [20]. Considering the importance of the relationship between language and thought, as the rhetoric with its persuasive intents had always in mind the interlocutor, his psychological mechanisms and the fact that the persuasiveness can be realized also by the implicit layer of the discourse. These skills will not be limited to climate change and environmental issues but are also generalizable to many other fields as they are transversal.

When what is communicated is not sufficiently clear, it often depends on how it was thought out, therefore improving critical thinking skills is crucial as it means creating a solid basis for improving reasoning and developing arguments [21]. Being aware of the tricks of the human mind and knowing how to think correctly implies knowing how to argue, how to convincingly put forward one's opinions, how to solve sometimes complex and counterintuitive problems, leading to positive effects on self-perception.

There are many concrete examples of how an improvement in critical thinking can enhance environmental literacy. For instance, critical thinking and psychorhetorical training, according to us, should lead to a better ability to recognize fake news and misinformation about climate change, or to formulate evidence-based arguments on rather controversial topics.

In summary, it is essential to develop critical thinking and argumentative skills to be able to:

(i) recognize the main reasoning biases in general and, in particular, in discussions about the environment:

- in logical-deductive reasoning, such as *confirmation bias* and *belief bias,*
- in decision-making (e.g. discounting effect and climate change);

(ii) become resistant to disinformation by learning how to recognize and deconstruct the arguments adopted by propaganda and fake news on climate change (see Fig. 2);

(iii) sustain an argumentative debate on a key topic (e.g., meat consumption and its environmental impact), by constructing discourses based on correct reasoning, facts and reliable sources, supporting opposing positions;

(iv) adopt behavioural interventions (nudge, boost) for oneself and one's family (e.g., providing *feedback* on one's energy consumption, decreasing emailing, decreasing water consumption in daily life, for instance by using the dishwasher, which consumes 10 times less water than washing by hand or by controlling the duration of showers).

4   Intervention

4.1   Brief Introduction of the Intervention

This section of the project aims to develop students' critical thinking and argumentative skills on climate change through three main parts. The first part focuses on explanation and comprehension of the climate phenomenon, the second one on biases of thought related to the issue of climate change and on the analysis of false reasoning and denialist information about climate change, while the third part consists in an argumentative debate between students on a key topic for climate change aimed at promoting their participation.

This intervention is targeted for secondary high school students who are not required to have previous knowledge about climate and environmental topics, as it will be provided in depth in the first part of the intervention. Moreover, no particular skills are requested as prerequisites to participate to this module.

The specificity of the Edus4EL intervention regards the fundamental consideration that the transmission of knowledge alone is not sufficient to improve Environmental Literacy, especially when cognitive and emotional biases that may hinder understanding and awareness are involved. These biases – such as the optimistic bias, confirmation bias, discounting effect and cognitive dissonance - must necessarily be addressed, as they discourage people from taking action. Within the present module, this aim has been pursued by focusing on the development of critical thinking and psychorhetorical skills of the participants.

Another key aspect of the intervention concerns the comparison of the group of participants who experienced the intervention with another group from the same school, with same grades and with homogeneous characteristics, who did not. This comparison makes it possible to measure the effects of the activities carried out with the intervention group and thus the intervention's effectiveness and impact.

4.2 Intervention Objectives

As previously mentioned, the intervention is divided into three sessions. The first session aims to educate students on the critical topics of environmental phenomena to enable them to understand climate change, particularly the distinction between Global warming, Climate change and Climate crisis. The second session aims to educate students on Critical Thinking, reasoning biases and rhetorical tricks to make them less prone to accept and promote false information and fake news through lessons and practical exercises. The last session included in this module is designed to develop and challenge the students' acquired skills through a debate on a critical topic about climate change. The specific procedure to be carried out in the three above-mentioned sessions is described in depth in the following paragraphs.

4.3 Sessions of the Intervention

1st Session: *The Earth Has Fever* - Lecture on climate change. The first session lasts two hours and is focused on the education of students on the key topics of environmental phenomena and climate change.

Lecture by Elisa Palazzi on "What is Climate Change?". The project began with an online lesson conducted by a climate physicist, designed to lay the basis for a solid and critical understanding of global warming and connected climate change. This lecture introduced key concepts and the physical basis of climate and climate change, as well as the appropriate terminology, as using the correct language facilitates a more accurate and informed discussion of climate issues.

The first part of the lesson focused on briefly introducing the two related but distinct concepts of global warming and climate change. The record-breaking year of 2023 was chosen as the starting point for the discussion about global warming. In 2023, temperatures were the highest on record since 1850, nearly 1.5 °C above pre-industrial levels (1850–1900). This year also marked the first time that every single day exceeded 1 °C above pre-industrial levels for that time of year. About 50% of the days were more than 1.5 °C warmer than the 1850–1900 average, with two days in November surpassing 2 °C (https://library.wmo.int/records/item/68835-state-of-the-global-climate-2023).

The example of 2023 was chosen for several reasons: it is the most recent year, making it the one we remember best, and it surpassed many climatic thresholds beyond just air temperatures. Moreover, it illustrates how 2023 fits into a long-term warming trend, which is more significant than breaking records in a single year. It's important to recognize that we don't need to chase records to understand that Earth, ecosystems, and human societies are facing a climate crisis.

As a next step, a series of compelling graphics, videos, images, and photographs was displayed to illustrate climate change (e.g. the "Warming Stripes" and "Biodiversity Stripes", see https://showyourstripes.info/, an example is shown in Fig. 1). These visuals also included maps showing sea level rise, photos of retreating glaciers, and pictures of extreme events that recently occurred in Italy, such as floods, droughts, and heat waves. The concept of extreme events was highlighted, noting that many types are becoming increasingly frequent or intense, raising the risk of damage to both territories and human lives.

When extreme events occur, they draw public attention to climate issues because they cause immediate, tangible damage and pose threats to human safety. For this reason, additional emphasis was placed on defining these events, including a statistical perspective. Discussing extreme events also provided the opportunity to mention those that seemingly contradict global warming, such as cold waves or heavy snowfalls. This allowed, on the one hand, to highlight that there can be counterintuitive aspects of climate when relying on personal perceptions; and on the other hand, to reinforce a fundamental concept, the difference between weather – the short-term conditions of the atmosphere, and climate - the average conditions of the atmosphere, hydrosphere, cryosphere, biosphere, i.e. the climate system components, for an extended period of time.

Subsequently, the lesson discussed the specificities of today's climate changes compared to those of the past, such as the last 2000 years or the Holocene, which began 11,700 years ago. One of the most significant differences is the speed at which changes are occurring today, which is also an indication of the cause generating them: the anthropogenic forcing. Natural forcings, such as the changes in the sun's energy output or the regular changes in Earth's orbital cycle, generate slower changes characterized by known cycles.

The current climate change is due to the increase in atmospheric concentration of greenhouse gases, such as carbon dioxide (CO_2), methane (CH_4), or nitrous oxide (N_2O), which originate from human activities such as fossil fuel combustion for energy production, agriculture, intensive farming, transportation, industry, etc. The lesson presented the data on the increase of greenhouse gases from around the mid-20th century to today, to highlight their unequivocal growth trend. The current warming trend induced by anthropogenic forcing is then amplified by a series of positive feedback within the climate system, which are part of the internal natural climate variability, such as the ice-albedo feedback. It is a process where a change in the extent of land/sea ice or snow areas modifies the albedo of the surface and by consequence the surface temperature of the planet. Global warming results in a decrease in the extent of snow- and ice-covered areas (reflective surfaces) and thus in an increase of darker and more absorbing surfaces, which amplifies surface warming. This feedback is particularly active in the coldest parts of the planet, such as the polar and high-altitude regions.

After briefly outlining possible future climate projections based on different greenhouse gas emission scenarios defined by the Intergovernmental Panel on Climate Change [23], the lesson saw its final part dedicated to illustrating what can and must be done to address the climate crisis through mitigation and adaptation, always keeping in mind that the climate we experience in the future depends on our decisions now.

The lesson made extensive use of metaphors – e.g. in explaining the difference between weather and climate, as well as to help understand how our emissions accumulate in the atmosphere when natural absorbers fail to remove them (through a simple bath analogy, see https://www.climateinteractive.org/ourwork/climate-bathtub-simulation/). Overall, metaphors have the potential to enrich the climate discourse by making it more inclusive, engaging, and actionable for individuals and the community.

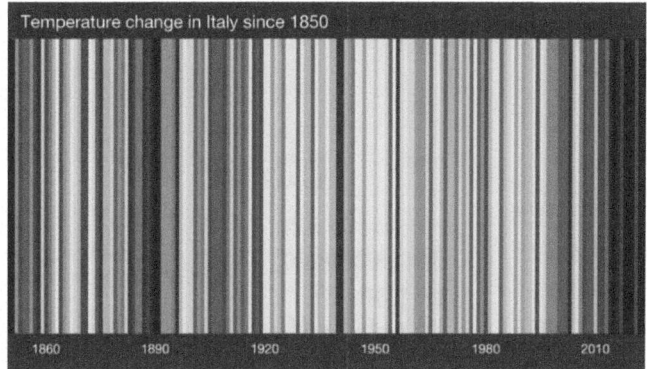

Fig. 1. Example of "Warming stripes" for Italy. They are graphical representations of the change in temperature as measured in Italy, from 1850 to 2023. Each stripe represents the temperature averaged over a year. The stripes turning from mainly blue to mainly red in more recent years, illustrate the rise in average temperatures in that country.

2nd Session: *Critical Thinking: Biases, reasoning and Communication about climate change.* This second session has the goal of educating students on Critical Thinking, reasoning biases and communication and also lasts two hours. The intervention is aimed at improving critical thinking through the provision of tools for forming an opinion and debating (this specific step is developed during the 3rd session) and is mainly based on (1) how to test an idea or hypothesis and establish whether it is *true or false?* (2) how to *draw conclusions* from facts. In particular, everything that has been covered in the two first sessions should implicitly contribute to the construction of the arguments for a fact-based debate in session 3 and correct reasoning.

Lecture by Laura Macchi on "Why is it difficult to communicate and reason about Climate Change?".

- Confirmation/verification bias: erroneous tendency to verify a hypothesis or idea considering only the cases that confirm it, without looking for cases that could potentially falsify it (counterexamples);
- Draw a conclusion from the premises: valid conclusions vs. factual true conclusions *belief bias;*

- Exercises - practical tests - inventions of false reasoning;
- Psychorhetorical intervention: in particular, our intervention was conceived for developing basic communicative skills for the understanding of the implicit layers of news and to discover rhetorical tricks in public communications on climate change. To develop basic skills for understanding information and reasoning about arguments with a scientific method, students were presented with several real newspaper articles reporting completely true news or news promoting climate denialism, which was followed by an in-depth analysis and explanation of its contents with identification of the potential reasoning biases and rhetorical tricks/fallacies. This stage is crucial since these news, which downplay the importance and severity of climate change, are real and young people are in contact with and thus influenced by them every day. Learning to disentangle the news into its possible biases can help youngsters not to take everything for true and critically review information. The skills developed are, moreover, generalizable and applicable to other domains;
- Newspaper articles

Fig. 2. Newspaper articles, in the original version (italian) used in our intervention.

In summary, to overcome bias, one can improve critical thinking and psychorhetorical tools → for debating and forming an opinion:

(i) How to test a statement, an idea and establish whether it is *true or false*?
(ii) How to *draw conclusions* from premises/facts?
(iii) How to detect *rhetorical tricks* in news and public communications

3rd Session: «*Taking the floor*». The third two-hours session, consists of "taking the floor". A few days before the debate, participants have been provided with material useful to organize and construct their argumentations.

The debate consists subsequently of an argumentative confrontation on a key topic for Climate Change in accordance with the following:

- the rules of civil debate (e.g., *no ad personam attacks*)
- the rules of reasoning

5   Methods

5.1   Participants

The intervention was carried out at the high school Liceo Claudio Cavalleri in the province of Milan (Parabiago). Students of 11th and 12th grade took part in the intervention. To investigate the effect of our intervention, we compared - using identical stimuli – the performance of a control group that had not done the intervention (n = 32) with the group that conversely had (n = 41) (climate physicist lecture, three syllogisms, two selection tasks and two newspaper articles).

5.2   Materials

After the climate physicist lecture, participants were presented with a series of stimuli consisting of three syllogisms, two selection tasks created on the topics of Climate Change, and two newspaper articles reporting misleading or false information about climate change. After each syllogism composed of two premises and a conclusion, participants were asked the following question: *"Does the conclusion necessarily follow from the premises?"*. One of the three syllogisms presented to participants is reported in Fig. 3.

Syllogism #2

Premise 1: All extreme weather phenomena are phenomena caused by global warming

Premise 2: All anomalous summers are caused by extreme weather phenomena

Conclusion: All anomalous summers are phenomena caused by global warming

---

Fig. 3. One of the syllogisms presented to participants

Similarly, after presenting the rules and some information about the four cards in the selection task, participants were asked to choose which cardboards to uncover to check whether the statement was true or false (see Fig. 4 in which an example of selection task is reported).

Lastly, for what concerns the psychorhetorical intervention, participants were asked to identify the implicit message of the two newspaper articles (see Fig. 2), and to say if they agreed with the implicit message - which reported arguments denying climate change – contained in each. All materials are reported in the Appendix.

Moreover, we presented to the participants who made the intervention the six images reported in Fig. 5, before and after the intervention, to investigate their opinion on the capacity to promote awareness, asking them *"Which of the following images do you think would most promote awareness of climate change?"*.

Selection task #3

Task: Below you are presented with a statement:

"If a bag is made of plastic, then you have to throw it away with the plastic waste".

followed by four cardboards referring to this statement, which are divided into two parts: on the left, there is information about the type of material (from which the bag is made), while on the right, there is information on the waste with which it has to be thrown. Only one part of each cardboard is visible (the other part is covered):

| Bag made of plastic | | Bag made of organic material | | | Plastic waste | | Biodegradable waste |

Which of these cardboards do you need to uncover to check if the statement is true or false?

Fig. 4. One of the selection tasks presented to participants

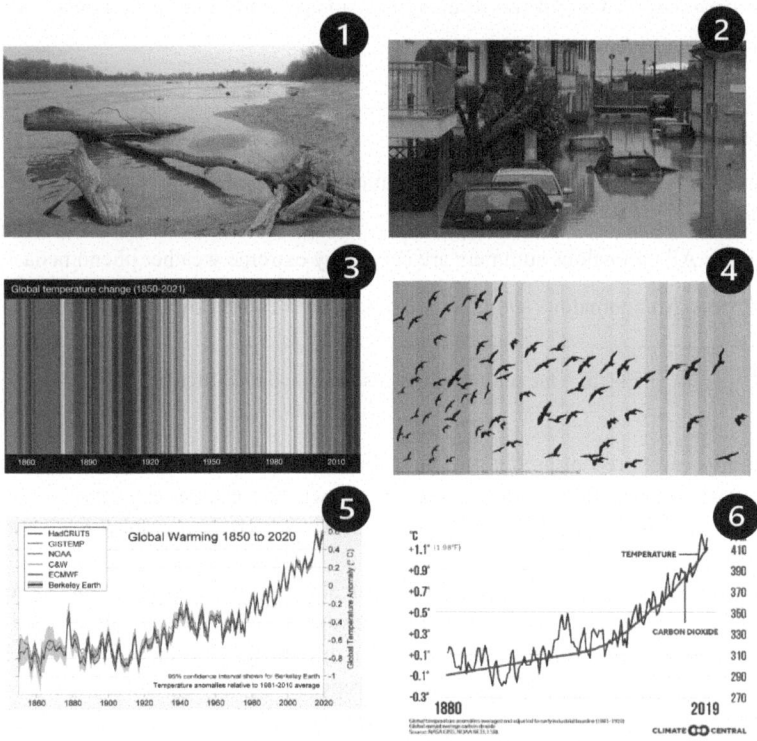

Fig. 5. Six images presented to participants who were asked to evaluate which, according to them, *would most promote awareness of climate change?*

5.3   Results

No significant differences emerged from the analysis of the number of correct responses in the evaluation of the syllogisms by the control and the intervention group. As reported in Fig. 6, both groups showed great capacity to understand whether a conclusion derived necessarily from the given premises or not, even if the intervention group outperformed the control group. The percentage of correct solutions shows that the first syllogism was evaluated correctly by 53% of the participants in the control group and 63,4% in the intervention group; similar results were obtained with the syllogism n.3 (respectively 53,1% vs 58,5%). The best performance emerged with the syllogism n.2, which brought to 68,8% of correct responses in the control group and 73,2% in the intervention group. These results will be discussed in the next section.

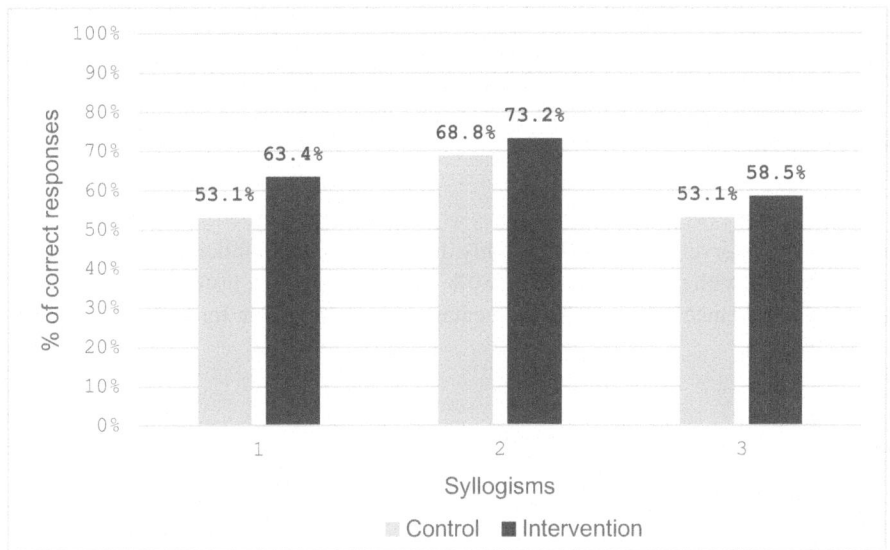

Fig. 6. Percentage of correct responses of the control and intervention groups in evaluating the syllogisms.

Statistically significant results have instead been found with the selection tasks. In fact, the differences between the control group and the intervention group were significant for selection tasks n°1 and n°3 ($\chi2 = 4,189, p = ,041$), showing a greater performance for the intervention group (from 7,3% for selection task n.2 to 12,2% for selection tasks n.1 and n.3) compared to the performance of the control group, - 0% of correct responses, indicating that not even one participant responded correctly to one of the selection tasks). Results are reported in Fig. 7 and discussed in the following section.

The most remarkable results emerged from the analysis of the article evaluation. The difference between the control and intervention groups in accepting the implicit message in the newspaper article was statistically significant for both articles, n.1 ($\chi2 = 10,264, p = ,001$) and n.2 ($\chi2 = 5,303, p = ,021$). As shown in Fig. 8, the group

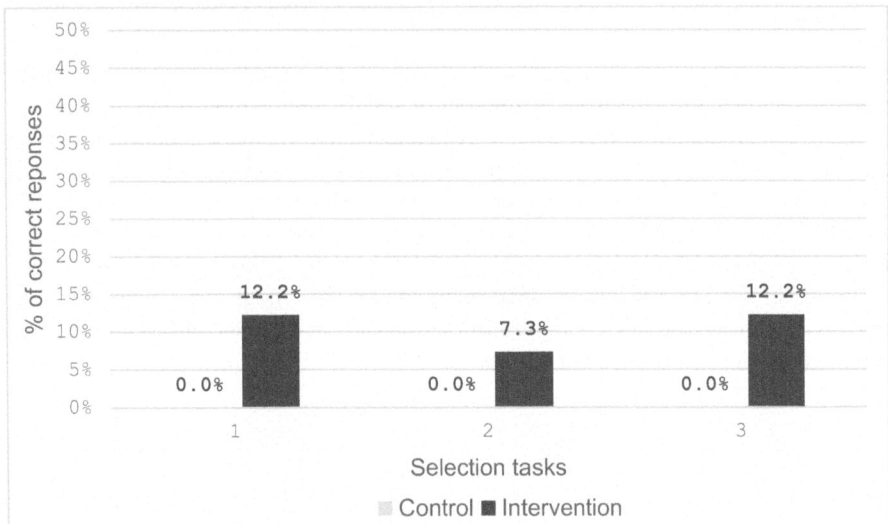

Fig. 7. Percentage of correct responses of the control and intervention groups in evaluating the selection tasks.

that made the intervention tended to highly disagree with the implicit message denying climate change, with percentages above 80%. The control group, instead, appeared to be mainly in accordance with it, as the percentages of disagreement for both articles were below 50%.

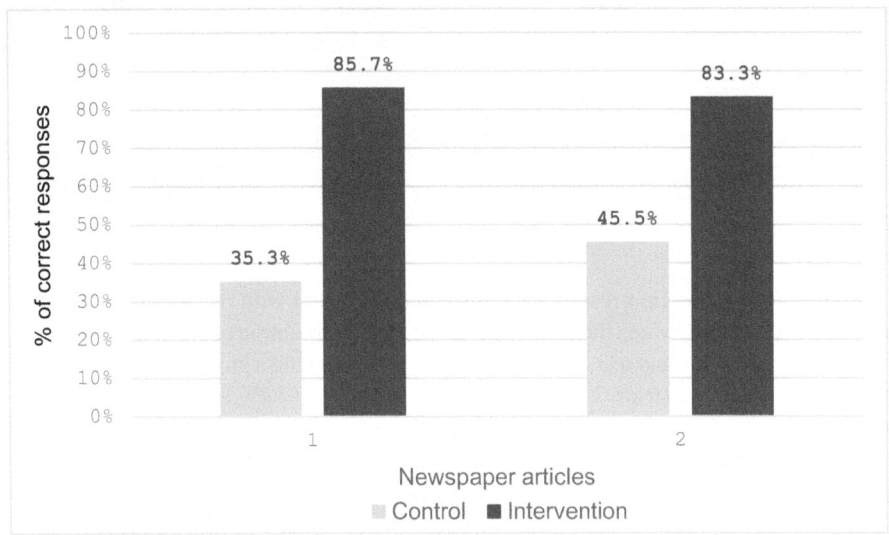

Fig. 8. Percentage of disagreement with the implicit messages contained in the newspaper articles, in the control and intervention groups.

The percentage of choice of each image concerning its ability to promote awareness is shown in Fig. 9.

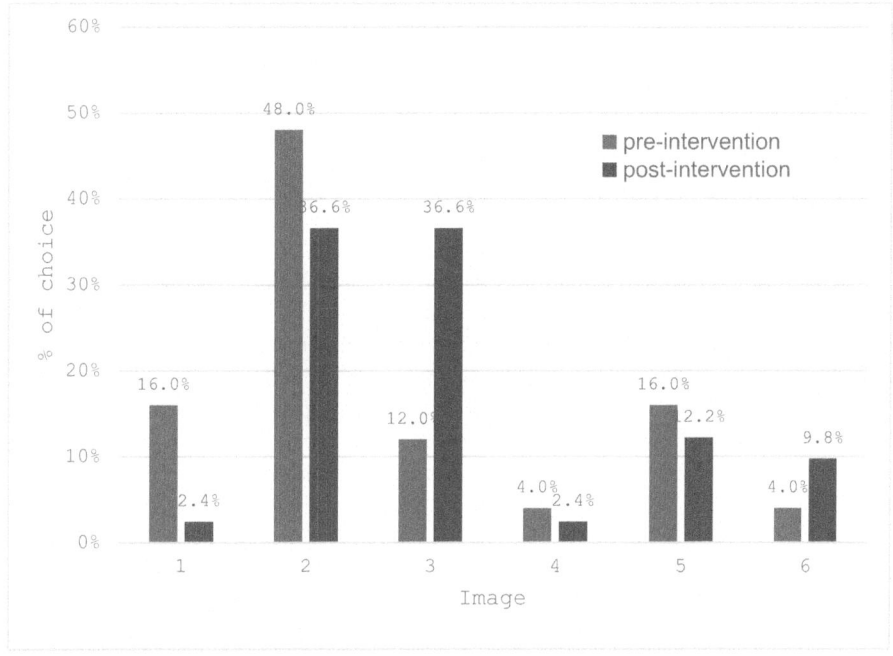

Fig. 9. Selection by the intervention group – before and after - of the image (reported in Fig. 5) considered more able to promote awareness of climate change.

6 Conclusions

The most important finding that seems to emerge from our intervention is that it appears to have been effective in promoting a critical approach among students who participated in the intervention. This is demonstrated by the very high percentage of disagreement with the implicit message contained in the articles denying climate change in the intervention group (approximately 85%) compared to the control group in which only around 35%–45% of the participants disagreed with it, showing, therefore, a propensity to accept the implicit message. Participants' disagreement with messages such as "the weather debunks global warming lies" or "it's so hot that it's snowing in Cortina" suggests that they are able to grasp the reality and immediacy of the phenomenon in time and space, potentially going beyond all those biases that lead to consider the effects as distant or irrelevant.

Other noteworthy results concern the selection task, with significant differences emerging between the control and the intervention group in two out of three tasks. Although the percentages of correct answers may seem low (at most around 12%), it is important to remember that in the literature, the solution rate for this type of task with adult university participants is less than 5%. The fact that the intervention group selected the correct cards (p and non-q) more frequently, which can lead to the falsification of the logical rule, suggests that the tendency to verify the rule (verification/confirmation bias) has been reduced, even if to a small extent. Being able to search for the counterexamples can be of great relevance when addressing climate change, as it prevents individuals from merely confirming their own and possibly erroneous ideas (such as denying climate change). Instead, it encourages the search for new information to develop a more complete and correct understanding of the phenomenon.

With the syllogisms, the already very high response percentages did not allow us to detect statistical significance. However, we note that they are much higher than those found in the literature, where they are less than 20%, and we attribute this to the inclusion of the term "necessarily" in the task instructions, which clarifies how the conclusion (logically valid vs factually true) is to be evaluated [24]. The high rate of correct evaluation of the syllogisms indicates that participants can distinguish between a syllogism's logical validity and its contents' credibility, thus avoiding belief bias.

Overall, our results indicate a better performance by participants who attended the critical thinking and psychorhetorical intervention compared to those who did not - even in cases where the results were not statistically significant – suggesting that our intervention has been successful.

Another interesting finding is the type of communication considered to be the most effective in promoting awareness about climate change. After the intervention, participants most often chose the image that visualizes the climate change and increasingly represents global warming (image 3 of Fig. 5, also reported in Fig. 1) in contrast to the preference before the intervention. The catastrophic images 1 and 2 (such as those depicting floods), reported in Fig. 5, were instead preferred before the intervention (in particular image 2) and less chosen after the intervention. This suggests that the image of global warming would represent the importance of conveying/transmitting in public communication the evolution of the phenomenon beyond isolated extreme weather events. Such events, when viewed as isolated events are often perceived as fatalistic and are less able to make the phenomenon understandable in its entirety and prospective nature. These results highlight the importance of the climatologist's lecture in explaining to the students how to interpret image 3 of Fig. 5, enabling them to grasp the fundamental aspect of climate change, which is its progression and worsening over the years, making salient the urgency to cope with it.

Funding. This project has been funded with support from the European Commission. Project reference: 2021-1-IT02-KA220-SCH-000027976.

Appendix

Syllogisms

Syllogism #1

Premise 1: If everyone stops using single-use plastics, the problem of plastic pollution will be solved

Premise 2: The problem of plastic pollution has been solved

Conclusion: Everyone has stopped using single-use plastics

Syllogism #3

Premise 1: If greenhouse gas emissions increase, then the earth's temperature rises

Premise 2: Earth's temperature rises

Conclusion: Greenhouse gas emissions increase

Selection Tasks

Selection task #1

Below, you are presented with a statement:

"To produce a 200-gram beef burger, at least 4 thousand litres of water are needed."

followed by four cardboards referring to this statement, which are divided into two parts: on the left, there is information about the type of burger, while on the right, there is information on the litres of water needed to produce it. Only one part of each cardboard is visible (the other part is covered):

 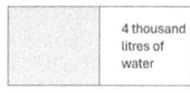 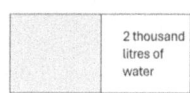

Which of these cardboards do you need to uncover to check if the statement is true or false?

Selection task #2

Task: Below you are presented with a statement:

"The incorrect disposal of used cooking vegetable oil is very serious.

If one kilo of used olive oil is disposed of incorrectly, then an amount of water equal to 1000m2 will be polluted".

followed by four cardboards referring to this statement, which are divided into two parts: on the left, there is information about the disposal of one kilo of used olive oil, while on the right, there is information on the amount of polluted water. Only one part of each cardboard is visible (the other part is covered):

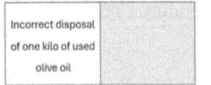 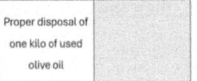 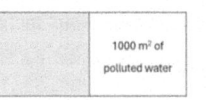 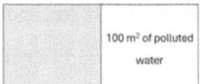

Which of these cardboards do you need to uncover to check if the statement is true or false?

Newspaper Article

Newspaper article #2

Il meteo smonta le balle sul surriscaldamento globale

C'è talmente caldo che nevica a Cortina

di **COSTANZA CAVALLI**

Il riscaldamento globale è una scarpa vecchia. Così scarpa che nevica a fine agosto. È successo la notte di sabato, a Cortina d'Ampezzo: le montagne intorno alla cittadina si sono ricoperte di neve, la temperatura è crollata a quattro gradi sopra lo zero e la gente ha tirato fuori il piumino mentre consumava il cellulare a riempire Instagram di foto fuori stagione.

La fissa del "global warming" (...)

segue a pagina 13

References

1. NEEF: Environmental Literacy in the United States: An Agenda for Leadership in the 21st Century. Washington, DC: National Environmental Education Foundation (2015). https://www.neefusa.org/resource/environmental-literacy-report-2015

2. Palazzi, E.: Narrate the Climate Crisis: Beyond the Language of Emergency (Original title: "Raccontare la crisi climatica: oltre il linguaggio dell'emergenza") (2021). https://www.you tube.com/watch?v=WDll5uXC63A
3. Weber, E.U.: Experience-based and description-based perceptions of long-term risk: why global warming does not scare us (yet). Clim. Change. Change **77**, 103–120 (2006). https://doi.org/10.1007/s10584-006-9060-3
4. Weinstein, N.D., Lyon, J.E.: Mindset, optimistic bias about personal risk and health-protective behaviour. Br. J. Health Psychol. **4**(4), 289–300 (1999). https://doi.org/10.1348/135910799 168641
5. Arpaia, B.: Perché non prendiamo abbastanza sul serio il cambiamento climatico [Why we do not take the climate change seriously enough]. Il Bo Live, 22 Dec 2018. https://ilbolive.unipd.it/it/cambiamento-climatico-perche-non-lo-prendiamo-sul-serio
6. Mazutis, D., Eckardt, A.: Sleepwalking into catastrophe: cognitive biases and corporate climate change inertia. Calif. Manag. Rev. **59**(3), 74–108 (2017). https://doi.org/10.1177/000 8125617707974
7. Drabek, T.E.: Human system responses to disaster: an inventory of sociological findings. Springer, New York, NY (1986)
8. Gupta, V.K., Saini, C., Oberoi, M., Kalra, G., Imran Nasir, M.: Semmelweis reflex: an age-old prejudice. World Neurosurgery **136**, 119–125 (2020). https://doi.org/10.1016/j.wneu.2019. 12.012
9. Jones, M., Sugden, R.: Positive confirmation bias in the acquisition of information. Theor. Decis.. Decis. **50**(1), 59–99 (2001). https://doi.org/10.1023/A:1005296023424
10. Kahneman, D., Knetsch, J.L., Thaler, R.H.: Anomalies: the endowment effect, loss aversion, and status quo bias. J. Econo. Perspect. **5**(1), 193–206 (1991). https://doi.org/10.1257/jep.5. 1.193
11. Festinger, L.: Cognitive dissonance. Sci. Am. **207**(4), 93–107 (1962). https://doi.org/10.1038/ scientificamerican1062-93
12. Seligman, M.E.P.: Helplessness: On Depression, Development, and Death. W.H. Freeman, San Francisco, New York (1992)
13. Stoknes, P. E.: What We Think About When We Try Not To Think About Global Warming: Toward A New Psychology of Climate Action. Chelsea Green Publishing, Chelsea, VT (2015)
14. Macchi, L.: La comunicazione pubblica. In: Viale, R., Macchi, L. (eds.), Analisi Comportamentale delle Politiche Pubbliche, Nudge e interventi basati sulle scienze cognitive (pp.271–293). Bologna: Il Mulino (2021)
15. Byerly, H., et al.: Nudging pro-environmental behavior: evidence and opportunities. Front. Ecol. Environ. **16**(3), 159–168 (2018). https://doi.org/10.1002/fee.1777
16. Grilli, G., Curtis, J.: Encouraging pro-environmental behaviours: a review of methods and approaches. Renew. Sustain. Energy Rev. **135**(12) (2021). https://doi.org/10.1016/j.rser.2020. 110039
17. Grüne-Yanoff, T., Hertwig, R.: Nudge versus boost: how coherent are policy and theory? Mind. Mach. **26**(1–2), 149–183 (2016). https://doi.org/10.1007/s11023-015-9367-9
18. Hertwig, R., Grüne-Yanoff, T.: Nudging and boosting: steering or empowering good decisions. Perspect. Psychol. Sci.. Psychol. Sci. **12**(6), 973–986 (2017). https://doi.org/10.1177/174569 1617702496
19. OECD: Thinking outside the box. The PISA 2022 Creative Thinking Assessment (2022). https://www.oecd.org/pisa/innovation/creative-thinking/
20. Macchi, L.: La psicoretorica. dall'arte del dire alla forma del pensiero. Raffaello Cortina Editore, Milano (2024)
21. Bagassi, M., Macchi, L.: The interpretative function and the emergence of unconscious analytic thought. In: Macchi, M.L., Bagassi, R.V. (eds.), Cognitive Unconscious and Human Rationality, pp. 43–76. MIT Press, Cambridge, MA (2016)

22. World Meteorological Organization (WMO): State of the Global Climate 2023. WMO: Geneve (2024). https://library.wmo.int/records/item/68835-state-of-the-global-climate-2023
23. IPCC, 2021: Climate Change 2021: The Physical Science Basis. Contribution of Working Group I to the Sixth Assessment Report of the Intergovernmental Panel on Climate Change [Masson-Delmotte, V., P. Zhai, A. Pirani, S.L. Connors, C. Péan, S. Berger, N. Caud, Y. Chen, L. Goldfarb, M.I. Gomis, M. Huang, K. Leitzell, E. Lonnoy, J.B.R. Matthews, T.K. Maycock, T. Waterfield, O. Yelekçi, R. Yu, and B. Zhou (eds.)]. Cambridge University Press, Cambridge, United Kingdom and New York, NY, USA, In press, https://doi.org/10.1017/978 1009157896
24. Macchi, L., Poli, F., Caravona, L., Vezzoli, M., Franchella, M.A., Bagassi, M.: How to get rid of the belief bias: boosting analytical thinking via pragmatics. Europe's J. Psychol. 15(3), 595 (2019)

The Effect of Screens on Children's Development: Concrete Action Taken in Schools, Closer to Families

Sabrina Reffad[1,2]([✉]) [iD] and Joelle Provasi[1,2] [iD]

[1] Ecole Pratique des Hautes Etudes, 4-14 rue Ferrus, 75014 Paris, France
sabrina.reffad@etu.ephe.psl.eu
[2] Laboratoire CHArt-EPHE, 4-14 rue Ferrus, 75014 Paris, France

Abstract. Evidence of the damaging effects of screens on all pillars of child development continues to emerge. The danger of screens lies above all in the behavior they adopt to prevent the conversations, games, vacations and family discussions through which children do most of their learning and form their character. In their pharmacological dimension, screens can also be beneficial when used in a healthy way at a certain age or for educational purposes, and can support learning sessions. However, early and excessive use has been associated with deterioration in language, attention, executive functions and learning delays. In health terms, it affects social maturity, increases the risk of eating and sleeping disorders, and alters mental and psychological health in general. Given the growing importance of screen use among very young children, at stages when brain plasticity is greatest, the cognitive and cerebral development of the current generation of children exposed to screens raises major questions. Official recommendations do not seem to be reaching everyone. Furthermore, some families are beginning to realize the consequences of the upheavals they are experiencing at school. It is possible to promote healthy behavior and children's well-being in the face of screens, and thus reduce their negative effects, by improving knowledge and proposing alternatives to children and their families. Concrete actions in the field, as well as at school, and accessible to all social classes, seem essential.

Keywords: Child development · Screen use · Health · Cognition · School · Socioeconomic context · Concrete actions

1 Introduction

Children's immersion in the world of screens seems inescapable in our hyper-digitised societies. Several studies have shown the early and excessive use of multiple digital media by preschool and school-age children (Bernard et al. 2023; Gauthe 2020; Ipsos 2022). Although health guidelines are attempting to regulate this digital and audiovisual consumption, parents underestimate the amount of

J. Baratgin et al. (Eds.): HAR 2024, LNCS 15504, pp. 345–365, 2025.
https://doi.org/10.1007/978-3-031-84595-6_21

time their children spend in front of screens (Arcom 2022; Beau 2022). However, over the last fifteen years, research has highlighted the harmful effects of excessive digital use on children's development. Yet exposure to and use of screens continues to grow, and at an increasingly early age, particularly since the recent pandemic (Bergmann et al. 2022). Official recommendations are not enough to raise awareness, as this digital phenomenon affects adults and children of all ages. Several initiatives undertaken by pediatrics authorities (Dieu-Osika 2020; Ponti et al. 2017), and recent measures taken by the public authorities, need to be relayed by other key contacts for families, and continued in a tangible way, i.e. in situ. While the importance of considering the various factors linked to exposure to screens, beyond the time of exposure, has been highlighted (Esseily et al. 2017; Guellai et al. 2022; Yang et al. 2023), the public needs to be widely informed and alternatives to screens need to be proposed. The issue of regulating children's biological rhythms, and the time devoted to each activity, therefore seems fundamental. It is rhythm that influences the ways in which we use the elements received in each experience. While the effects of television and screens on children's development are multifactorial, and depend, among other things, on the characteristics of the child, the family and social context, and the characteristics of the content watched (Guellai et al. 2022; Kotstyrka-Allkorne et al. 2017), it must be borne in mind that these effects, whatever the conditions, affect the child's health, academic success and future (Domingues-Montanari 2017; Muppalla et al. 2023). The time spent exposed to screens, once considered to be the common denominator of harmful effects, now appears to be one factor among others. The correlates studied on the issue of screens in children remain insufficient (Veldman et al. 2023), and proves the absolute need for more in-depth and better-quality research, with greater social cohesion. The role played by all the adults who contribute to a child's development is undoubtedly essential. However, regulating the right amount of time still raises several questions, particularly as regards the means of action to be taken with this audience to ensure that it is kept under control. Does school have a role to play?

The purpose of this article is to review the effects of screens on children's development, and to set out the practical measures being taken in the field to help families, particularly those most at risk. At a time of heated debate following the announcement of legislation to protect children from overexposure to screens, it is important to propose practical solutions in addition to those that already exist. At the initiative of educational teams in several primary school Priority Education Networks (PEN), and in partnership with early childhood professionals, initiatives have been set up to explore strategies for raising awareness among parents of children from birth to age 12 and reducing screen time. The choice of this age group reflects both our desire to initiate a preventive approach to the use of screens, and our professional sector, that of early childhood. Schools, often the first place where children are socialized, remain an institution that is close to the population in disadvantaged neighborhoods. Furthermore, the views of school psychologists and teachers, as co-contributors to the success of pupils' educational careers, resonate strongly within homes, with both children and their parents.

2 The Effects of Screens on Children's Development

2.1 Cognition

Digital technology has both beneficial and deleterious effects on children's cognitive skills. Used wisely, it can be used as a teaching aid by teachers in reading or writing sessions with dyslexic pupils (Poirel 2020; Ruiz et al. 2018), or in some cases, in the use of educational software (Anderson and Subrahmanyam 2017). Benefits can also be observed in watching even entertaining TV with a parent to stimulate the joint attention mechanism (Wass et al. 2018). However, several studies have highlighted negative associations between the duration of exposure to screens and cognitive development and learning in young children (Gastaud et al. 2023; Harlé and Desmurget 2012; Ottemer 2021; Prieur 2020; Rocha et al. 2021). Furthermore, television viewing is directly linked to neurocognitive functioning and therefore to the neuroanatomy of the brain (Desmurget 2011, 2022). In both cross-sectional and longitudinal analyses, television viewing was positively correlated with changes in medial frontopolar and prefrontal regions, the first contributing to intellectual functioning (Takeuchi et al. 2015). The negative correlation between television viewing and IQ was confirmed. Rodriguez-Ayllon et al. [2020] have shown that increased screen time use during childhood has also been linked to changes in global white matter microstructure.

Several studies have shown that watching television, or watching it in the background, is a predictor of hyperactivity or attention disorders (Barr 2013; Christakis et al. 2004; Takai et al. 2005), and each hour spent watching television increases the risk of developing an attention disorder in adolescence (Landhuis et al. 2007). Similar negative effects have also been demonstrated for the frequent use of video games (Masi 2020; Swing et al. 2010). Furthermore, the notion of "multitasking" induced by having several devices in the home implies a reduction in attention to other visual and sound fields (Gueron-Sela et al. 2020). Long-term exposure to touch screens from the age of 12 months is associated with faster exogenous attention and a concomitant reduction in the control of endogenous attention (Portugal et al. 2021). The problem with multiple tasks lies in the difficulty of developing a fixed attentional focus that is effective and beneficial to learning behavior, particularly in the classroom (Pagani et al. 2013, 2010). Executive functions (EF) also play an important role in learning academic subjects and regulate behavior in school life (Lussier et al. 2018). Children's use of screens before the age of 2 affects EF-related cortical activity and executive development longitudinally (Law et al. 2023; McHarg et al. 2020). Excessive exposure to fast-paced cartoon content in preschoolers undermines executive functioning (Liliard and Peterson 2011). Children under the age of 3 learn less from exposure to screens than from actual demonstration, and age-related constraints on memory flexibility contribute to transfer deficits (Barr 2010, 2013).

Inappropriate use of digital media may also play a role in the onset of developmental disorders (Chen et al. 2020; Madigan et al. 2019). The researchers showed that excessive screen time at age 1 was associated with a statistically higher likelihood of ASD at age 3 (Kushima et al. 2022). Chonchaiya et al. [2015] reported that infants aged 6–18 months who had been exposed to television since the age of 6 months had higher levels of pervasive developmental disorders (using the Child Behavior Checklist). A team of pediatricians has highlighted the effects of this inappropriate use (hyperconnected parent, still face, early excessive exposure, background TV...) on children's development (Dieu-Osika et al. 2020). She describes this new pathology called "early media overexposure syndrome", which affects young children and combines language delay with symptoms similar to those of autism. However, this syndrome can be reversed by completely ceasing exposure to media and screens.

Furthermore, this early exposure to screens can be problematic for cognitive and social functions years later (Desmurget 2020; Nathanson et al. 2014). Exposure to television between the ages of 5 and 15, regardless of gender, has been associated with poor school results, behavioral difficulties, a high risk of leaving school without qualifications, and low chances of obtaining a university degree by the age of 26 (Hancox et al. 2005). While stories read or audio books used at an early age promote socio-cognitive skills and knowledge in children, adolescents and adults (theory of mind, empathy, etc.)(Lenhart and Richter 2024). Early exposure to screens, television in the background, or in the bedroom impoverish the child's socio-cognitive skills. However, when the child is accompanied by an adult who discusses the content viewed, the development of theory of mind is then encouraged (Nathanson et al. 2013).

2.2   Language

Language is the major issue at school, and the role of the child's surroundings and environment influences his or her learning (Florin 2020). Although there may be evidence that specific vocabulary items are learned through exposure to television programs (Mares and Pan 2013; Wright et al. 2001), more complex aspects of language, such as phonetics and grammar, are not acquired through exposure to television in infants (Barr et al. 2007; Kuhl et al. 2003). Up to the age of 2, infants have difficulty understanding speech on screen without adult interaction (Pempek et al. 2010). This deficit in language expression is explained by a lack of social communicative exchanges, as well as transmodal difficulties, which prevent babies from transferring auditory or visual information, as they would in a real communication situation with a partner (Barr 2010, 2013). One study also found an association between early use of touch screens and language delay (Van Den Heuvel et al. 2019). Children exposed to TV in the background or first thing in the morning have been associated with risks of developing primary language disorders (Collet et al. 2019; Martinot et al. 2021). Similarly, interactions between children and parents, and the latter's involvement in media use, can have a negative influence on language development and literacy skills (Sundqvist et al. 2021; Tremblay et al. 2021).

Screen use is most often linked to watching adult programs, or to passive or recreational use, which is harmful to language (Anderson and Subrahmanyam 2017; Gueron-Sela et al. 2020; Kotstyrka-Allkorne et al. 2017; Nathanson et al. 2014).

Although it is imperative to explore the links between early and excessive screen exposure, in light of context (Yang et al. 2023), most research suggests a strong link between screen time and language delay (Bhutani et al. 2024).

2.3 Health and Socio-Affective Skills

In the first few years of life, children learn to forge meaningful links with the world around them. If social and emotional skills are compromised, this is obviously detrimental to the child's development and entry into school. A negative association between media use and the quality of life of young children aged 2 to 6 has been assessed by questionnaire (Hcsp 2020). Lin et al. [2020] showed that exposure to television viewing is correlated with depressive disorders in children. It has also been repeatedly documented that heavy television viewing at the age of 2–3 years leads to an increased risk of peer victimisation in the early years of school (Pagani et al. 2013). Recent studies have found that screen time, as well as background TV, in early childhood is negatively associated with later self-regulation, while fantasy or recreational content appears to have immediate negative effects on children's self-regulatory abilities (Cliff et al. 2018; Huang et al. 2024; Lawrence and Choe 2021; Uzundağ et al. 2022). Furthermore, access to social networks has been democratised among primary school-age children.

However, for several years now, harassment and cyberbullying at school have been on the rise and continue outside the school walls, with pupils being harassed 24 hours a day (Bellon and Gardette 2018). The legal age for authorising access to social networks in France, set at 15, was only recently passed, by the law of 7 July 2023. The effects of high screen time accumulate with age, with clinical symptoms more evident from early adolescence onwards (Domingues-Montanari 2017), as with the association of screen communication with lower self-esteem and psychological well-being in this audience (Twenge et al. 2018). Self-regulatory abilities are also closely linked to sensorimotor development and health. The study by (Pagani et al. 2013) is the first to suggest a prospective association between excessive early exposure and motor skills in kindergarten. Excessive television viewing at an early age is also prospectively associated with lower explosive leg strength, impaired gross motor skills, impaired coordination and impaired locomotor skills (Edelson et al. 2015; Pagani et al. 2013).

In addition, a recent study found a statistically significant negative association between weekly exposure to screens and an alteration in the quality of graphic skills in children aged 5 to 6 (André and Cochetel 2023). Regarding children's fundamental rhythms, studies have shown a reduction in sleep in infants and young children associated with screen viewing (Diler and Başkale 2022; Li et al. 2020; Ribner et al. 2019), as well as an alteration in sleep quality (Cheung et al. 2017). Studies show the association between eating in front of a screen and junk food consumption in children aged 2 and over (Jusienė et al.

2019; Lutz et al. 2024). Several links have been found between an increase in screen time and an increase in sedentary behavior and obesity (Khadilkar et al. 2023; Li et al. 2020; Robinson et al. 2017; Skrede et al. 2019; Tremblay et al. 2011), notably because screen time replaces the physical activity associated with the development of white matter microstructure (Rodriguez-Ayllon et al. 2020). Exposure to screens therefore disrupts sleep patterns and eating habits, with boys predominating (Zink et al. 2024). The study by (Small et al. 2020), reports the importance of recognise the effects of screen time on sleep, considered as a moderator of various negative effects on cognition and brain function, such as ADHD (Cao et al. 2018). Finally, a recent study highlighted an association between increased screen time and increased myopia in children aged 6 to 8 years, after lockdown (Wang et al. 2021).

2.4   The Family and Socio-Economic Context

Although inequalities in equipment persist according to standard of living, more and more households are equipped with electronic goods. An INSEE survey reveals that in 2019, 83% of households will own a computer and 96% a smartphone (Gleizes et al. 2021). Digital tools have become indispensable and accessible to all. In qualitative interviews, many stay-at-home mothers said they use digital technology to 'escape' the boredom or frustrations of raising children, or to regulate their own emotions or excitement (Radesky et al. 2015). The use of digital tools as a means of calming, (re)compensating or occupying a child (Fondard 2014), can be detrimental to their social-emotional skills (Huang et al. 2024; Radesky et al. 2015). Today's research therefore explores more contextual aspects, particularly those linked to parental digital engagement and practices. Children's behavioral problems are thought to be linked to technological interruptions or "technoference", a phenomenon that hinders parents' interactions with their children (Devouche et al. 2023; McDaniel 2015; McDaniel and Radesky 2017). Psychosocial factors also influence digital use, such as the role played by parents and the level of stress in the child's environment (Prieur 2020). Other influences need to be considered, such as environmental and economic variables and social fractures, which have an impact on academic success and are aggravating factors (Zaouche-Gaudron 2017; Zaouche Gaudron 2021). Furthermore, studies have shown that children from families with low socio-economic status were almost universally exposed to mobile devices (Kabali et al. 2015); tended to be exposed to higher levels of electronic media; and showed poorer self-regulation (Corcos and Bergman 2020). The higher the parents' average level of education, the lower the level of screen use among children aged 2 to 12 (Combes et al. 2023; Gassama et al. 2018). Several studies have shown that early exposure to screens is greater for children from low-income families (Barr 2013; De Lepeleere et al. 2018; Rideout 2011), and is also correlated with the mother's age, parents' low level of education, and immigrant families (Poncet et al. 2022).

Lockdown during the Covid-19 pandemic exacerbated the health problems associated with exposure to screens in these same families (Kurz et al. 2023; Molleri et al. 2023; Thompson et al. 2022). Early exposure for more than 2 hours a day in families with low socio-economic status results in fewer verbal exchanges with their parents, poorer cognitive performance, poorer language skills later, shorter school careers and less access to healthcare in general (Arnold 2011; Chardon et al. 2015; Chonchaiya and Pruksananonda 2008; Combes et al. 2023; Hancox et al. 2005; Tomopoulos et al. 2010; Tremblay et al. 2021). It should be remembered that the influence of the environment on the neuroanatomy of the brain has been highlighted in scientific research. Genetic effects explain only a small part of the variance in intelligence (Davies et al. 2007). Longitudinal studies have shown improvements in children's verbal IQ, correlated with changes in regional grey matter volume (Konrad et al. 2012; Ramsden et al. 2011). This research shows the importance of conducting psycho-educational action programs on the ground to achieve positive changes in children and their families.

3   Professional Action with Families and Their Children

3.1   Understanding How to Implement Official Recommendations in Sensitive Neighbourhoods

Recommendations that encourage parents to set consistent limits on screen time are beneficial. Indeed, research shows that children of parents who control the amount of time spent in front of a screen are more inclined to take part in physical activities and spend time reading (Carlson et al. 2010). However, exposure to screens is multifactorial, and is correlated at least in part with the level of education of the mother, who is sometimes isolated, and with socioeconomic status (Combes et al. 2023; Gassama et al. 2018; Prieur 2020). Specific parental practices therefore need to be closely explored, to better understand the parental correlates of children's screen time. The parent's role in managing screen time is crucial. One study examined the paradigms explaining the influence of parents, who by limiting the time they spent in front of the television in the vicinity of their child aged 6 to 9, helped to improve their children's screen time, and thus reduce socio-economic disparities (De Lepeleere et al. 2018). Avoiding negative parental practices regarding time spent watching television seems to be an important parental practice to target, as children's self-regulation skills depend on parental commitment (Uzundağ et al. 2022). Schools must be able to contribute to reducing social inequalities and become involved in prevention processes and actions on the effects of screens, with children and their families (Muppalla et al. 2023).

3.2   Concrete Actions in the Field: Proof of the Effectiveness of Our Commitment on the Ground

Several concrete actions programs on the ground, as close as possible to vulnerable groups, have already proved their effectiveness over a few years. Both to reduce the socio-cognitive differences linked to low socio-economic living conditions, and more recently to regulate the use of screens, which is one of the aggravating factors.

From a constructivist and interactionist perspective, the "parler bambin" program by (Zorman et al. 2011) proposed language workshops to stimulate children aged between 18 and 30 months in nurseries located in sensitive urban areas. The children were selected based on their language level and the results showed positive long-term effects, particularly on their school performance. The study by (Berger et al. 2021) confirmed the importance of nurseries, compared with other forms of childcare, in preserving the language skills of children from disadvantaged socio-economic backgrounds. School readiness has pre-school and socio-emotional components that depend on the stimulation of the child from birth, and of the child's carer. The Head Start program in the USA, made up of healthcare professionals, works with families to promote quality stimulation of their children's cognitive development and learning (Crotty et al. 2023). A family training program improved brain function, cognition and behavior in preschool children from lower socio-economic backgrounds (Neville et al. 2013).

Interventions aimed at informing and raising awareness among parents during discussions about screens (Escobar-Chaves et al. 2010), or via advice given by midwives to parents when their babies were three, sixteen, twenty-eight and forty weeks old. Adams et al. [2018] have demonstrated awareness of the importance of the recommendations and compliance with them. As part of a coherent diagnostic approach, and with the aim of avoiding over-medicalisation of the intensive use of screens in the long term, healthcare professionals in pediatric or psychological consultations offer families food for thought on how to adapt their practices and remain in control of digital tools, to safeguard their children's development (Dieu-Osika 2020; Duflo 2018). Initiatives in the professional field can therefore be beneficial in the short, medium and long term.

3.3   Why and How Can PEN Schools Foster Healthy Digital Habits in a Constantly Changing Society?

Schools and their various institutions should work in harmony with the other components of society, offering the younger generations a privileged center for instruction, education and training in autonomy and critical thinking. Despite all the criticisms that can be levelled at it, the school, or the French education system, still echoes in many people's minds the ultimate opportunity to gain knowledge, and not just a certain social status that meets the economic requirements of the day. In this sense, it still has a role to play through educational psychology professionals and their actions in situ, with families who place their trust in them.

3.4   Professionals Dealing with the Complex Realities on the Ground

The early and excessive use of screens can be seen in all environments, but it is an aggravating factor among disadvantaged populations. Difficult living conditions have an impact on children's development. The consequences can be seen within the school. Professionals in charge of early childhood, as well as teaching teams, sometimes find themselves at a loss when faced with the ubiquitous use of screens by underprivileged families. For several years now, interviews with families and children at school, with the school psychologist, have been corroborating the research findings, and highlighting the massive use of screens by preschool and primary school children. The issue raises concerns for children's future and their progress as pupils. In particular, when adapting to school proves difficult.

Solidarity, commitment, high-quality teaching relationships and cohesion within the teams foster trust with families from disadvantaged neighborhoods. However, we also must consider the emotional impact that these situations, with which they are regularly confronted, can have on professionals. Although the national education psychologist is trying to strengthen partnerships between the social, socio-educational and health services, to encourage cross-fertilisation, the objective of strengthening prevention remains difficult due to a lack of time and logistical and human resources. Yet teachers need time for analysis, training and support from specialists. Their proximity to children and their families, as professionals in the field, helps to shed light on parents' beliefs. They reflect their typical behavior within a naturalistic framework, and account for the complex and changing processes that occur daily in the lives of young children. Parents may ambivalently use screens to occupy, distract, calm or (re)compensate their little ones, while at the same time raising concerns about the "hypnotic" or "addictive" effects they observe.

Generally speaking, economic deprivation in disadvantaged neighborhoods means less access to certain material goods or cultural and artistic activities, which exposes children to dangerous situations that their parents would like to avoid. For some parents, the screen represents access to culture and knowledge. For others, the predominant use of the screen becomes an activity, or a way of compensating for shortcomings and/or escaping the reality of difficult living conditions. Lastly, it is essential to avoid distilling guilt-inducing information or disguising injunctions with infantilising advice. On the contrary, the aim is to build a genuine partnership with families, who will become ambassadors to their peers, as well as all the professional teams involved in early childhood, thus enabling the emergence of deep-seated convictions that will change mentalities.

Rapid, tailored and targeted prevention and health promotion initiatives can have a real positive impact. There are 3 main strands to these initiatives, in the form of conferences for parents at school, training for professionals, and educational and psycho-pedagogical workshops for kindergarten school and their parents.

3.5   Conferences-Coffee: a Friendly Place to Get Information and Exchange Ideas

From the start of the new school year in September 2021, conferences for families were organised in all the school groups in the school psychologist's sector in the city of Saint-Denis, in partnership with school nurses, head teachers and teachers from nursery to elementary school. In a friendly setting familiar to parents, a presentation on the effects of screens on children's development was given in 2 parts: information and exchanges between professionals and families, then in 3 parts. The first focused on the importance of the developmental needs of babies aged 0–3 and the need to protect them from exposure to screens; the second showed the effects of screens on cognition and language and interpersonal skills at school; and the third highlighted the effects on children's physical and physiological health (sleep, diet, motor skills, etc.). We were able to reach parents whose children were most exposed, their participation was active in the proposed projects.

3.6   The Importance of Training Professionals in Early Childhood and Pedagogy

All concerned by the issue of screens, several conferences have been given by the school psychologist to early childhood professionals within local structures (neighborhood centers, parents' centers, etc.)

Training sessions on the effects of screens at school have been organised for teachers by the school psychologist, as part of the educational activities run by the National Education Inspector for the Saint-Denis 3 district.

3.7   A Booklet on Screens: the Fruit of a Multi-Professional Partnership

A partnership was formed to produce a booklet for parents that would be both informative and provide practical tips (tips from parents) and multiple alternatives to screens[1]. The project is part of the Saint-Denis city policy, which is committed to promoting sporting, family, artistic and cultural events and activities that are accessible to all. The working group was made up of professionals and parents (project leader for risk and addictive behavior, nursery nurses, school psychologist, teacher-researcher, mothers, etc.), in partnership with the mother and child protection (PMI), GPs and the IRI[2], all of whom are committed to combating overexposure to screens.

---

[1] Bien grandir dans un monde d'ecrans_Guide à l'usage des parents d'enfants de 0 à 6 ans, 2024, available here: https://ville-saint-denis.fr/actualites/bien-grandir-dans-monde-ecrans.

[2] Institute of Research and Innovation.

3.8   Prevention and Health Promotion in Kindergarten Classes with Parents

In close contact with families and children, it has been observed that, depending on the context, knowledge about early-learning games, motor skills, language and stimulation in general is sometimes not readily available to parents in their immediate family environment. Some parents, often from low-income backgrounds or newly arrived, rely entirely on the school.

This observation led to the organisation of 'psycho-educational' workshops from the start of kindergarten class, with the teachers, the educational psychologist and the parents. The families actively participated in the games workshops to stimulate their children's language, cognitive and psychomotor skills. The list of materials and games used is circulated to teams and parents to stimulate ideas. Simple language games can be reproduced at home with equipment that is already available, inexpensive and accessible.

3.9   In-Class Workshops with Elementary School Children

It's essential to take stock of the situation, regarding several factors that are worsening in schools as a direct result of the use of screens. At the start of the new school year in September 2020, in a post COVID context, the observations by educational teams made in our schools in the sector during the first term related to a disruption of changes in daily activities and sleep-wake rhythms. These observations were put into practice by scientific research (Bergmann et al. 2022; Molleri et al. 2023). The effects of this homeostatic disruption were caused by physical inactivity, lack of exposure to daylight, heavy exposure to screens late at night, irregular mealtimes, lack of social interaction, etc. The testimonies of families and children and the observations of teaching teams have highlighted symptoms of irritability, emotional and behavioral problems, daytime sleepiness, weight gain and even obesity, and a deleterious impact on cognitive performance and decision-making. Observation and interviews have enabled us to assess and identify the psychosocial criteria that alert us to the cases of the most vulnerable pupils, whose symptoms correspond to the clinical picture of depressive and anxiety disorders with addictive behavior, particularly eating disorders (Bantuelle and Demeulemeester 2008; Cohen 2008; Doré 2017; Shankland 2020). A therapeutic project led by the school psychologist and the school nurse has been set up within a school group, in collaboration with the teaching teams, the neighborhood center, the secondary school, the families, the central kitchen, the educational success program manager and the town's dietician. Two groups of children from cycle 2 and cycle 3 (aged 6–12) were formed: one focusing on the cooperative and educational game "psychogame", looking for ways of expressing themselves to resolve intrapsychic and interpersonal conflicts; and the second on culinary therapy "foodstyle". The children revisited food by creating a balanced menu served by the canteen to all the pupils in the school group. These actions were carried out with the aim of promoting the well-being of the pupils, using methods that enable them to learn to express, solve and confront problems in a fun and playful way. By becoming involved in the project, children fully identify themselves as "autonomous subjects", able to grapple with the world, the world of objects as much as the world of others, capable of learning and building with others.

Since 2023, we conducted a survey on 366 pupils on the use of screens and its evolution among children aged 3 to 12 (a different paper dedicated to those results is in print). This use is evolving in parallel with age, with increasing use of certain types of media, depending on gender, and at an increasingly early age.

Screen awareness days were then organised for all classes from CP[3] to CM2[4]. The mornings were devoted to presenting the effects of early and excessive use of screens, followed by a debate. Later, teachers used a questionnaire designed by the national educational psychologist to assess pupils' knowledge of the issue. Sporting activities and board game workshops were offered in the afternoons, to show alternatives to screens, as well as the many benefits of playing together. The children received a magnet with Sabine Duflo's 4 steps. These are four simple and effective rules to regulate the use of screens in the home: no screens in the morning; no screens during meals; no screens before going to bed; no screens in the bedroom.

The issue of cyberbullying was addressed during the awareness campaign. However, several initiatives for CM1[5] and CM2 classes were stepped up, in the form of theatre sketches, debates and a poster competition for a campaign to prevent harassment within the school. Some classes were also able to take part in activities organised by outside associations, where these were funded by the town.

4   Discussion and Conclusion

The use of screens can have both beneficial and negative effects on children's development. When used in a healthy way or as an educational aid, they can potentially improve learning (Poirel 2020). In fact, digital tools are tending to be widely used in schools, and their use in a school setting, in a manner designed and controlled by an educational professional, can be effective. Nevertheless, questions remain about the relevance of this tool in kindergarten school. Early, multitasking and excessive use of this tool has consequences for language, attention and EF, for socio-affective skills, for the quantity and quality of interactions between peers and with adults, and for cognition and academic success. They can also have an impact on sleep, eating, vision and sensorimotor skills. Research has highlighted protective factors such as co-viewing, control over the content and availability of screens for the child, and depending on the family, socio-economic and cultural background, the child's leisure activities, or nursery attendance for language, for example. This highlights the crucial importance of the role of parents in managing digital technology. Children's surroundings and environment

---

[3] first grade.
[4] fifth grade.
[5] fourth grade.

influence their learning (Florin 2020). Family educational practices and those of the adults who support the child's development contribute to the construction of social and cognitive skills and language skills.

These conferences, held in a friendly setting within the school, helped to bring out parents' beliefs on the issue of screens and to provide concrete information. Exchanges between peers and multi-professional teams make it possible to: change the representations distorted by unfounded beliefs (DeLoache et al. 2010), share tips and strategies for reducing and better managing the time spent in front of screens, but above all, become aware of the issues, and nurture solid convictions for healthy use of digital tools. These school-based initiatives to counter the effects of screens on children's development have also helped to provide children with real information. It has also led to recognition of the legitimate discourse on potentially harmful media practices put forward by education professionals.

So, schools still have a role to play. Other bodies contribute to socialisation and support for parenthood from the moment a child is born, which is essential. But while only some children attend the (PMI) or nursery or are in contact with other players in the medico-social sector, all children (with a few exceptions) attend school. Schools alone will not be able to carry out one more of the many missions' incumbent upon them, but a well-understood partnership approach will include a single, coherent discourse. Schools must bear witness to the deleterious effects of the use of screens on children's schooling. The tangible impact of these digital uses, which can be seen in what a child's first social environment is more often than not, can raise awareness and enlighten conviction.

These actions can easily be extended to teachers, families and pupils. From the post-COVID period onwards, the various initiatives have made it possible to systematically raise the issue of screens during discussions with parents and pupils. This made it possible to reach a wide audience, thanks to the availability of motivated teams who agreed to give up their time in the mornings and evenings. However, there are still a few limitations, particularly when it comes to mobilising parents with major psycho-social or cultural difficulties and language barriers. In addition, the teams, which are often mobile within the city, must be re-mobilised each year, which can be a heavy burden for novice teachers and organisers. A pilot school with a school psychologist and a team of teachers/principal could be appointed in each district, to disseminate information and propose a program of actions to be carried out with pupils and their families. Furthermore, the long-term beneficial effects can be measured using a methodology already in place.

In this paper, our therapeutic interventions are essentially qualitative and do not include a controlled evaluation of the effect. Despite these limitations, this work can make a modest contribution to inspiring concrete action in the field, using a systemic approach. The issues of socio-cognition and language at school depend not only on environmental conditions, but also on the practices of well-informed adults and their convictions. Educational habits that are strongly linked to the family, environmental and socio-economic context are also correlated with the school institution, often the child's first place of socialisation. To combat the worsening social inequalities in health, and contribute to the proper cognitive, psychosocial and relational development of children, it is vital that the issue of screens takes its rightful place in schools. The most vulnerable families need to be informed by means other than recommendations that are far removed from the reality of their daily lives, and often experienced as injunctive. Engaging the public in making significant changes to their behavior requires a systemic approach that considers all the parameters that influence human development. Informing and training educational teams, as well as all those involved in education and pediatrics authorities, remains essential.

Acknowledgments. We would like to thank all the participating children from the schools, in particular the *Pierre Sémard* school group, the families, the educational teams, Mr. Delmond, the National Education Inspector of Saint-Denis 3, the *Sémard* community center and one of its pillars Jérome (rip), the partners and the early childhood professionals.

Disclosure of Interests. The authors have no competing interests to declare that are relevant to the content of this article.

References

Adams, E.L., Marini, M.E., Stokes, J., Birch, L.L., Paul, I.M., Savage, J.S.: INSIGHT responsive parenting intervention reduces infant's screen time and television exposure. Int. J. Behav. Nutr. Phys. Activ. **15**(1), 24 (2018). https://doi.org/10.1186/s12966-018-0657-5

Anderson, D.R., Subrahmanyam, K.: Digital screen media and cognitive development. Pediatrics **140**(Supplement_2), S57–s61 (2017). https://doi.org/10.1542/peds.2016-1758C

André, A., Cochetel, O.: Effet du temps d'exposition aux écrans sur le graphisme des enfants de 5 à 6 ans: Une étude transversale conduite au cours de l'année scolaire 2019-2020 chez des enfants âgés de 5 à 6 ans, en grande section de maternelle, dans sept écoles d'Auvergne. La nouvelle revue - Éducation et société inclusives , N° **95**(3), 191–214 (2023). https://doi.org/10.3917/nresi.095.0191

Arcom. (2022, January). Observatoire de l'équipement audiovisuel des foyers de France métropolitaine - résultats des 1er et 2e trimestres 2021 pour la télévision(Enquête)

Arnold, I., Hattie, J.: Visible Learning: A Synthesis of Over 800 Meta-Analyses Relating to Achievement. Routledge, Abingdon (2011, August). 2008, 392 pp, ISBN 978-0-415-47618-8 (pbk). International Review of Education, 57(1-2), 219–221. https://doi.org/10.1007/s11159-011-9198-8

Bantuelle, M., Demeulemeester, R.: Comportements à risque et santé: agir en milieu scolaire, programmes et stratégies efficaces référentiel de bonnes pratiques. Éd. INPES, Saint-Denis (2008)

Barr, R.: Transfer of learning between 2D and 3D sources during infancy: informing theory and practice. Dev. Rev. **30**(2), 128–154 (2010). https://doi.org/10.1016/j.dr.2010.03.001

Barr, R.: Memory constraints on infant learning from picture books, television, and touchscreens. Child Dev. Perspect. **7**(4), 205–210 (2013). https://doi.org/10.1111/cdep.12041

Barr, R., Muentener, P., Garcia, A.: Age-related changes in deferred imitation from television by 6- to 18-month-olds. Dev. Sci. **10**(6), 910–921 (2007). https://doi.org/10.1111/j.1467-7687.2007.00641.x

Beau, A.: COVID-19 les enfants passent plus de temps devant les écrans depuis le début de la crise, selon un.webarchive (2022, February)

Bellon, J.-P., Gardette, B.: Harcèlement et cyberharcèlement à l'école: une souffrance scolaire en réseau (3e éd. actualisée ed.). Paris: ESF sciences humaines (2018)

Berger, L. M., Panico, L., Solaz, A.: The impact of centerbased childcare attendance on early child development: evidence from the French elfe cohort. Demography **58**(2), 419–450 (2021). https://doi.org/10.1215/00703370-8977274

Bergmann, C., et al.: Young children's screen time during the first COVID-19 lockdown in 12 countries. Sci. Rep. **12**(1), 2015 (2022). https://doi.org/10.1038/s41598-022-05840-5

Bernard, Y.-J., et al.: Temps D'écran De 2 À 5 Ans Et Demi Chez Les Enfants De La Cohorte Nationale Elfe/Screen Time Among Children Aged 2 To 5-and-a-half Years In The French Nationwide Cohort Elfe (2023)

Bhutani, P., Gupta, M., Bajaj, G., Deka, R.C., Satapathy, S.S., Ray, S.K.: Is the screen time duration affecting children's language development? - A scoping review. Clin. Epidemiol. Glob. Health **25**, 101457 (2024). https://doi.org/10.1016/j.cegh.2023.101457

Cao, H., et al.: Prevalence of attention-deficit/hyperactivity disorder symptoms and their associations with sleep schedules and sleep-related problems among preschoolers in mainland China, no. 8 (2018)

Carlson, S.A., Fulton, J.E., Lee, S.M., Foley, J.T., Heitzler, C., Huhman, M.: Influence of limit-setting and participation in physical activity on youth screen time. Pediatrics **126**(1), e89–e96 (2010). https://doi.org/10.1542/peds.2009-3374

Chardon, O., Guignon, N., De Saint Pol, T.: La santé des élèves de grande section de maternelle en 2013 - des inégalités sociales dès le plus jeune âge (2015). Dress

Chen, J.-Y., Strodl, E., Huang, L.-H., Chen, Y.-J., Yang, G.-Y., Chen, W.-Q.: Early electronic screen exposure and autistic-like behaviors among preschoolers: the mediating role of caregiver-child interaction, sleep duration and outdoor activities. Children **7**(11), 200 (2020). https://doi.org/10.3390/children7110200

Cheung, C.H.M., Bedford, R., Saez De Urabain, I.R., Karmiloff-Smith, A., Smith, T.J.: Daily touchscreen use in infants and toddlers is associated with reduced sleep and delayed sleep onset. Sci. Rep. **7**(1), 46104 (2017). https://doi.org/10.1038/srep46104

Chonchaiya, W., Pruksananonda, C.: Television viewing associates with delayed language development, no. 6 (2008). https://doi.org/10.1111/j.1651-2227.2008.00831.x

Chonchaiya, W., Sirachairat, C., Vijakkhana, N., Wilaisakditipakorn, T., Pruksananonda, C.: Elevated background tv exposure over time increases behavioural scores of 18-month-old toddlers. Acta Paediatr. **104**(10), 1039–1046 (2015). https://doi.org/10.1111/apa.13067

Christakis, D.A., Zimmerman, F.J., DiGiuseppe, D.L., McCarty, C.A.: Early television exposure and subsequent attentional problems in children. Pediatrics 8 (2004). https://doi.org/10.1542/peds.113.4.708

Cliff, D.P., Howard, S.J., Radesky, J.S., McNeill, J., Vella, S.A.: Early childhood media exposure and self-regulation: bidirectional longitudinal associations. Acad. Pediatr. **18**(7), 813–819 (2018). https://doi.org/10.1016/j.acap.2018.04.012

Cohen, D.: Vers un modèle développemental d'épigenèse probabiliste du trouble des conduites et des troubles externalisés de l'enfant et de l'adolescent. Neuropsychiatrie de l'Enfance et de l'Adolescence 56(4-5), 237–244 (2008). https://doi.org/10.1016/j.neurenf.2007.07.003

Collet, M., Gagnière, B., Rousseau, C., Chapron, A., Fiquet, L., Certain, C.: Case–control study found that primary language disorders were associated with screen exposure. Acta Paediatrica 108(6), 1103– 1109 (2019). https://doi.org/10.1111/apa.14639

Combes, C., Guerra, A., Létang, C., Roy, A.: Usages des écrans: état des lieux auprès d'une cohorte d'enfants français. Anae (2023)

Corcos, M., Bergman, B.: Le parent hyperconnecté à son portable risque de se déconnecter de son enfant. Journal le Monde (2020, January)

Crotty, J.E., Martin-Herz, S.P., Scharf, R.J.: Cognitive development. Pediatr. Rev. 44(2), 58–67 (2023). https://doi.org/10.1542/pir.2021-005069

Davies, W., Isles, A.R., Humby, T., Wilkinson, L.S.: What are imprinted genes doing in the brain? Epigenetics 2(4), 201–206 (2007). https://doi.org/10.4161/epi.2.4.5379

De Lepeleere, S., et al.: Parenting practices as a mediator in the association between family socio-economic status and screen-time in primary schoolchildren: a feel4diabetes study. Int. J. Environ. Res. Public Health 15(11), 2553 (2018). https://doi.org/10.3390/ijerph15112553

DeLoache, J.S., et al.: Do babies learn from baby media? Psychol. Sci. 21(11), 1570–1574 (2010). https://doi.org/10.1177/0956797610384145

Desmurget, M.: TV Lobotomie. La vérité scientifique sur les effets de la télévision. (J'ai lu ed.) (2011)

Desmurget, M.: La Fabrique du crétin digital (Points ed.) (2020)

Desmurget, M.: Temps d'écran : "On atteint des niveaux extravagants", dénonce le chercheur Michel Desmurget (2022, February)

Devouche, E., Morange-Majoux, F., Lebouc, M.: Écran et technoférence chez le bébé de 6 à 12 mois:. Contraste, N° 57(1), 261–285 (2023). https://doi.org/10.3917/cont.057.0261

Dieu-Osika, S.: Aborder la place des écrans en consultation- une démarche indispensable quel que soit l'âge de l'enfant. Médecine & enfance (2020, April)

Dieu-Osika, S., Bossière, M.-C., Osika, E.: Early media overexposure syndrome must be suspected in toddlers who display speech delay with autism-like symptoms. Glob. Pediatr. Health 7, 2333794x2092593 (2020). https://doi.org/10.1177/2333794x20925939

Diler, F., Başkale, H.: The influence of sleep patterns and screen time on the sleep needs of infants and toddlers: a cross-sectional study. J. Pediatr. Nurs. 67, e201–e207 (2022). https://doi.org/10.1016/j.pedn.2022.07.014

Domingues-Montanari, S.: Clinical and psychological effects of excessive screen time on children. J. Paediatr. Child Health 53(4), 333–338 (2017). https://doi.org/10.1111/jpc.13462

Doré, C.: L'estime de soi: analyse de concept: Recherche en soins infirmiers, No 129(2), 18–26 (2017). https://doi.org/10.3917/rsi.129.0018

Duflo, S.: Quand les écrans deviennent neurotoxiques. (Marabout ed.) (2018)

Edelson, L.R., Mathias, K.C., Fulgoni, V.L., Karagounis, L.G.: Screen-based sedentary behavior and associations with functional strength in 6–15 year-old children in the United States. BMC Public Health 16(1), 116 (2015). https://doi.org/10.1186/s12889-016-2791-9

Escobar-Chaves, S.L., Markham, C.M., Addy, R.C., Greisinger, A., Murray, N.G., Brehm, B.: The fun families study: intervention to reduce children's TV viewing. Obesity **18**(S1), S99–s101 (2010). https://doi.org/10.1038/oby.2009.438

Esseily, R., Guellai, B., Chopin, A., Somogyi, E.: L'écran est-il bon ou mauvais pour le jeune enfant ?: Une revue de la littérature sur la prévalence de l'écran et ses effets sur le développement cognitif précoce. Spirale, N° **83**(3), 28–40 (2017). https://doi.org/10.3917/spi.083.0028

Florin, A.: Le développement du langage (Dunod 2° ed.) (2020)

Fondard, F.: Étude sur les attitudes des parents. Union nationale des associations familiales, p. 12 (2014)

Gassama, M., Bernard, J., Dargent-Molina, P., Charles, M.-A.: Activités physiques et usage des écrans à l'âge de 2 ans chez les enfants de la cohorte Elfe, no. 24 (2018)

Gastaud, L.M., et al.: Screen time: implications for early childhood cognitive development. Early Hum. Dev. **183**, 105792 (2023). https://doi.org/10.1016/j.earlhumdev.2023.105792

Gauthe, M.: L'utilisation des écrans par les enfants de 0 à 6 ans dans le cadre familial. Étude quantitative à partir de 375 enfants (2020)

Gleizes, F., Legleye, S., Pla, A.: Ordinateur et accès à Internet : les inégalités d'équipement persistent selon le niveau de vie. INSEE focus (2021)

Guellai, B., Somogyi, E., Esseily, R., Chopin, A.: Effects of screen exposure on young children's cognitive development: a review. Front. Psychol. **13**, 923370 (2022). https://doi.org/10.3389/fpsyg.2022.923370

Gueron-Sela, N., Gordon-Hacker, A.: Longitudinal links between media use and focused attention through toddlerhood: a cumulative risk approach. Front. Psychol. **11**, 569222 (2020). https://doi.org/10.3389/fpsyg.2020.569222

Hancox, R.J., Milne, B.J., Poulton, R.: Association of television viewing during childhood with poor educational achievement. Arch. Pediatr. Adolesc. Med. **159** (2005)

Harlé, B., Desmurget, M.: Effets de l'exposition chronique aux écrans sur le développement cognitif de l'enfant. Archives de Pédiatrie **19**(7), 772–776 (2012). https://doi.org/10.1016/j.arcped.2012.04.003

Hcsp. (2020). Analyse des données scientifiques- effets de l'exposition des enfants et des jeunes aux écrans.pdf

Huang, P., et al.: Screen time, brain network development and socio-emotional competence in childhood: moderation of associations by parent–child reading. Psychol. Med. 1–12 (2024). https://doi.org/10.1017/s0033291724000084

Ipsos. (2022). Etude_open_unaf_ipsos_parents_enfants_numerique.pdf

Jusienė, R., Urbonas, V., Laurinaitytė, I., Rakickienė, L., Breidokienė, R., Kuzminskaitė, M., Praninskienė, R.: Screen use during meals among young children: exploration of associated variables. Medicina **55**(10), 688 (2019). https://doi.org/10.3390/medicina55100688

Kabali, H.K., et al.: Exposure and use of mobile media devices by young children. Pediatrics **136**(6), 1044–1050 (2015). https://doi.org/10.1542/peds.2015-2151

Khadilkar, V., et al.: Indian academy of pediatrics revised guidelines on evaluation, prevention and management of childhood obesity. Indian Pediatrics **60**(12), 1013–1031 (2023). https://doi.org/10.1007/s13312-023-3066-z

Konrad, A., Vucurevic, G., Musso, F., Winterer, G.: VBM–DTI correlates of verbal intelligence: a potential link to Broca's area. J. Cogn. Neurosci. **24**(4), 888–895 (2012)

Kotstyrka-Allkorne, K.K., Cooper, N.R., SImpson, A.: The relationship between television exposure and children's cognition and behaviour- A systematic review_. Elsevier , vol. 44, pp. 19–58 (2017). https://doi.org/10.1016/j.dr.2016.12.002

Kuhl, P. K., Tsao, F.-M., Liu, H.-M.: Foreign-language experience in infancy: Effects of short-term exposure and social interaction on phonetic learning. Proc. Natl. Acad. Sci. **100**(15), 9096– 9101 (2003). https://doi.org/10.1073/pnas.1532872100

Kurz, D., Braig, S., Genuneit, J., Rothenbacher, D.: Trajectories of child mental health, physical activity and screen-time during the COVID-19 pandemic considering different family situations: results from a longitudinal birth cohort. Child Adolescent Psychiatry Ment. Health **17**(1), 36 (2023). https://doi.org/10.1186/s13034-023-00581-3

Kushima, M., et al.: Association between screen time exposure in children at 1 year of age and autism spectrum disorder at 3 years of age: the Japan environment and children's study. JAMA Pediatr. **176**(4), 384 (2022). https://doi.org/10.1001/jamapediatrics.2021.5778

Landhuis, C.E., Poulton, R., Welch, D., Hancox, R.J.: Does Childhood Television Viewing Lead to Attention Problems in Adolescence? Results From a Prospective Longitudinal Study, no. 6 (2007)

Law, E.C., et al.: Associations between infant screen use, electroencephalography markers, and cognitive outcomes. JAMA Pediatr. **177**(3), 311 (2023). https://doi.org/10.1001/jamapediatrics.2022.5674

Lawrence, A., Choe, D.E.: Mobile media and young children's cognitive skills: a review. Acad. Pediatr. **21**(6), 996–1000 (2021). https://doi.org/10.1016/j.acap.2021.01.007

Lenhart, J., Richter, T.: Media exposure and preschoolers' socialcognitive development. Br. J. Dev. Psychol. **42**(2), 234–256 (2024). https://doi.org/10.1111/bjdp.12478

Li, C., Cheng, G., Sha, T., Cheng, W., Yan, Y.: The relationships between screen use and health indicators among infants, toddlers, and preschoolers: a meta-analysis and systematic review. Int. J. Environ. Res. Public Health **17**(19), 7324 (2020). https://doi.org/10.3390/ijerph17197324

Liliard, A.S., Peterson, J.: The Immediate Impact of Different Types of Television on Young Children's Executive Function (2011)

Lin, H.-P., et al.: Prolonged touch screen device usage is associated with emotional and behavioral problems, but not language delay, in toddlers. Infant Behav. Dev. **58**, 101424 (2020). https://doi.org/10.1016/j.infbeh.2020.101424

Lussier, F., Chevrier, E., Gascon, L.: Neuropsychologie de l'enfant et de l'adolescent. Troubles développementaux et de l'apprentissage. (DUNOD 3ème édition ed.) (2018)

Lutz, M.R., et al.: Television time, especially during meals, is associated with less healthy dietary practices in toddlers. Acad. Pediatr. **24**(5), 741–747 (2024). https://doi.org/10.1016/j.acap.2023.09.019

Madigan, S., Browne, D., Racine, N., Mori, C., Tough, S.: Association between screen time and children's performance on a developmental screening test. JAMA Pediatr. **173**(3), 244 (2019). https://doi.org/10.1001/jamapediatrics.2018.5056

Mares, M.-L., Pan, Z.: Effects of sesame street: a meta-analysis of children's learning in 15 countries. J. Appl. Dev. Psychol. **34**(3), 140–151 (2013). https://doi.org/10.1016/j.appdev.2013.01.001

Martinot, P., et al.: Exposure to screens and children's language development in the EDEN mother–child cohort. Sci. Rep. **11**(1), 11863 (2021). https://doi.org/10.1038/s41598-021-90867-3

Masi, L.: TDAH et usage addictif des jeux vidéo chez les enfants (2020)

McDaniel, B.T.: Technoference: Everyday Intrusions and Interruptions of Technology in Couple and Family Relationships (2015)

McDaniel, B.T., Radesky, J.S.: Technoference: parent distraction with technology and associations with child behavior problems. Child Dev. 1–10, 10 (2017). https://doi.org/10.1111/cdev.12822

McHarg, G., Ribner, A.D., Devine, R.T., Hughes, C.: The NewFAMS study team. Infant screen exposure links to toddlers' inhibition, but not other EF constructs: a propensity score study. Infancy **25**(2), 205–222 (2020). https://doi.org/10.1111/infa.12325

Molleri, N., Gomes Junior, S.C., Marano, D., Zin, A.: Survey of the adequacy of Brazilian children and adolescents to the 24-hour movement guidelines before and during the COVID-19 pandemic. Int. J. Environ. Res. Public Health **20**(9), 5737 (2023). https://doi.org/10.3390/ijerph20095737

Muppalla, S.K., Vuppalapati, S., Reddy Pulliahgaru, A., Sreenivasulu, H.: Effects of excessive screen time on child development: an updated review and strategies for management. Cureus (2023, June). https://doi.org/10.7759/cureus.40608

Nathanson, A.I., Aladé, F., Sharp, M.L., Rasmussen, E.E., Christy, K.: The relation between television exposure and executive function among preschoolers. Dev. Psychol. - Am. Psychol. Assoc. **50**(5), 1497–150 (2014). https://doi.org/10.1037/a0035714

Nathanson, A.I., Sharp, M.L., Aladé, F., Rasmussen, E.E., Christy, K.: The relation between television exposure and theory of mind among preschoolers. J. Commun. **63**(6), 1088–1108 (2013). https://doi.org/10.1111/jcom.12062

Neville, H.J., Stevens, C., Pakulak, E., Bell, T.A., Fanning, J., Klein, S., Isbell, E.: Family-based training program improves brain function, cognition, and behavior in lower socioeconomic status preschoolers. Proc. Natl. Acad. Sci. **110**(29), 12138–12143 (2013). https://doi.org/10.1073/pnas.1304437110

Ottemer, O.: Conséquences de l'exposition aux écrans sur le développement neurocognitif des enfants de moins de sept ans: revue de la littérature. HAL Open Sci. 115 (2021)

Pagani, L.S., Fitzpatrick, C., Barnett, T.A.: Early childhood television viewing and kindergarten entry readiness. Pediatr. Res. **74**(3), 350–355 (2013). https://doi.org/10.1038/pr.2013.105

Pagani, L.S., Fitzpatrick, C., Barnett, T.A., Dubow, E.: Prospective associations between early childhood television exposure and academic, psychosocial, and physical well-being by middle childhood. Arch. Pediatr. Adolesc. Med. **164**(5) (2010). https://doi.org/10.1001/archpediatrics.2010.50

Pempek, T.A., Kirkorian, H.L., Richards, J.E., Anderson, D.R., Lund, A.F., Stevens, M.: Video comprehensibility and attention in very young children. Dev. Psychol. **46**(5), 1283–1293 (2010). https://doi.org/10.1037/a0020614

Poirel, N.: Votre enfant devant les écrans: ne paniquez pas: Ce que disent vraiment les neurosciences. (1er ed.). De Boeck Sup (2020)

Poncet, L., et al.: Sociodemographic and behavioural factors of adherence to the no-screen guideline for toddlers among parents from the French nationwide Elfe birth cohort. Int. J. Behav. Nutr. Phys. Activity **19**(1), 104 (2022). https://doi.org/10.1186/s12966-022-01342-9

Ponti, M., et al.: Le temps d'écran et les jeunes enfants : promouvoir la santé et le développement dans un monde numérique. Paediatr. Child Health **22**(8), 469–477 (2017). https://doi.org/10.1093/pch/pxx121

Portugal, A.M., Bedford, R., Cheung, C.H.M., Mason, L., Smith, T.J.: Longitudinal touchscreen use across early development is associated with faster exogenous and reduced endogenous attention control. Sci. Rep. **11**(1), 2205 (2021). https://doi.org/10.1038/s41598-021-81775-7

Prieur, C.: Exposition des enfants de 0 à 3 ans aux écrans : résultats des cohortes de naissance sur les déterminants et les conséquences en termes de développement, no. 7 (2020)

Radesky, J.S., Schumacher, J., Zuckerman, B.: Mobile and interactive media use by young children: the good, the bad, and the unknown. Pediatrics **135**(1), 1–3 (2015). https://doi.org/10.1542/peds.2014-2251

Ramsden, S., et al.: Verbal and non-verbal intelligence changes in the teenage brain. Nature **479**(7371), 113–116 (2011). https://doi.org/10.1038/nature10514

Ribner, A.D., McHarg, G.G.: Why won't she sleep? Screen exposure and sleep patterns in young infants. Infant Behav. Dev. **57**, 101334 (2019). https://doi.org/10.1016/j.infbeh.2019.101334

Rideout, V.: Zero to eight children's media use in America (Tech. Rep.). San Francisco (2011)

Robinson, T.N., et al.: Screen media exposure and obesity in children and adolescents. Pediatrics **140**(Supplement_2), S97–s101 (2017)

Rocha, H.A.L., et al.: Screen time and early childhood development in Ceará, Brazil: a population-based study. BMC Public Health **21**(1), 2072 (2021)

Rodriguez-Ayllon, M., et al.: Associations of physical activity and screen time with white matter microstructure in children from the general population. NeuroImage **205**, 116258 (2020). https://doi.org/10.1016/j.neuroimage.2019.116258

Ruiz, J.-P., Lassault, J., Sprenger-Charolles, L., Richardson, U., Lyytinen, H., Ziegler, J.C.: GraphoGame: un outil numérique pour enfants en difficultés d'apprentissage de la lecture (2018)

Shankland, R.: Les troubles du comportement alimentaire - Prévention et accompagnement thérapeutique (2e ed.). Dunod (2020)

Skrede, T., Steene-Johannessen, J., Anderssen, S.A., Resaland, G.K., Ekelund, U.: The prospective association between objectively measured sedentary time, moderate-to-vigorous physical activity and cardiometabolic risk factors in youth: a systematic review and metaanalysis. Obesity Reviews **20**(1), 55–74 (2019). https://doi.org/10.1111/obr.12758

Small, G.W., et al.: Brain health consequences of digital technology use. Dialogues Clin. Neurosci. **22**(2), 179–187 (2020). https://doi.org/10.31887/DCNS.2020.22.2/gsmall

Sundqvist, A., Koch, F.-S., Birberg Thornberg, U., Barr, R., Heimann, M.: Growing up in a digital world – digital media and the association with the child's language development at two years of age. Front. Psychol. **12**, 569920 (2021). https://doi.org/10.3389/fpsyg.2021.569920

Swing, E.L., Gentile, D.A., Anderson, C.A., Walsh, D.A.: Television and video game exposure and the development of attention problems. Pediatrics **126**(2), 214–221 (2010)

Takai, Y., Sato, M., Tan, R., Hirai, T.: Development of stereoscopic acuity: longitudinal study using a computer-based randomdot stereo test. Jpn. J. Ophthalmol. **49**(1), 1–5 (2005). https://doi.org/10.1007/s10384-004-0141-4

Takeuchi, H., et al.: The impact of television viewing on brain structures: cross-sectional and longitudinal analyses. Cereb. Cortex **25**(5), 1188–1197 (2015). https://doi.org/10.1093/cercor/bht315

Thompson, S.F., Shimomaeda, L., Calhoun, R., Moini, N., Smith, M R., Lengua, L.J.: Maternal mental health and child adjustment problems in response to the COVID-19 pandemic in families experiencing economic disadvantage. Res. Child Adolesc. Psychopathol. **50**(6), 695–708 (2022). https://doi.org/10.1007/s10802-021-00888-9

Tomopoulos, S., Dreyer, B.P., Berkule, S., Fierman, A.H., Brockmeyer, C., Mendelsohn, A.L.: Infant media exposure and toddler development. Arch. Pediatr. Adolesc. Med. **164**(12), 7 (2010)

Tremblay, M.S., et al.: Systematic review of sedentary behaviour and health indicators in school-aged children and youth. Int. J. Behav. Nutr. Phys. Act. **8**(1), 98 (2011). https://doi.org/10.1186/1479-5868-8-98

Tremblay, T., Gagné, A., Bigras, N.: Family literacy activities mediate the effects of recreational screen time on children's language development. Psychol. Behav. Sci. Int. J. (2021). https://doi.org/10.19080/pbsij.2021.17.555951

Twenge, J.M., Martin, G.N., Campbell, W.K.: Decreases in psychological well-being among American adolescents after 2012 and links to screen time during the rise of smartphone technology. Emotion **18**(6), 765–780 (2018). https://doi.org/10.1037/emo0000403

Uzundağ, B.A., Altundal, M.N., Keşşafoğlu, D.: Screen media exposure in early childhood and its relation to children's selfregulation. Hum. Behav. Emerg. Technol. **2022**, 1–34 (2022). https://doi.org/10.1155/2022/4490166

Van Den Heuvel, M., Ma, J., Borkhoff, C.M., Koroshegyi, C., Dai, D.W.H., Parkin, P.C.: On behalf of the TARGet Kids! Collaboration. Mobile media device use is associated with expressive language delay in 18-month-old children. J. Dev. Behav. Pediatr. **40**(2), 99–104 (2019)

Veldman, S., Altenburg, T., Chinapaw, M., Gubbels, J.: Correlates of screen time in the early years (0–5 years): a systematic review. Prevent. Med. Rep. **33**, 102214 (2023)

Wang, J., et al.: Progression of myopia in school-aged children after COVID-19 home confinement. JAMA Ophthalmol. **139**(3), 293 (2021). https://doi.org/10.1001/jamaophthalmol.2020.6239

Wass, S.V., et al.: Parental neural responsivity to infants' visual attention: how mature brains influence immature brains during social interaction. PLOS Biol. **16**(12), e2006328 (2018). https://doi.org/10.1371/journal.pbio.2006328

Wright, J.C., et al.: The relations of early television viewing to school readiness and vocabulary of children from lowincome families: the early window project. Child Dev. **72**(5), 1347–1366 (2001). https://doi.org/10.1111/1467-8624.t01-1-00352

Yang, S., et al.: Associations of screen use with cognitive development in early childhood: the ELFE birth cohort. J. Child Psychol. Psychiatry (2023). jcpp.13887. https://doi.org/10.1111/jcpp.13887

Zaouche-Gaudron, C.: Enfants de la précarité (Érès ed.) (2017)

Zaouche Gaudron, C.: Les inégalités de l'enfance. Approche en conditions de vie (2021)

Zink, J., et al.: Longitudinal associations of screen time, physical activity, and sleep duration with body mass index in U.S. youth. Int. J. Behav. Nutr. Phys. Activity **21**(1), 35 (2024). https://doi.org/10.1186/s12966-024-01587-6

Zorman, M., Duyme, M., Kern, S., Pouget, G.: << Parler bambin >> un programme de prévention du développement précoce du langage. Anae(112-113), 8 (2011)

Can We Really Learn in the Metaverse?
A Discussion on Learning Through Immersive Technology

Raphaël Bompy[1,2](✉)

[1] Sorbonne Université, Paris, France
rbompy@ipc-paris.fr
[2] IPC – Facultés Libres de Philosophie et de Psychologie, Paris, France

Abstract. Will the metaverse enable us to learn, and learn better? Matthew Ball, one of the leading experts on the metaverse, defends this position, arguing that immersive technology will, for instance, enable students to understand the laws of physics or budding surgeons to acquire mechanisms by operating on a virtual patient. Meta's advertising campaign takes a similar approach, featuring students in the middle of dinosaurs or showing professionals practising technical movements. It is this perspective that we will attempt to question. What are the conditions under which there is learning? And what qualifies a good and successful learning? Learning cannot be reduced to the repetition of technical gestures or cognitive reflexes that could easily be reproduced by an artificial agent. Nor is it a stream of information received and memorised. Yet both elements are part of all learning. Learning in a modelled and simulated environment can undoubtedly modify our reflexes and practices, but learning means foremost acquiring knowledge through the exercise of intelligence. If we assume that all knowledge begins with the experience of the things to be known, then not everything can be learned in the metaverse, because the metaverse transforms the way we experience things by imposing a partially or totally simulated reality on our senses.

Keywords: Metaverse · Learning · Immersive Technology

1 Introduction

The recent development of so-called immersive technologies could revolutionise and even optimise the way we learn, as Matthew Ball[1] suggests (Ball, 2022). The new Apple Vision Pro[2] headset, released at the beginning of 2024, makes it possible to project indicators or educational interfaces in augmented reality (AR). Meta offers a

---

[1] Matthew Ball is a Canadian investor and theorist, pioneer in the field of the metaverse: the exhaustive definition that he proposes in his work has been taken up by most of the major expert reports on the subject (from large international companies like Meta to public authorities like certain governments or the European Union).

[2] https://www.apple.com/apple-vision-pro/ [accessed 07/07/2024].

© The Author(s), under exclusive license to Springer Nature Switzerland AG 2025
J. Baratgin et al. (Eds.): HAR 2024, LNCS 15504, pp. 366–381, 2025.
https://doi.org/10.1007/978-3-031-84595-6_22

virtual reality (VR) simulator for acquiring skills based on realistic customer interaction scenarios[3], and the company LuminousXR gives workers the opportunity to train on virtual oil machines in perfectly safe conditions[4].

The aim of this contribution is to examine the possibility of learning by means of immersive technology of the metaverse type, in other words to question whether it is possible to learn in the metaverse[5]. Although these technologies are not yet widely used, particularly in education, the issues involved and their potential uses make them sufficiently relevant to be considered authoritative. The exploratory mission report on metaverses submitted to the French government in October 2022 (Basdevant et al., 2022) is a milestone in this direction, urging political players to take up this subject and encourage students at all levels to take an interest in these issues (Basdevant et al., 2022, 16).

Immersive technologies allow us to experience things from a first-person perspective. Since practice makes perfect, isn't it through an immersive experience very close to reality that we can learn better? In other words, to optimise learning, isn't it better to move from the status of a passive, external observer – as when we interact with screens or books – to that of an active[6] participant in an immersive universe? We will try to uphold that accepting this hypothesis leads necessarily to a framework of thought that makes learning partly impossible. If learning is the acquisition of knowledge or skills through the exercise of intelligence[7], then the metaverse proposes only an improvement in learning *conditions*, not an enhanced learning in itself.

To this end, we shall first elucidate the theses put forward by Matthew Ball. From his perspective, learning something is always an act external to the subject, and is no longer a

---

[3] https://meta-skills.io/ [accessed 07/07/2024].

[4] https://www.luminousxr.com/vr-oil-gas-energy-landing-page/ [accessed 07/07/2024].

[5] The term metaverse, contraction of the Greek μετά "in the middle (of), with, after, beyond" and "universe", makes its appearance in the science fiction novel *Snow Crash*, written by Neal Stephenson in 1992 to name a fictional universe entirely computer-generated. Today, such a metaverse does not exist, but there is a plurality of metaverses allowing all kinds of applications. Three types of technologies actually provide access to what experts call metaverse: virtual reality (VR), augmented reality (AR), and mixed reality (MR). The World Economic Forum defines VR as a fully digital world, while AR technology is an overlay of a digital layer on the physical world, with limited interaction – for example information or indications, and MR is a mixture of reality and digital – in particular to manipulate or move digital 3D objects in the physical world. All three technologies are commonly called "extended" reality (XR). The term metaverse is therefore broader than Stephenson imagined, because it today involves many means of access: not only VR, but also all other XR technologies, whether used with 3D or not – one can in principle enter the metaverse simply via a smartphone or other immersive 2D technologies (Ball, 2022, 33–35 and 306). For these reasons, the title of this article specifies that the notion of learning will be discussed as related to immersive technology, that is to say any existing technology which contributes to the metaverse as envisaged by Matthew Ball.

[6] In the words used by Louis B. Rosenberg, CEO & Chief Scientist of Anonymous AI.

[7] The Oxford English Dictionary provides us with this definition of "learning": "To acquire knowledge of (a subject) or skill in (an art, etc.) as a result of study, experience, or teaching". This "acquisition" supposes an understanding of the content, which is precisely the etymological definition of intelligence: https://www.etymonline.com/word/intelligence [accessed 07/07/2024].

matter of knowledge internalised by a subject. Thus, is it still possible to know things, or has the age of new technologies relinquished the subject's ability to access knowledge on his own? What lies behind these questions is one of the most fundamental philosophical issues, namely the possibility of attaining some form of truth and transmitting it – whether it be learned knowledge, an acquired skill or disposition, or retained information.

2   Learning in the Metaverse Era

After proposing a definition of the metaverse[8] and describing how it can be implemented, Matthew Ball presents the key sectors concerned by the metaverse and whose transformation will revolutionise the way we live; education comes first (Ball, 2022, 250–254), being of critical importance in its contribution to society and the economy. The Economist Impact 2023 report[9] supports this claim, identifying four high-potential application areas: learning and education, work and collaboration, social interaction and entertainment.

Ball starts from the observation that while labour productivity has increased significantly in most areas over the last 50 years, it has stagnated or increased very little in the field of education, resulting in wages being dragged down. He puts forward two reasons for this: on the one hand, neither the time spent per unit of education[10] has been reduced, nor can the number of students be increased indefinitely[11]. On the other hand, the nature of the educational world leads to very high costs[12], which have increased over the years due to the introduction of new computer or technical equipment – projectors, screens, etc. It is therefore clear that the author's concern about education is foremost economic.

This preliminary statement helps us introduce the central thesis put forward by Ball: traditional means[13] of education are no guarantee of good learning[14] insofar as they do not provide a conducive environment for students, i.e. they lack the layout and immersion necessary to cultivate engagement through a sense of 'presence'. Apart from the clear

---

[8] "A massively scaled and interoperable network of real-time rendered 3D virtual worlds and environments. These can be experienced synchronously and persistently by an effectively unlimited number of users with an individual sense of presence, and with continuity of data, such as identity, history, entitlements, objects, communications and payments": https://www.matthewball. co/all/forwardtothemetaverseprimer [accessed 07/07/2024].

[9] Economist Impact: Towards a successful metaverse: the case for measuring enabling factors (2023). https://impact.economist.com/perspectives/technology-innovation/towards-a-suc cessful-metaverse [accessed 07/07/2024]. This research and development report supported by the company Meta aims to establish the basis for setting up a tool for countries wishing to participate in the construction of the metaverse.

[10] We can illustrate it in the following way: to teach the same module x (for example an introductory course in philosophy of science), the time devoted in number of hours in the 1970s and today (2024) is identical, providing that all other conditions are equal.

[11] If for the same teaching unit we increase the number of students in a class, there is a limit beyond which the quality of what is taught is drastically reduced.

[12] In particular those of the premises or the maintenance of the premises, school materials, etc.

[13] Computer screens included.

[14] In Matthew Ball's sense, "learning well" therefore designates both a qualitative dimension (learning quickly) and a quantitative dimension (learning a lot or to many).

benefits of being physically present for students[15], there are a number of activities that cannot be carried out online via a screen, unlike in a face-to-face situation in a classroom. Matthew Ball therefore suggests that the gap between virtual simulations and physical situations[16] be reduced as much as possible in order to enhance learning.

In his book, Ball points out that students who learn in the metaverse will no longer simply be spectators of outer knowledge, but actors in knowledge that they will construct or experience themselves. In other words, students will be required to build things as they have been thought of in the past, in other words to retrace the steps taken by those who came before us: "build Rome in a semester", or to experiment for themselves: "visit the inside of a volcano and be ejected with the lava" and "test gravity on the Moon"[17] (Ball, 2022, 253). Aware of the difference between the physical universe and the simulation offered by immersive technology, his ambition remains to reduce as far as possible the gap between the universe and the meta-universe.

This imitation of nature by art raises the following question: in order to learn, immersive technologies give us access to the thing to be learned in a way that is as faithful as possible to reality. In this way, we can see the similarity between learning in the metaverse and learning in the physical world, insofar as learning possesses the same functional aspect in both cases. However, going back to the origin of the word "to learn"[18] tells us that it bears at least two aspects: that of acquiring knowledge or information, and that of being able to teach or communicate it. Learning multiplication tables, for instance, meets this definition: a child first acquires knowledge, in this case the correspondence between multiplied numbers and their results, and then can recite the tables to prove that he has learned. If the child can learn the tables, it is also important to note that he or she did not have this knowledge beforehand, otherwise we would rather say that he or she is revising the tables, not learning them.

Here we have a twofold problem with the definition: to what extent am I able to learn what a thing is if I am myself its designer (in the metaverse)? What do we grasp that does not come from us or that we do not already possess? And on the other hand, do we still have the possibility of transmitting something if the only way to learn is through first-person experience? Is it not in everyone's interest then to simply "experience" things without relying on the experience of others – with the risk of being isolated from reality? In *Les liens artificiels* (Devers, 2022), the main character Julien gradually distances himself from reality as he immerses himself in the virtual world[19], with a shocking ending that we will not divulge: the fact that he experiences everything virtually and

---

[15] The benefit of having students physically present is thus stated as obvious by the author: research shows remote learning can be in some cases less engaging and more challenging than in-person learning (Phoutopoulos et al., 2022).

[16] In particular through the use of virtual or augmented reality tools, haptic technologies, etc.

[17] In order to observe that two objects fall at the same speed on the Moon.

[18] *Apprehendere* in its latin etymology, which means to grasp, to take (dictionary Larousse: https://www.larousse.fr/dictionnaires/francais/apprendre/4746 [accessed 07/07/2024]).

[19] "In the days that followed, Julien never took off his suit. Barely leaving his home, he immersed himself in Vangel's body, learned to discover life with his eyes, to feel at home at the top of the Eiffel Tower. Little by little, he forgot that his name was Julien Libérat, that he was a former pianist, that he lived in Rungis and other biographical data: nothing existed except his new identity, the best, that of his anti-Me" (Devers, 2022, 255–256) [personal translation].

by himself, gradually cutting himself off from his physical life and his relationships, teaches him nothing but rather leads him to unlearn about his own existence. Doesn't the metaverse isolate us from reality, thus preventing us from learning? Don't we learn when we receive a feedback from experience, that is to say from reality? In the same vein, this saying is granted to Nelson Mandela: "I never lose. I either win or learn". While the metaverse allows us to win on many levels, it is not certain that it allows us to learn from our mistakes when it becomes a bulwark against reality.

We need to rigorously define what learning means from the perspective of the development of immersive technologies and the metaverse, and the acceleration of technological progress. From this point of view, there seems to be an equivalence between learning and assimilating a certain mechanism: one of the experts in the Harvard Business Review (Purdy, 2022) concentrates under this notion everything that corresponds to training or acquiring skills, with the same ambition as Matthew Ball to reduce learning time or increase the number of learners. We can see that learning becomes equivalent to a kind of problem-solving that requires certain methodological steps to be followed, what Purdy calls "how to guides", rather like nudges (Thaler, Sunstein, 2008): "[Virtual-world training provides greater opportunity for] learning by doing, and overall higher engagement through immersion in games and problem-solving through "quest-based" methods. Virtual-world learning can also make use of virtual agents, AI-powered bots that can assist learners when they get stuck, provide nudges, and set scaled challenges" (Purdy, 2022).

Learning with the help of immersive technologies thus comes down to following a process similar to that of a video game, in other words immersing oneself in scripted environments where the achievement of a quest is stimulated by a system of nudges, i.e. rewards and reprimands (with a time limit, inciting colours or sounds, etc.). Mathieu Triclot also points out that "for the player, it's no longer a question of interpreting a world, but of transforming it"[20] (Triclot, 2011, 113). Conceiving "learning" in the metaverse therefore implies identifying the acquisition of knowledge or information with the execution of a series of operations in an actionable universe[21].

The principle of learning in the metaverse, if we are to believe the elements put forward by Purdy, lies in the fact of experiencing a world in which the subject can act. In a sense, it is beneficial to consider learning this way, because the playful aspect and mechanics of the game can reinforce certain cognitive dispositions (Rahman et al., 2024) or facilitate involvement in a given task or environment (Reeves, Read, 2009). It should be noted, however, that the Economist Impact report[22] does not mention an improvement in learning as such, but rather in the sensory experience of the learner, in other words the *learning conditions*: "Interactive technologies, such as brain technology interaction, VR and AR promise sensory stimulation that could facilitate and enhance

---

[20] [personal translation].

[21] Even if the objective of a virtual world is not necessarily the same as in a game, the principle remains similar (Ball, 2022, 30).

[22] Economist Impact: Towards a successful metaverse: the case for measuring enabling factors (2023). https://impact.economist.com/perspectives/technology-innovation/towards-a-successful-metaverse [accessed 07/07/2024].

the learning experience for this population"[23]. While there may be a correlation between better *learning conditions* and better *learning*, a causal link is by no means necessary.

While the promoters of immersive technologies are up in arms against what they denounce as scepticism about these technologies[24], the doubts expressed since the emergence of the World Wide Web in the 90s are not without interest in terms of the digital innovations we have to deal with in 2024. Clifford Stoll's famous article[25] denounced then the impossibility of perceiving what is *trustworthy*, literally what is worth trusting. The stated aim of immersive technologies is to reduce the difference between physical reality and virtual worlds, so that the subject *trusts* what is projected before his or her eyes (Ball, 2022, 30)[26] and behaves towards virtual objects *in the same way* as towards a physical object. To agree with Matthew Ball that we can learn anything by means of immersive reality is therefore necessarily to accept that we can no longer distinguish between what belongs to the physical universe and what belongs to the metaverse. This inability to distinguish between the physical and the digital appears to be a questionable presupposition that impedes the very possibility of learning, the main objective of which is always to develop a critical thinking (Walter, 2024)[27].

3 An Experiential Philosophy

Matthew Ball's conception of learning considers the learner and the act of teaching from a purely experiential angle. Ball's three seemingly innocuous examples (Ball, 2022, 253) of how the metaverse can be applied to education are very telling:

- "Constructing Rome in a semester": a 3D taste of life in Rome as it was under the Roman Empire aims to bring students closer to historical reality. They will also be able to remake the activities of that time (i.e. building aqueducts) to better understand the internal logic behind it.
- "Recreating the conditions of gravity": like a scientist in a laboratory, a student will be able to reconstruct the behaviour of objects in different environments to which they do not have easy access (on the Moon, for example) or in such a way that these experiments can be repeated by others.
- "Being ejected with lava from a volcano": instead of recreating an unconvincing chemical reaction with vinegar in class, students will be able to follow the path that lava runs through during a volcanic eruption, sensations guaranteed.

From these opportunities that the metaverse will bring to the field of education, several characteristics of a good learning can be extracted. We formulate them as follows:

1. Live a thrilling experience

---

[23] *Ibid.*, p. 17.

[24] *Ibid.*, p. 7.

[25] Stoll, C.: "Why the Web won't be Nirvana". Newsweek, February, 25th, 1995. https://www.newsweek.com/clifford-stoll-why-web-wont-be-nirvana-185306 [accessed 07/07/2024].

[26] Regarding the desire to reduce the gap between the physical world and virtual simulation.

[27] The article in question warns in particular of the risk of hallucinations caused by AI, which can generate untruths while appearing authoritative. As a solution, it proposes strengthening critical thinking capable of distinguishing what is "real" from what is proposed by AI.

2. Test for yourself under realistic conditions
3. Reveal the underlying arrangement of things
4. Engage in a favourable environment
5. Expand the perimeter of what is possible in this environment

These theses underpin Matthew Ball's discourse and perfectly reflect the development logic of immersive technologies. We will now review and discuss them.

Live a Thrilling Experience. Just like at a funfair or at a Disneyland attraction, it is assumed that students derive added value from the emotions and sensations they experience during the learning process. Matthew Ball transposes this added value to learning as a whole, as if living a thrilling experience necessarily implied better learning. The feeling of "presence" mentioned above is reinforced by sensory stimuli that facilitate motivation and involvement (Parong, Mayer, 2021).

Therefore, learning is partly an emotional and sensitive experience. In fact, most metaverse applications contain a playful dimension through the immersive digital environment they offer[28]. In the case of learning, however, the playful aspect does not always seem to be appropriate. In that case learning would be reduced to a simple alternation of sensitive challenges and rewards; teaching would simply be a matter of stimulating or supervising the sensory experience. However, studies on reinforcement learning (Tolman, Honzik, 1930) have long shown that learning can be much faster in the case of delayed gratification than in the case of immediate gratification.

While immersive reality may in some cases have a positive influence on the quality of learning, there is no necessity for this (Piccione et al., 2019): students may even in some cases turn away from the utilitarian value of what is being learned – the content, in favour of the entertainment value – i.e. an aspect of entertainment that is only incidental (van der Heijden, 2004) to the learning in question. The feeling of presence can even be negatively correlated with learning (Makransky et al., 2019), insofar as the overload of sensory stimuli has the effect of atrophying cognitive capacity through an excess of secondary information (Pimentel et al., 2022, 6)[29].

Test for Yourself Under Realistic Conditions. If sensory experience is central to immersive reality, it is because the metaverse puts us in near-real conditions that allow us to experience things in the first person. Dropping objects on the Moon is something I can realise by recreating a similar situation in virtual reality, rather than watching a video of the single event that initially enabled us to understand its behaviour.

Two aspects are worth mentioning. Firstly, this philosophy urges observation, in other words to base the acquisition of knowledge on experience. This axiom seems to have

---

[28] The pleasant aspect of immersive technologies helps to overcome the arduousness of certain tasks, for example household chores: https://80.lv/articles/here-s-how-vr-can-make-house-chores-fun-engaging/ [accessed 07/07/2024].

[29] On a personal note of no scientific value here, the immersive exhibitions Eternelle Notre-Dame (https://www.eternellenotredame.com/en/ [accessed 07/07/2024]) or L'Horizon de Kheops (https://eclipso-entertainment.com/en/paris/horizon-de-kheops/ [accessed 07/07/2024]) put wonder at immersive technology ahead of educational content. In each case, the members of the group left with the memory of a good sensory experience, but without having acquired any new knowledge or retained much information.

been fairly consensual in philosophy for a long time, notably with Kant (1975, 31)[30], even though he did not identify experience with its purely sensory dimension. But it is clear that there can also be experience, in its emotional or sensitive meaning, without any real observation (Hacking, 1983, 167). Secondly, students are not simply receptacles of knowledge, but real actors in their own learning and at the centre of the teaching process: they are the explorers, and the teacher becomes their guide. We can draw a parallel here with connectivism in psychology: "[Educators] need to create conducive environments that promote networking, guiding learners to form "successful" connections [...] In essence, the world is now a classroom, and every connection, a potential lesson" (Kurt, 2023). Whatever the case, we can agree with connectivists that learning requires the acquisition of knowledge, in other words, there's nothing immediate about it: learning to play the piano does indeed require the painful practice of scales for a time, rather like a muscle that you strengthen: "When we learn, when we know, what we're doing is more like growing and developing ourselves the way we might build a muscle" (Downes, 2022, 63).

All learning implies a certain kind of readiness, which can be reinforced by regular and precise practice. It is by no means instantaneous. This is undoubtedly the strongest argument in favour of learning in the metaverse: recreating a workspace similar to a classroom, just as secure and stimulating, as long as we remember that the classroom is not quite equivalent to the real world.

Reveal the Underlying Arrangement of Things. Learning requires an understanding of the underlying mechanisms of things, says Matthew Ball, in other words an understanding of how things are the way they are, and how they came to be. In short, learning is successful when the student has followed, memorised and internalised the stages of the learning process. After all, learning something means knowing and understanding the steps involved in making it. For Ball, I can only understand reality if I can identify, grasp and reproduce the underlying patterns, in other words what the mechanics of things are.

Fig. 1. Illustration taken from the article by S. Downes (Downes, 2022, 71).

---

[30] "That all knowledge begins with experience, this raises no doubt" [personal translation].

The image above (Fig. 1) has been taken from Stephen Downes' article. It is entitled "Emergence" and is accompanied by the following quote: "The image is just a series of circles. And yet, when people look at this image, they see a woman wearing a hat. This common phenomenon makes it possible to watch a sporting event on a dot matrix screen on television. But the woman is not actually in the picture, nor is the sporting event in the dot matrix screen. It is only because of the way they are organised that they are perceived as they are" (Downes, 2022, 71). This example is fundamental to our understanding of immersive realities and of contemporary technological epistemology in general.

In 2005, Ray Kurzweil[31] was already depicting himself as a patternist (Kurzweil, 2005, 5 and 371): according to him, a global model is first understood in terms of the fundamental units that make it up. Fractal design helps to visualise this assertion: in the figure below (Fig. 2), there is a single pattern that is repeated potentially ad infinitum. While the total structure may appear complex at first glance, the fundamental pattern is very simple. By grasping the fundamental pattern that serves as a principle, it is possible to understand the entire structure.

Fig. 2. Example of a fractal design

We find the same type of perspective with Downes: "In connectivism as precisely understood, by "knowledge" we are not alluding to cognitive or representational structures, but rather, specifically and only to patterns of connectivity between entities in such physical structures" (Downes, 2022, 65). From this perspective that we may label as Kurzweilian, knowledge is the grasp of underlying patterns. What we know indeed always depends to a greater or lesser extent on our previous experience; knowledge is cumulative. For example, learning to solve complicated equations necessarily involves mastering multiplication tables and other operating rules.

Thus, it is by learning simple elements that we can grasp more complex wholes. However, the patternist approach undoubtedly raises a difficulty: does the intelligibility of

---

[31] American author, engineer and futurologist, director of engineering at Google, teacher at the prestigious MIT, "pope" of transhumanism, and member of the Army Science Advisory Board (committee of experts responsible for scientific and technological advice for the United States).

things emerge only from an understanding of the elements of which they are composed? Is it possible to know things beyond their simple mechanical aspect? If the answer is no, then knowing what a chocolate cake is, for example, would simply mean carrying out each stage of the chocolate cake-making process oneself and deriving a representation of it; learning would simply mean following a recipe and memorising it. But isn't it possible to say what a chocolate cake is without mentioning the steps involved in making it? For instance, by mentioning its flavour, its smell, the person who made it, or even its usefulness and purpose? Because understanding the *how* is not necessary to our grasp of a certain causality (Cheng, Buehner, 2012, 212), we can define what a thing is in several ways without reducing its definition to its material causality.

If we go even further, the skill I acquire in correctly executing a programme – following a certain process in immersive reality that simulates a real situation – is not necessarily a sign of successful learning, but rather of functional mastery of a certain combination of "reflex" gestures[32]. One could happen to no longer be in certain conditions, in a fully equipped kitchen with certain ingredients, etc., therefore being unable to make a cake. In that case, one will have to use language for purposes other than simply listing the steps in the recipe. One might be able to understand what a chocolate cake is without having gone through the steps oneself.

Engage in a Favourable Environment. Given that learning conditions are paramount, the environment will play a key role in immersive technology, since it is precisely its immersive nature that makes it innovative. From the learner's perspective, just as in a video game, immersive reality seeks a form of total adhesion of the subject to the content to which he or she is subjected. Mathieu Triclot refers to the "shift of the place of the game from the physical space of the table to the virtual space of the screen" (see footnote 20) (Triclot, 2011, 45–46): in order to engage a user of a digital product, we need to look at it from the perspective of the attention market, in other words, consider activities from a purely informational angle (Reeves, Read, 2009). However, even if immersion can generate greater enjoyment and develop pleasure in the educational content (Makransky et al., 2019), the risk is that the learner's motivation be artificialised: he or she may lose all interest once he or she has mastered the workings of the programme, just as a player tires of a game once he or she has perfectly mastered the winning mechanics (Koster, 2013).

That said, we understand the importance of creating the conditions of engagement for people learning in the metaverse: again from a connectivist perspective, learning takes place not at the level of an individual subject, but at the level of a network: "For me, connectivism is the thesis that knowledge is distributed across a network of connections, and therefore that learning consists of the ability to construct and traverse those networks […] When a person learns, or when something learns, a connection is physically created between two nodes or two entities in a network […] I say a connection exists between two entities when a change of state in one entity can cause or result in a change of state in the second entity" (Downes, 2022, 59–60). The possibility of learning at the individual

---

[32] The Kurzweilian presupposition consists of identifying qualitative improvement (better learning) with quantitative increase (the quantity of information circulating in a system). A critique of this presupposition would require more detailed argumentation to show that the quantity of information is not equivalent to better learning.

level seems to be rejected here. It is rather in the relationship between one entity and another that knowledge is increased and learning reinforced.

The metaverse fits perfectly into this logic, since it is defined fundamentally as being a network[33], or emerging from a network: "It's most likely the Metaverse emerges from a network of different platforms, bodies, and technologies working together (however reluctantly) and embracing interoperability" (Ball, 2020). If engagement, taken from the world of games, is a fundamental characteristic of immersive realities, it is because collective performance is perceived to be far superior to that of the individual: a highly intelligent person will never be as intelligent as society as a whole. Does this mean that we are incapable of knowing individually? Or that it is never better to know as individuals than to know collectively? This is a question that merits further elaboration. But at this stage we can elaborate the following conclusion: if knowledge is only possible at the level of a network, then learning is neither desirable nor necessary, nor even possible for the individual alone. The human being is no longer capable of knowledge as an individual, but only as a member of a larger whole whose elements are interrelated.

Expand the Perimeter of What is Possible in This Environment. Once plunged in an immersive environment, one needs to push back the boundaries of what is possible in order to avoid boredom. The important element is no longer the means used to achieve a result, but rather the scale of the result itself. As we mentioned at the beginning of this article, Ball does not seem to be any more interested in the quality of education, the virtue of teachers or the follow-up of students than in the number of students per class or the speed with which a course is being assimilated. For him, the quality of education is measured by these criteria. In the same vein, connectivism is less concerned with the way in which an individual acts to achieve a goal than with the speed with which he or she achieves a result: "the knowledge isn't the map. The knowledge consists of whatever neural network structures are developed by using the map and exploring the city, or perhaps conducting activities in the city. 'Knowing', in this sense, isn't remembering the map; it is successful navigation through the city" (Downes, 2022, 80).

With this main objective of achieving results, we can better understand Driscoll's definition of learning: "a persisting change in performance or performance potential that results from experience and interaction with the world" (Driscoll, 2000, 1). But if learning comes down to achieving a result – a certain performance – how and according to what indicators should we define this result? What benchmark should we rely on to say that learning has been successful? If a student does not achieve a sufficient or expected result, but makes significant progress, has he learnt nothing or learnt badly? Purdy shows us that the metaverse can enable us to learn more quickly by compressing the time needed to develop and acquire new skills (Purdy, 2022). From a pedagogical point of view, what should we do with those who do not learn as quickly or who find themselves in difficulty when faced with a frantic pace of learning? Is performance the goal of education? It is to be feared that the purely economic measurement of learning (more students or faster learning) will not provide answers to these questions.

---

[33] See definition in Footnote 8.

4   Limits of This Philosophy

The emergence of immersive technologies profoundly transforms the notion of learning. This framework of thought, which externalizes learning, lies at the crossroads between connectivist psychology and the information sciences, of which Kurzweil is one of the most emblematic contemporary authors. But learning is defined first and foremost as an acquisition of knowledge[34] – which supposes a certain exercise of intelligence that cannot be reduced to accumulating information, since memory is not the only measure of intelligence (Coane et al. 2023). Following our analysis of the presuppositions underlying Matthew Ball's thesis, we can identify a number of resolutive elements that summarise the limits of his philosophy, and which will enable us to gather further details on what a good learning is.

A Simulation or Virtual Object Cannot Be Identified with the Real Thing. In other words, the image of the woman that appears on my screen is not the person as such but her digital representation[35] which I imagine to be that person. In this sense, the image can be a certain representation that emanates from a set of simple elements. But the pitfall would undoubtedly be to transpose this to the physical world and identify the thing with its representation, or a representation with a real thing.

The prudent thing to do would be to raise questions: an image created by generative artificial intelligence does not necessarily mean that what is represented corresponds to physical reality[36]. In other words, one should not believe everything one see, especially in the age of fake news. Despite the emotions we feel or the impressions we form, critical intelligence always requires us to take a step back from images and our first impressions.

Not Everything Can Be Reduced to Information. The information science paradigm assumes that meaning, words and definitions emanate from an agglomeration of simple patterns. If we accept this hypothesis, which is by the way open for debate, then there is effectively no difference between a simulation and a physical experiment, to the extent that we can sometimes no longer distinguish them appreciably. If we accept that there is equivalence between the physical and virtual universes, it is because we look at them from the point of view of the subject's experience of them, which has become the measure of reality and of our knowledge of things. We are thus forced to accept Chalmers' thesis: "The world we're living in could be a virtual world. I'm not saying it is. But it's a possibility we can't rule out" (Chalmers, 2022, Introduction, xvii). But that's not what Matthew Ball himself supports, since he concedes that there is a difference between the physical and digital worlds.

Nevertheless, it would be unwise to believe that we can gain no understanding of reality other than by looking at it through the prism of the fundamental elements that make it up – in this case, their informational aspect. Bergson argued that the structure of the elements is more important than knowledge of the elements themselves – in our

---

[34] See definition in Footnote 7.

[35] René Magritte, with his famous painting representing a pipe and bearing the words "Ceci n'est pas une pipe", was already keen to differentiate the image from the object represented.

[36] https://news.sky.com/story/fake-ai-images-keep-going-viral-here-are-eight-that-have-caught-people-out-13028547 [accessed 07/07/2024].

case patterns: "Do you know a building when all its stones have been shown to you in advance? And yet there are only stones in the building. This is because the art lies in the arrangement, and the important thing is not to know the stone, but the place it will occupy" (see footnote 20) (Bergson, 2011, 41). The importance of a form of causality that is not patternist but that accounts for the formal dimension of things – the principle of their arrangement or structure – and thus guarantees their intelligibility, is one of the major challenges for the philosophy of knowledge.

One Can Have An Understanding of Things Without Having Experienced Them. A first element of understanding is this. If I have to experiment and test on my own in order to acquire knowledge, then very little knowledge is attainable for me. We can even put forward the hypothesis that most of what we know, starting with the rotation of the Earth or the history of the dinosaurs, is simply an adherence on our part to theses that cannot be verified empirically.

But it seems that, even if learning something requires first and foremost an experience of it, not every understanding of things necessarily requires a self-experimentation: in this respect, we can make better decisions by drawing on the experience of others (Läpple, Barham, 2019). Moreover, there is no need to kill someone to learn that it is an evil act, but it is possible to understand it relying both on ethical principles that have been transmitted through generations, and on a philosophical consideration of the nature of a human person. Likewise, it would not be possible to give advice or to teach, and we would be confining ourselves to pure, incommunicable experience. If we take advice, or deliberate, is this not a sign that the experience of others can be transmitted in a different way than through a sensory vector?

Not all learning is conditioning. Conditioning by nudges encourages the development of reflex mechanisms that can facilitate a decision or an action. However, not all learning is conditioning, or at least not all of the learning process is conditioning, in the sense that we are always learning *something*. This means that learning is not simply a process that a person goes through in multiple iterations in order to acquire the appropriate configuration. There is a *content* to learning: *what is learned*. It seems that this point is totally neglected by connectivists, for example Siemens: "The pipe is more important than the content within the pipe" (Siemens, 2005, 7). or when they talk about students using MOOCs[37]: "We did not care what they learned. There was no content we wanted them to learn (in the sense of acquiring knowledge and representing it accurately) and we even said to them, "You determine what counts as success for you; we don't have a body of content here. What's important to us is that you can function in a network and learn things from the network, and that's connectivism"" (Downes, 2022, 81–82).

Of course, learning conditions can always be improved. That in no way means that learning will be better as a result. Let us take an analogy with GPS: having a GPS in my car doesn't make me a better driver, even if it makes my driving easier. One will always prefer to get into the car of a good driver with limited driving conditions, rather than that of a bad driver with optimal and comfortable driving conditions[38]. The same applies to the metaverse: while immersive technologies facilitate learning conditions

---

[37] https://www.downes.ca/post/57911 [accessed 07/07/2024].

[38] Unless the "driving conditions" completely replace the driver: in the case of a perfectly autonomous car, this problem supposedly disappears.

thanks to their immersive and playful aspect, these are only accidental conditions that are secondary to the content of what is being learned.

5 Conclusion

Marching in the footsteps of connectivists, proponents of the metaverse focus on the conditions for learning and the realism of stimulating immersive environments.

In the metaverse, knowledge is no longer something internalised, but seems on the contrary to be externalised outside the individual and entrusted to the performance of an artificial network – today, we no longer memorise certain information that we hasten to search for on our browser as soon as we need it.

The very definition of the term "learning" actually has two dimensions: that of the *subject* – who receives something – and that of the *object* – which is received. Learning can therefore mean acquiring or transmitting knowledge; in both cases, a *subject* and a *content* come into play. This is the reason why an in-depth reflection on learning must ask two questions: who learns? And what do we learn?

If, in immersive reality, the individual learner is merely an instrument of an ever more powerful neural network, and if the content of what is learned is of no importance to the designers of these technologies as long as the performance criteria are met (the speed of teaching and the large number of students reached), then the metaverse may offer a fun and stimulating experience, but it won't teach us much.

Acknowledgments. This contribution was prepared and written without any help from an artificial agent (generative artificial intelligence, conversational agent, etc.).

References

Ball, M.: The Metaverse. And How it Will Revolutionize Everything. Liveright Publishing, W.W. Norton & Company, New York (2022)

Ball, M.: "The Metaverse: What It Is, Where to Find it, and Who Will Build It". Mat-thewball.co (2020). https://www.matthewball.co/all/themetaverse

Basdevant, A., François, C., Ronfard, R.: Mission exploratoire sur les métavers. Rapport commandé le 14 février 2022 par une lettre de mission du gouvernement français pour les compte du Ministre de l'Économie, des Finances et de la Relance, de la Ministre de la Culture et du Secrétaire d'État chargé de la Transition numérique et des Communications électroniques (2022). https://cnnumerique.fr/nos-travaux/mission-exploratoire-sur-les-metavers

Bergson, H.: Ecritsphilosophiques. 2011, 1ère éd., PUF, coll. Quadrige (1882)

Chalmers, D.: Reality+: Virtual Worlds and the Problems of Philosophy. W. W. Norton, New York (2022)

Cheng, P.W., Buehner, M.J.: Causal learning. In: Holyoak, K.J., Morrison, R.G. (eds.) The Oxford Handbook of Thinking and Reasoning, pp. 210–233. Oxford University Press (2012)

Coane, J.H., Cipollini, J., Barrett, T.E., Kavaler, J., Umanath, S.: Lay definitions of intelligence, knowledge, and memory: inter- and independence of constructs. J. Intell. 11(5), 84 (2023). https://doi.org/10.3390/jintelligence11050084

Devers, N.: Les Liens Artificiels. Albin Michel, Paris (2022)

Downes, S.: Connectivism. Asian J. Dist. Educ. **17**(1), 58–87 (2022). https://doi.org/10.5281/zen odo.6173510

Downes, S.: Buntine oration: learning objects. Int. J. Instr. Technol. Distance Learn. **1**(11), 3–15 (2004)

Driscoll, M.: Psychology of Learning for Instruction. Allyn & Bacon, Needham Heights (2000)

Economist Impact: Towards a successful metaverse: the case for measuring enabling factors (2023). https://impact.economist.com/perspectives/technology-innovation/towards-a-suc cessful-metaverse

Georgiou, Y., Tsivitanidou, O., Ioannou, A.: Learning experience design with immersive virtual reality in physics education. Educ. Technol. Res. Dev. **69**, 3051–3080 (2021). https://doi.org/10.1007/s11423-021-10055-y

Hacking, I.: Representing and Intervening: Introductory Topics in the Philosophy of Natural Science. Cambridge University Press, Cambridge (1983). https://doi.org/10.1017/CBO978051 1814563

van der Heijden, H.: User acceptance of hedonic information systems. MIS Q. **28**(4), 695–704 (2004). https://doi.org/10.2307/25148660

Jovanović, A., Milosavljević, A.: VoRtex metaverse platform for gamified collaborative learning. Electronics **11**, 317 (2022). https://doi.org/10.3390/electronics11030317

Kant, E.: (trad. Tremesaygueset Pacaud, préf. Charles Serrus): Critique de la raison pure. 1975, 8e éd. PUF, coll. "Bibliothèque de Philosophie contemporaine" (1781)

Koster, R.: A Theory of Fun for Game Design, 2nd edn. O'Reilly Media (2013)

Kurt, S.: Connectivism Learning Theory. Educational Technology (2023). https://educationaltech nology.net/connectivism-learning-theory/

Kurzweil, R.: The Singularity is Near. Londres, Duckworth (2005)

Läpple, D., Barham, B.: How do learning ability, advice from experts and peers shape decision making? J. Behav. Exp. Econ.Behav. Exp. Econ. **80**, 92–107 (2019). https://doi.org/10.1016/j. socec.2019.03.010

Makransky, G., Terkildsen, T.S., Mayer, R.E.: Adding immersive virtual reality to a science lab simulation causes more presence but less learning. Learn. Instr. **60**, 225–236 (2019). https://doi.org/10.1016/j.learninstruc.2017.12.007

Parong, J., Mayer, R.E.: Cognitive and affective processes for learning science in immersive virtual reality. J. Comput. Assist. Learn.Comput. Assist. Learn. **37**(1), 226–241 (2021). https://doi.org/10.1111/jcal.12482

Photopoulos, P., Tsonos, C., Stavrakas, I., Triantis, D.: Remote and in-person learning: utility versus social experience. SN Comput. Sci. **4** (2022). https://doi.org/10.1007/s42979-022-015 39-6

Piccione, J., Collett, J., De Foe, A.: Virtual skills training: the role of presence and agency. Heliyon **5**(11), e02583 (2019). https://doi.org/10.1016/j.heliyon.2019.e02583

Pimentel, D., Fauville, G., Frazier, K., McGivney, E., Rosas, S., Woolsey, E.: An Introduction to Learning in the Metaverse. Meridian Treehouse (2022)

Purdy, M.: How the Metaverse Could Change Work. Harvard Business Review (2022)

Rahman, H., Wahid, S.A., Ahmad, F., et al.: Game-based learning in metaverse: virtual chemistry classroom for chemical bonding for remote education. Educ. Inf. Technol. (2024). https://doi.org/10.1007/s10639-024-12575-5

Reeves, B., Read, L.: Total Engagement: Using Games and Virtual Worlds to Change the Way People Work and Businesses Compete. Harvard Business Press (2009)

Siemens, G.: Connectivism: a learning theory for the digital age. Int. J. Instr. Technol. Distance Learn. **2**, 1–9 (2005)

Stephenson, K.: (Internal communication, no. 36): what knowledge tears apart, net-works make whole. http://www.netform.com/html/icf.pdf. Accessed 10 Dec 2004

Stoll, C.: "Why the Web won't be Nirvana". Newsweek (1995). https://www.newsweek.com/cli fford-stoll-why-web-wont-be-nirvana-185306

Surowiecki, J.: The Wisdom of Crowds: Why the Many are Smarter than the Few and How Collective Wisdom Shapes Business, Economies, Societies, and Nations. Doubleday & Co. (2004)

Thaler, R., Sunstein, C.: Nudge: Improving Decision About Health, Wealth and Happiness. Penguin Books, Londres (2008)

Tolman, E.C., Honzik, C.H.: Degrees of hunger, reward and non-reward, and maze learning in rats. Univ. Calif. Publ. Psychol. **4**, 241–256 (1930)

Triclot, M.: Philosophie des jeux vidéo. Zones, Paris (2011)

Walter, Y.: Embracing the future of artificial intelligence in the classroom: the relevance of AI literacy, prompt engineering, and critical thinking in modern education. Int. J. Educ. Technol. High. Educ. **21**(1), 15 (2024). https://doi.org/10.1186/s41239-024-00448-3

Wu, B., Yu, X., Gu, X.: Effectiveness of immersive virtual reality using head-mounted displays on learning performance: a meta-analysis. Br. J. Edu. Technol. **51**(6), 1991–2005 (2020). https://doi.org/10.1111/bjet.13023

Experimental Procedures in Cognition

Emotion-Enhanced Pain Assessment Protocol

Bruna Alves[1]([✉])[iD], Ana Almeida[1,2][iD], Catarina Silva[1,2], Daniela Pais[1][iD],
Rita P. Ribeiro[3,4][iD], João Gama[3,5][iD], José Maria Fernandes[1][iD],
Susana Brás[1][iD], and Raquel Sebastião[1,6][iD]

[1] IEETA, DETI, LASI, University of Aveiro, 3810-193 Aveiro, Portugal
{bruna.alves,acha,catarinavilar,danielapais,jfernan,susana.bras,
raquel.sebastiao}@ua.pt
[2] Department of Physics, University of Aveiro, 3810-193 Aveiro, Portugal
[3] INESC TEC, University of Porto, 4200-465 Porto, Portugal
[4] Faculty of Sciences, University of Porto, 4169-007 Porto, Portugal
rpribeiro@fc.up.pt
[5] Faculty of Economics, University of Porto, 4200-464 Porto, Portugal
jgama@fep.up.pt
[6] ESTGV, Polytechnic Institute of Viseu, 3504-510 Viseu, Portugal

Abstract. Pain is a highly subjective phenomenon that depends on multiple factors. The common methods used to evaluate pain require the person to be awakened and cooperative, which may not always be possible. Moreover, such methods are subject to non-quantifiable influences, namely the impact of an individual's emotional state on how pain is perceived or how negative emotions may exacerbate pain perception, while positive emotions may attenuate it. The goal of this study was to conduct a novel protocol for pain induction with emotional elicitation and assess its feasibility. In this protocol, the physiological responses were monitored, and collected, through Electrocardiogram, Electrodermal Activity, and surface Electromyogram signals. Along the protocol, the pain perception was evaluated using a 0–10 numerical rating scale and by registering the time from the pain stimulus beginning to the Pain and Tolerance Thresholds. This study comprised three emotional sessions, negative, positive, and neutral, which were performed through videos of excerpts of terror, comedy, and documentary films, respectively, followed by pain induction using the Cold Pressor Task (CPT). A total of 56 participants performed the study, with a CPT mean time of about 91.70 ± 39.64 s among all the sessions. The conducted protocol was considered feasible and safe as it allowed the collection of physiological data, pain, and questionnaires' reports from 56 participants, without any harm to them. Moreover, the collected data can be further used to assess how emotional conditions influence pain perception and to provide better emotion-calibrated pain recognition systems based on physiological signals.

Keywords: Emotion · Pain · Cold Pressor Task · Physiological Signals

J. Baratgin et al. (Eds.): HAR 2024, LNCS 15504, pp. 385–400, 2025.
https://doi.org/10.1007/978-3-031-84595-6_23

1   Introduction

Pain assessment is mainly performed through patient self-report, using Numerical Rating Scales (NRS) or visual analogue scales [15]. However, patients with limited communication skills cannot report their pain experiences [16]. Thus, it is important to create methods that use objective measurements to evaluate pain, such as automated pain recognition systems that are capable of detecting human body characteristics affected by pain, for example, facial expressions, sounds, gestures and physiological signals.

Analysing physiological changes is a feasible approach to detect pain. These changes may include variations in skin conductance due to an increased sympathetic outflow that leads to sweat secretion, therefore altering the electrical properties of the skin, and Electrodermal Activity (EDA), increasing the electrical conductance [16]. Heart rate is also affected by an increase which can be observed in the Electrocardiogram (ECG) [16]. Furthermore, the muscle tone also increases, showing this effect in the electrical muscle activity revealed by the Electromyogram (EMG) [16]. The facial muscles like the corrugator and zygomaticus are expected to be active during pain experience, and also the trapezius which is an indicator of high stress related to pain [15].

It has been disclosed that negative emotions with high arousal levels may increase pain sensitivity. On the contrary, in a positive emotional condition, the pain perception may be decreased [9].

This work presents the design, implementation and feasibility's assessment of a novel protocol that was designed to study the role of emotions in pain, by collecting emotional and physiological data, such as ECG, EDA, and EMG. This proposed protocol resources emotion-inducing videos (film excerpts) to assess the influence of emotional response on pain. Pain was induced through a Cold Pressor Task (CPT). The CPT involves placing a hand or forearm in cold water. After immersing the limb, the subject perceives a slowly mounting pain [2]. Cold-water immersion activates the sympathetic nervous system (SNS) through nociceptors in the skin, causing a physiological response. This task is also known as a stress induction protocol since pain response activates the hypothalamic-pituitary-adrenal (HPA) axis and releases glucocorticoids, thus preparing the body to cope with noxious stimuli, serving as warning signals to prevent injury or tissue damage [5]. Rainville et al. [10] compared four different methods for experimental pain induction and found that the CPT was suitable to mimic the effects of chronic pain.

Advancing the current state of the art, to better describe and characterize the pain, efforts to disclose the key role that emotions play in pain perception and in physiological responses to pain should be endeavored. To this aim, this work contributes with a full description of a protocol for pain assessment related to the emotional condition of the individuals by comprising the elicitation of three emotional conditions (negative, positive, and neutral) prior to pain induction through exposure to a cold stimulus. Furthermore, this work also assesses the feasibility of the protocol, showing that emotional elicitation was effectively achieved.

It is expected that, under the same scenario, the experienced pain from the same cold stimulus triggers similar physiological responses, rendered by similar information in the physiological signals. The feasibility of this protocol enables the construction of a database comprising physiological signals labelled with pain and corresponding questionnaire reports. Thus, the data collected under this protocol will allow the study of the hypothesis that emotions play a key role in pain perception, attaining it by the physiological characterization of pain induced by the same stimulus under different emotional contexts and by comparing the self-reported pain by the same participants across different sessions. Moreover, these data may contribute to the development of pain prediction models that take emotional conditions into account, thus filling some current gaps that exist in pain assessment. Additionally, the collected data will allow for the evaluation of how emotion modulates pain patterns, which will be a focus of future research.

The remaining of the present document is structured as follows: Sect. 2 highlights related works, Sect. 3 presents a detailed description of the methodology, from participants recruitment to the experimental procedure, Sect. 4 discusses the implementation of the protocol and assesses its feasibility, and Sect. 5 exposes the main conclusions and future research perspectives.

2   Related Works

Zhang *et al.* [17] investigated the different pain perceptions between males and females and how this is related to negative emotions. To quantify pain sensitivity, they used a CPT, and to assess the negative emotions they used several questionnaires and Magnetic Resonance Image (MRI) data. The authors found statistically significant gender differences in Pain Threshold and tolerance, and in two questionnaires: the males presented higher pain and Tolerance Thresholds and lower scores in two questionnaires. Further analysis showed that the two questionnaires' scores were negatively correlated with Pain Threshold and tolerance, showing that differences in pain sensitivity were mediated by pain-related negative emotions, specifically pain-related fear and anxiety.

Ensuing, relying on the participant's active affective state may not be sufficient to study emotion's influence on pain, therefore inducing different valences of emotions using emotional stimulation may be key to solving so. Studies in this area have already put these types of stimuli into practice, including via pictures [14], videos [6,18], virtual reality, games, music/sound [7,8], words, and recall [4]. However, emotional elicitation through videos has the advantage of being a multichannel method.

Velana *et al.* [14] collected the SenseEmotion database, a multimodal database for pain and emotion recognition. Pain was induced through hot stimuli using a thermal stimulator at the forearm in two phases (one per forearm) while 180 digital images were selected for affective stimuli (highly pleasant, neutral, and highly unpleasant). During each phase, the authors calibrated three individual pain stimuli for each participant, which were delivered during the experimental phase. During each experimental phase, 30 images of each type were presented under the three pain levels in a randomized order. Additionally, 30 images were presented without any pain stimulus. The participants rated their reaction to each image using the Self-Assessment Manikin questionnaire, measuring valence and arousal. Throughout the entire protocol, EDA, ECG, EMG of the trapezius muscle, and the respiration signal (RSP) were collected, as well as video and audio signals.

Zhang et al. [18] used videos for emotional elicitation to collect the BioVid Emo DB, a multimodal database for emotion recognition. The authors included 15 film clips to induce 5 different emotions while recording ECG, EDA, EMG from the trapezius, and video signals. Self-assessment of arousal, valence, and discrete emotions showed that sadness and anger were reported almost equally in the anger clip and that the emotion elicitation was well performed.

In another study, Zhang et al. [19] designed a protocol of ten tasks related to different emotions. The tasks include interpersonal conversations, watching film clips, a CPT and physical experiences. As the CPT intended to target physical pain, the other nine tasks intended to target several emotions such as sadness, happiness, anger, and fear, among others. Between two consecutive tasks, participants reported the feelings they experienced. During the protocol, the authors collected high-resolution 3D dynamic imaging, high-resolution 2D video, thermal (infrared) sensing, and physiological sensors that included EDA, respiration rate, blood pressure, and heart rate.

Silva and Sebastião [12] studied the ECG signal during pain induction under different emotional conditions. The participants were subjected to a CPT while watching an emotional elicitation video. The protocol consisted of two sessions, one using a fear-emotion-elicitation video and a second using a neutral one. The statistical analysis did not show a significant difference between the pain levels reported from both sessions. The implemented protocol was unsuitable for emotional elicitation since the participant focused only on the CPT instead of paying attention to the emotion-inducing videos.

3   Methodology and Protocol

The implemented protocol followed several procedures, from the recruitment of participants from the local community to the experimental procedure itself.

3.1   Participants Recruitment

Participants were recruited from the local community, through the dissemination on social networks and indirect access means. The inclusion criteria comprised:

- Age between 18 and 50;
- Not having been diagnosed with any severe mental illness and not suffering from any disease that causes chronic pain;
- Ability to understand and respond to measures that need to be self-reported;
- Agree to sign the informed consent to participate in the study.

Individuals with a clinical history of previous cardiovascular or cerebrovascular events, severe arterial insufficiency, convulsions and cuts, wounds or fractures on the hand and forearm to be immersed were excluded.

It is important to mention that this study was approved by the Ethical Committee Ethics and Deontology Council of the University of Aveiro (CED-UA - 24-CED/2021).

3.2 Protocol

After informed consent and detailed explanations of the entire procedure, data were collected in three distinct temporal sessions, with an interval of approximately 1 week. Figure 1 illustrates the data collection timeline for each session.

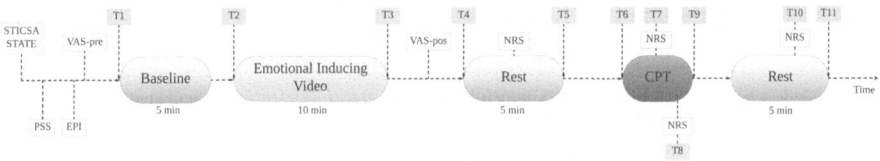

Fig. 1. Implemented protocol trail. The scheme includes emotional elicitation and pain-inducing phases, triggers (T1 to T11), and pain report moments.

Participants were exposed to an emotion elicitation task by watching videos suggestive (and already validated) of each emotion (positive, negative, and neutral) and a pain induction task referred to as the Cold Pressor Task (CPT), without compromising their health and well-being. Simultaneously, physiological signals, namely surface EMG, EDA, and ECG, were continuously monitored and collected through minimally invasive equipment. The EMG was collected in the non-dominant trapezius and triceps muscles, the EDA was collected in the dominant hand and the ECG was collected at the rib cage. Figure 2 schematises the placement of the electrodes on the participant. The white electrodes represent the reference electrodes for the ECG and the EMG signals, while the red and black electrodes represent the positive and negative ones, respectively.

Before participating in the emotion elicitation and pain induction task, participants responded to various questionnaires adapted and validated for the Portuguese language. These questionnaires address psychological traits and emotional states of the participants and include the short version of the Eysenck Personality Inventory (EPI) (48 items) [1], the Perceived Stress Scale (PSS) (10 items) [13], and the State-Trait Inventory for Cognitive and Somatic Anxiety (STICSA) (21 items + 21 items) [3,11]. Before and after the emotion elicitation task through the viewing of videos, participants responded to a Visual Analog Scale (VAS) questionnaire to self-report their emotional state.

The procedure begun with a 5-min collection of baseline physiological signals while the participant is at rest. This was followed by the visualization of a

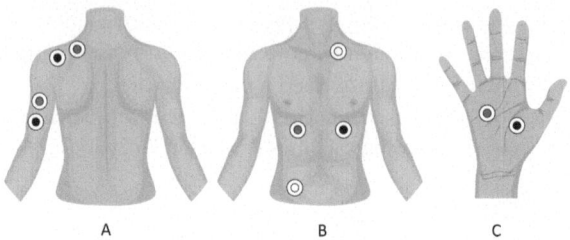

A   B   C

Fig. 2. Scheme of the placement of the electrodes: A) EMG electrodes on the trapezius and triceps muscles; B) ECG electrodes plus reference electrode of the EMG and C) EDA electrodes.

10-min video excerpt sequence with emotional content. A short description of each excerpt is presented in Table 1.

Table 1. Short description of the video segments used for emotional elicitation.

Emotion	Film Title	Length (s)	Segment Description
Neutral	The Secret Life of Birds	188	A documentary about birds
	Nature Documentary 1	188	A documentary about Portugal's landscapes
	Nature Documentary 2	217	A document about Antarctica
Positive (Happiness)	What Women Want	228	A man dancing, singing, driking wine, while dressing himself as a woman while his daughter walks in with her new boyfriend
	Three Fugitives	144	A normal day at a bank, untill a silly thief enters to rob it. However, the money gets stuck in the chandelier
	Sister Act	150	A musical scene where a choir of nuns sings and dances a religious song
	Madagascar	102	A group of speaking animals singing and dancing "I like to move it, move it"
Negative (Fear)	The Descendent	162	A group of girls in a dark cave where they find animal bones and a terrifying creature
	The Grudge	127	A man enters his home to find everything in disarray. His wife appears to be possessed and a horrifying child appears
	Insidious	89	A man is searching for paranormal activity with a machine, while a creepy lady appears
	A Tale Of Two Sisters	215	A girl wakes up from a nightmare to find a creepy girl that heads to her

These video excerpts were used in previous work to validate their ability for emotional elicitation [6]. Moreover, the order of the emotional sessions was randomised, obtaining 6 different possible order series: P-Nt-N, P-N-Nt, Nt-P-N, Nt-N-P, N-Nt-P, N-P-Nt, with P, N and Nt standing for Positive, Negative and Neutral, respectively. Each participant was allocated to one of the sequences, guaranteeing that the number of participants in each sequence was approximately the same.

After watching the video, there was a 5-min collection of physiological signals while the participant is at rest. Then, the CPT procedure was performed, followed by another 5-min resting period. In the CPT, participants were asked to submerge their non-dominant hand into a tank of cold water (approximately 7 °C) for 2 min (although participants could remove their forearm if they found the pain intolerable). This way, all participants were exposed to a standardized cold stimulus task to induce pain.

During this procedure, the moment at which participants indicated the onset of pain (Pain Threshold) was registered, and the level of pain was self-reported at four distinct time points (pain levels), using a NRS ranging from 0 (no pain) to 10 (maximum pain).

During data collection, several triggers were considered:

- T1: Beginning of the physiological data acquisition;
- T2: Beginning of the Emotional Inducing Video and end of the Baseline;
- T3: End of the Emotional Inducing Video;
- T4: Beginning of the first Rest epoch;
- T5: End of the first Rest epoch;
- T6: Beginning of the CPT epoch (the participant immerses their hand in the water tank);
- T7: Pain Threshold (the participant begins to feel pain);
- T8: Tolerance Threshold (the participant can no longer tolerate the pain);
- T9: End of the CPT epoch (the participant removes the hand from the water tank);
- T10: 3 min after taking the hand of the water tank;
- T11: End of the physiological data acquisition.

These triggers pointed out crucial moments of the protocol and facilitated the division of the signal according to the different epochs of the protocol. The triggers also allowed the computation of the time of each segment a posteriori. If the participant could tolerate the maximum time of 2 min with the hand immersed in the water T8 and T9 were equal.

3.3 Experimental Setup

The experimental procedure was implemented at the Institute of Electronics and Informatics Engineering of Aveiro, University of Aveiro, in a room with dim light and prepared specifically for this purpose, with the equipment placed in the face-rear of the windows to minimize the visual external stimuli.

For the signals collection, a 4-channel Biosignalsplux[1] was used. This device simultaneously collects and digitalizes the signals from the sensors connected and transmits them via Bluetooth to a computer. Two EMG sensors, one EDA sensor, and one ECG sensor were connected to the device.

For the CPT, a stainless-steel tank of 45 litres with an immersion thermostat (Termotronic-100) was used. It includes a circulation pump to improve the

---

[1] https://www.pluxbiosignals.com/pages/biosignalsplux (Accessed 22nd April 2023).

homogeneity within the bath and to perform a closed liquid circulation circuit. The water was initially cooled with ice and the control panel allowed the adjustment of temperature. Also, two notebooks were used: one for the acquisition of the sensors' reading and the respective monitorization (through the software OpenSignals from Biosignalsplux) and the other for displaying the videos with an external monitor (used with headphones for audio). This notebook was also used for completing the questionnaires. Figure 3 presents the experimental setup used.

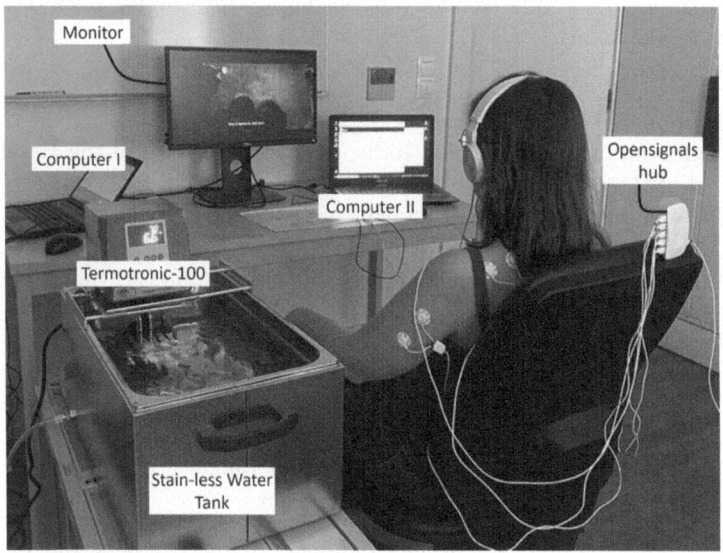

Fig. 3. Experimental setup, showing the Thermotronic-100, the computers, the monitor, and the Plux equipment for physiological data acquisition.

3.4 Experimental Procedure

The experimental procedure regards the preparation of the protocol before the participant's arrival, the implementation of the protocol, and organising the material for the next session/day. The following steps were taken before the participant's arrival:

– Turn on Termotronic-100 and cool the water;
– Make sure the temperature is between 6 and 6.5 Celsius degrees;
– Turn on the computers and monitor;
– Open Opensignals on computer I and Excel for data annotation;
– Open questionnaires and the movie excerpt on computer II;
– Place electrodes on sensors and connect sensors' cables to the hub (if necessary);
– Connect headphones to computer II;

At participant arrival, the subsequent steps were performed:

- Thank the participant and reinforce that they can give up on the study at any time;
- Detailed inform how the experiment will be conducted and how the data will be used, clarify any doubt or question of the participant, and make sure the participant does not have their smartphone in its pockets and it is in silent mode;
- Make sure STICSA-Trait was answered and the informed consent was signed;
- Lead the participant to sit on the chair near the monitor, and ask them to answer the initial questionnaires by order: PSS, EPI, STICSA-State, and VAS-Pre;
- Make sure that the participant understands all questions and if there is any doubt, clarify it;
- Place the sensors on the participant according to Fig. 2;
- Turn on the Biosignalsplux hub;
- Check if all sensors are well placed;
- Update the sensors on the Opensignals, define 1000 Hz for the sampling rate, and start recording (trigger - T1);
- Check if the acquisition track is ok;
- After 5 min, open the movie excerpts' sequence and adjust the sound volume to 50 (on computer II scale);
- Ask the participant to relax and be attentive to the video, press trigger (T2);
- After the movie is over, press trigger (T3);
- Open the VAS-Pos questionnaire;
- Ask the participant to answer the questionnaire and explain that the assessment of emotional state must be done using memory and to think about the most intense moment of each set;
- As soon as the participant submits the questionnaire press trigger (T4);
- After 5 min, press trigger (T5) and register water temperature;
- Before hand immersion, ask and register the participant's pain level;
- Ask the participant to introduce their hand in the cold water tank, and press the trigger (T6) as soon as their hand is immersed;
- When the patient hits the pain (sensation threshold) press trigger (T7) and register the pain level;
- When the patient hits the Tolerance Threshold (before taking its hand out of the water) press trigger (T8) and register pain level and time;
- Press trigger (T9) as soon as the participant takes the hand out of the water and ask the participant to relax;
- After 3 min of resting, register pain level and press trigger (T10);
- After 2 min press trigger (T11) and end the experimental procedure;

As soon as the experimental procedure finished, these steps were fulfilled:

- Stop the acquisition of physiological signals;
- Turn off the Biosignalsplux hub;
- Ask the participant to get up and remove the electrodes carefully;

– Thank the participant and schedule the next session, if necessary;
– Save the acquired signals.

The final steps were concluded at the end of the day:

– Put Biosignalsplux equipment to charge;
– Freeze water for the next day;
– Turn off computers and monitors;
– Store Biosignalsplux hub, sensor, and headphones.

4   Implementation and Discussion

A total of 56 volunteers (28 females) within the age ranges of 18 to 30 years old (mean of 22.46 and standard deviation of 2.04 years old) participated in the study. Only 2 participants did not undergo the last session due to personal reasons. Therefore, a total of 166 sessions were performed. This proves that the protocol was well-tolerated, as the vast majority of participants completed the three sessions without dropping out.

Among all the sessions the mean time of the CPT was about 91.70 ± 39.64 s (mean $\pm$ standard deviation). Within the 166 sessions, there were 9 sessions where the participant kept the hand in the cold-water tank for less than 30 s. On the opposite, there were 96 sessions where the participants endured the maximum time (2 min) with the hand immersed.

Moreover, Table 2 presents Tolerance and Pain Thresholds, in a measure of time, and NRS, showing that, in general, Pain Threshold and Tolerance times were higher for positive emotion elicitation compared to the negative condition.

Table 2. Descriptive evaluation for Pain and Tolerance Thresholds divided by gender and elicited emotion both in time (in seconds) and in NRS scores.

Subjects	Stimuli				
	Emotion	Pain			
		Pain Threshold (s)	Pain Threshold (NRS)	Tolerance Threshold (s)	Tolerance Threshold (NRS)
All (n = 54)	Neutral	15.08 ± 14.59	4.83 ± 2.04	92.32 ± 40.98	7.74 ± 1.66
	Positive	19.74 ± 21.94	4.71 ± 2.05	93.23 ± 38.71	7.78 ± 1.77
	Negative	15.76 ± 9.29	4.70 ± 2.20	92.11 ± 39.76	7.94 ± 1.77
Females (n = 27)	Neutral	16.15 ± 19.18	4.96 ± 2.13	90.23 ± 39.43	7.89 ± 1.74
	Positive	14.70 ± 11.56	4.88 ± 2.23	89.20 ± 38.60	7.74 ± 2.03
	Negative	13.39 ± 9.99	4.42 ± 2.19	85.70 ± 40.53	8.11 ± 1.87
Males (n = 27)	Neutral	14.01 ± 7.99	4.70 ± 1.98	94.41 ± 43.13	7.59 ± 1.60
	Positive	24.78 ± 28.19	4.54 ± 1.88	97.25 ± 39.12	7.81 ± 1.49
	Negative	18.14 ± 8.02	4.96 ± 2.21	98.52 ± 38.67	7.78 ± 1.69

Figure 4 presents the violin plots of the scores reported by the participants and the times for Pain Threshold and tolerance.

Regarding Pain Threshold time (top left Fig. 4) the Gaussian formed from the distribution in a Positive condition presents a higher density for females rather

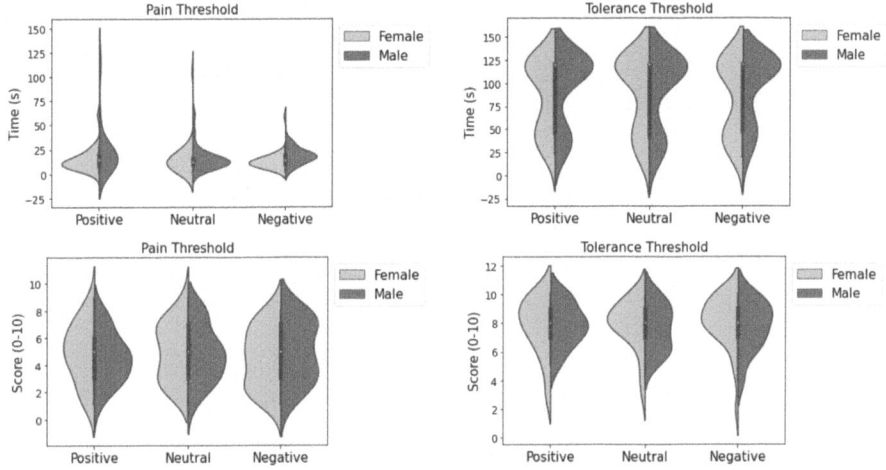

Fig. 4. Violin plots regarding pain (left) and Tolerance Thresholds (right) in a measure of time (top) and NRS (bottom), divided by gender.

than males, as the males present higher threshold times, which indicates a later pain report. A similar observation is found for the Negative condition, while for the Neutral, the opposite occurs. As for Tolerance time (top right Fig. 4) the violin plots are symmetric, indicating that the distributions for the two genders appear to be similar among all the emotional conditions, and hinting at the possibility of emotions not influencing Tolerance Threshold. Regarding the NRS scores (bottom Fig. 4), in general, females tended to report higher scores than males.

Figure 5 presents the four physiological signals acquired for a random participant in the different stages of the protocol. Within these, it is possible to observe some signal fluctuations, namely in the video and CPT epochs. Nevertheless, the EDA signal of this participant presents a similar behavior at the beginning of the emotional elicitation (Video) epoch and during the CPT, showcasing that a single physiological signal may not be able to disclose a proper measurement of pain, and therefore supporting the need for a multimodal physiological analysis.

Table 3 presents the results (mean $\pm$ standard deviation) of the VAS questionnaire items, which is intended to evaluate the emotional elicitation, before and after watching the emotional inducing video. Along with the questionnaires' responses, the results from statistical tests to evaluate the significance of the responses are also presented.

As can be seen, the results from VAS-pre are very similar across emotions, which was expected since the participants did not know the content of the video at this stage. Concerning the results from VAS-pos some differences can be highlighted, namely Anxiety, Stress, and Fear scores were higher for the negative emotional session, while Happiness and Valence scores were higher for the positive session, and Arousal score was higher for the positive and negative sessions.

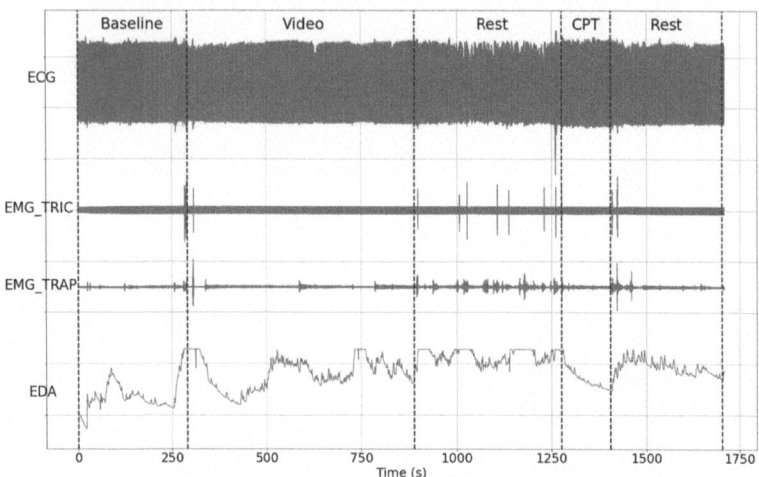

Fig. 5. ECG, EMG from triceps (EMG_TRIC) and trapezius muscles (EMG_TRAP), and EDA signals of a random participant during the different stages of the protocol.

To assess if emotional elicitation was achieved, a statistical analysis was performed comparing the differences between the VAS_pos and VAS_pre scores, of the several items, across the different emotional conditions. At first, the normality of all the items' scores was tested through the Shapiro-Wilk test. Thereafter, as the assumption of normality was rejected for all the items, the Friedman test was applied, followed by the application of the Tukey's post hoc tests to those items that presented statistically significant differences between different emotional conditions.

Concerning all the items, the assumption of normality was rejected with a significance of 5%. Regarding the Friedman test, the null hypothesis was rejected for all the items, meaning that there were significant differences among emotional conditions. The results for the Friedman test are presented in Table 3. The Anxiety was significantly different between the Negative and Positive conditions (Tukey's p-Value: 1.23e−10; Confidence Interval (CI): [0.76, 1.60]); and between the Negative and Neutral conditions (Tukey's p-value: 1.95e−6; CI: [0.47, 1.31]). Regarding the Happiness scores, differences were found between the Negative and Positive (Tukey's p-value: 3.92e−6; CI: [−1.28, −0.44]) and the Positive and Neutral conditions (Tukey's p-value: 0.003; CI: [0.163, 0.996]). Fear scores could distinguish the Negative condition from Positive (Tukey's p-value: 2.55e−11; CI: [0.77, 1.57]) and from the Neutral one (Tukey's p-value: 2.81e−10; CI: [0.71, 1.51]). Similarly, the Stress could distinguish the Negative condition from the Positive (Tukey's p-value: 9.49e−5; CI: [0.33,1.19]) and Neutral sessions (Tukey's p-value: 7.55e−6; CI: [0.43,1.29]). Arousal was significantly different regarding the Negative and Neutral sessions (Tukey's p-value: 0.023; CI: [0.05, 0.87]) and the Positive and Neutral session (Tukey's p-value: 1.46e−4; CI: [0.30, 1.12]). Finally, the Valence differentiated between the Positive sessions

Table 3. Results of the VAS questionnaires answers before (VAS-pre) and after (VAS-pos) the emotional inducing video and their statistical results.

Items	Emotion			Friedman Test	
	Neutral (Nt)	Positive (P)	Negative (N)	p-Value	χ^2
Anxiety	3.81±2.35	4.16±2.35	3.91±2.27	5.46e-11	47.26
	2.91±1.89	2.77±1.60	5.02±2.62	▼, •	
Hapiness	7.13±1.84	7.00±1.99	6.94±1.97	5.14e-6	24.35
	7.04±2.00	7.89±1.87	6.02±2.30	▲, •	
Fear	2.04±1.78	1.91±1.59	1.85±1.47	1.50e-13	59.06
	1.62±1.44	1.41±0.80	3.80±2.46	▼, •	
Stress	4.25±2.52	4.36±2.58	4.11±2.83	1.77e-06	26.49
	3.11±2.22	3.16±2.03	4.70±2.46	▼, •	
Arousal	7.68±2.05	7.88±1.62	7.59±2.03	2.08e-4	16.95
	7.04±2.06	8.34±1.37	7.96±1.76	▼, ▲	
Valence	8.19±1.49	7.96±1.66	7.87±1.69	1.81e-08	35.66
	8.06±1.47	9.00±1.14	7.30±1.64	▲, •	

Grey: VAS-pre; White: VAS-pos; Tukey's post-hoc test p-Value<0.05: ▲: Nt vs P; ▼: Nt vs N; •: P vs N

from the Negative (Tukey's p-value: 2.60e−8; CI: [−1.41, −0.59]) and the Neutral ones (Tukey's p-value: 6.11e−5; CI: [0.33, 1.14]). These results indicate that emotions were elicited properly.

As far as it is known by the authors, this is the first time a complete protocol for pain induction with previous emotional elicitation has been published. In the protocol implemented by Silva and Sebastião [12], emotional elicitation and pain induction were performed simultaneously. The authors concluded that this approach was not suitable as the participant while suffering the pain induction task, is distracted from the video contents. This protocol overcame this limitation since the video for emotional elicitation was visualized before the pain induction task, avoiding distracting the participant. Regarding the study of Velana et al. [14], the pain was also induced while watching the affective images. Zhang et al. [19] included physical pain as an emotion, not considering the influence of other emotions over pain.

The protocol was considered feasible, as emotional conditions were properly elicited, and safe. However, it is essential to acknowledge that the protocol has some limitations. It may be perceived as tiring by participants since it requires them to sit for about 30 min. During the Baseline and Rest periods, participants must remain seated in a comfortable position. Some individuals exhibited restlessness while seated in the chair. Also, the protocol was thoroughly explained to the participants and all the doubts were addressed, however, some participants did not understand that they were required to report the time they first perceived pain. Consequently, they only reported pain when specifically asked to do so. This may lead to increased values of pain perception in some cases. Additionally, some participants reported being confused when evaluating the pain felt using the NRS, struggling to choose the appropriate score.

Another circumstance that may have hindered emotional elicitation is the fact that some participants had already watched the movies from which the excerpts were taken. Consequently, in this case, the video was unable to elicit emotion, especially fear.

5   Conclusions

The contribution of this work is the design of a feasible protocol for pain induction with emotional elicitation to allow further studies on how emotional conditions influence pain perception. The first and main result is a fully anonymized and annotated dataset, which includes pain reports and biomedical signals, ensuring the safety and well-being of the participants without causing any harm or injury.

It would also be interesting to include a reevaluation of the emotional condition of the participants before the CPT to ensure that the emotional elicitation remained consistent before the pain induction since there is a 5-min interval between the two tasks.

Regarding the physiological data, different analyses can be performed in further research endeavours to assess the influence of pain and emotions over physiological response. Moreover, the data can be used to evolve better emotion-calibrated pain recognition systems/scores.

Furthermore, the authors intend to make the dataset available after the conclusion of the Emotional Modulation in Pain Assessment (EMPA) project (DOI: 10.54499/2022.05005.PTDC). It is worth noting that a comprehensive analysis of this dataset must be pursued. Moreover, the goal of this work was achieved from which the constructed dataset presents a promising avenue for forthcoming investigations, promising valuable insights and further advancements in the field.

Acknowledgments. This work was funded by national funds through FCT - Fundação para a Ciência e a Tecnologia, I.P., under the grant UI/BI62/10827/2023 (B.A.), UI/BI62/10828/2023 (A.A.) and the Scientific Employment Stimulus CEECINST/00013/2021 (DOI: 10.54499/CEECINST/00013/2021/CP2779/CT0001, R.S.). S. B. is funded by (DOI: 10.54499/DL57/2016/CP1482/CT0096) national funds, European Regional Development Fund, FSE through COMPETE2020, through FCT, in the scope of the framework contract foreseen in the numbers 4, 5, and 6 of the article 23, of the Decree-Law 57/2016, of August 29, changed by Law 57/2017, of July 19. This work is also supported by the FCT through national funds, under unit 00127-IEETA and under the project EMPA (2022.05005.PTDC, DOI: 10.54499/2022.05005.PTDC). We present our sincere appreciation to all the volunteers who participated in the study.

References

1. Almiro, P.A., Moura, O., Simões, M.R.: Psychometric properties of the European Portuguese version of the eysenck personality questionnaire-revised (EPQ-R). Pers. Individ. Differ. **88**, 88–93 (2016). https://doi.org/10.1016/j.paid.2015.08.050
2. von Baeyer, C.L., Piira, T., Chambers, C.T., Trapanotto, M., Zeltzer, L.K.: Guidelines for the cold pressor task as an experimental pain stimulus for use with children. J. Pain **6**(4), 218–227 (2005). https://doi.org/10.1016/j.jpain.2005.01.349
3. Barros, F., Figueiredo, C., Brás, S., Carvalho, J.M., Soares, S.C.: Multidimensional assessment of anxiety through the state-trait inventory for cognitive and somatic anxiety (STICSA): from dimensionality to response prediction across emotional contexts. PLoS ONE **17**(1), e0262960 (2022). https://doi.org/10.1371/journal.pone.0262960
4. Bota, P.J., Wang, C., Fred, A.L., Da Silva, H.P.: A review, current challenges, and future possibilities on emotion recognition using machine learning and physiological signals. IEEE Access **7**, 140990–141020 (2019). https://doi.org/10.1109/ACCESS.2019.2944001
5. Fanninger, S., Plener, P.L., Fischer, M.J., Kothgassner, O.D., Goreis, A.: Water temperature during the cold pressor test: a scoping review. Physiol. Behav. **271**, 114354 (2023). https://doi.org/10.1016/j.physbeh.2023.114354
6. Ferreira, J., Brás, S., Silva, C.F., Soares, S.C.: An automatic classifier of emotions built from entropy of noise. Psychophysiology **54**(4), 620–627 (2016). https://doi.org/10.1111/psyp.12808
7. Hsu, Y., Wang, J., Chiang, W., Hung, C.: Automatic ECG-based emotion recognition in music listening. IEEE Trans. Affect. Comput. **11**(01), 85–99 (2020). https://doi.org/10.1109/TAFFC.2017.2781732
8. Kim, J., André, E.: Emotion recognition based on physiological changes in music listening. IEEE Trans. Pattern Anal. Mach. Intell. **30**(12), 2067–2083 (2008). https://doi.org/10.1109/TPAMI.2008.26
9. Lumley, M.A., et al.: Pain and emotion: a biopsychosocial review of recent research. J. Clin. Psychol. **67**(9), 942–968 (2011). https://doi.org/10.1002/jclp.20816
10. Rainville, P., Feine, J.S., Bushnell, M.C., Duncan, G.H.: A psychophysical comparison of sensory and affective responses to four modalities of experimental pain. Somatosensory Motor Res. **9**(4), 265–277 (1992). https://doi.org/10.3109/08990229209144776
11. Ree, M.J., French, D., MacLeod, C., Locke, V.: Distinguishing cognitive and somatic dimensions of state and trait anxiety: development and validation of the state-trait inventory for cognitive and somatic anxiety (STICSA). Behav. Cogn. Psychother. **36**(3), 313–332 (2008). https://doi.org/10.1017/S1352465808004232
12. Silva, P., Sebastião, R.: Using the electrocardiogram for pain classification under emotional contexts. Sensors **23**(3) (2023). https://doi.org/10.3390/s23031443
13. Trigo, M., Canudo, N., Branco, F., Silva, D.: Estudo das propriedades psicométricas da perceived stress scale (PSS) na população portuguesa. Psychologica (53), 353–378 (2010). https://doi.org/10.14195/1647-8606_53_17
14. Velana, M., et al.: The SenseEmotion database: a multimodal database for the development and systematic validation of an automatic pain- and emotion-recognition system. In: Schwenker, F., Scherer, S. (eds.) MPRSS 2016. LNCS (LNAI), vol. 10183, pp. 127–139. Springer, Cham (2017). https://doi.org/10.1007/978-3-319-59259-6_11

15. Walter, S., et al.: Automatic pain quantification using autonomic parameters. Psychol. Neurosci. **7**(3), 363–380 (2014). https://doi.org/10.3922/j.psns.2014.041
16. Werner, P., Lopez-Martinez, D., Walter, S., Al-Hamadi, A., Gruss, S., Picard, R.W.: Automatic recognition methods supporting pain assessment: a survey. IEEE Trans. Affect. Comput. **13**(1), 530–552 (2022). https://doi.org/10.1109/taffc.2019.2946774
17. Zhang, H., Bi, Y., Hou, X., Lu, X., Tu, Y., Hu, L.: The role of negative emotions in sex differences in pain sensitivity. Neuroimage **245**, 118685 (2021). https://doi.org/10.1016/j.neuroimage.2021.118685
18. Zhang, L., et al.: "BioVid Emo DB": a multimodal database for emotion analyses validated by subjective ratings. In: 2016 IEEE Symposium Series on Computational Intelligence (SSCI). IEEE (2016). https://doi.org/10.1109/ssci.2016.7849931
19. Zhang, Z., et al.: Multimodal spontaneous emotion corpus for human behavior analysis. In: 2016 IEEE Conference on Computer Vision and Pattern Recognition (CVPR). IEEE (2016). https://doi.org/10.1109/cvpr.2016.374

Additional Paper From HAR-2023

Discharge of Responsibility as an Enhancer of Utilitarian Choices

Maxime Bourlier[1], Cassandra Leroux[1], Hirofumi Hashimoto[2], Kaede Maeda[3], Hiroshi Yama[2], and Jean Baratgin[1,4(✉)]

[1] Cognition Humaine et Artificielle (CHArt) Laboratory, Paris 8 University, Saint-Denis, France
jean.baratgin@paris-reasoning.eu
[2] Graduate School of Literature and Human Sciences, Department of Human Behavioral Sciences, Osaka Metropolitan University, Osaka, Japan
[3] Department of Psychology, College of Contemporary Psychology, Rikkyo University, Tokyo, Japan
[4] Probability, Assessment, Reasoning and Inferences Studies (P-A-R-I-S) Association, Paris, France

Abstract. The tramway problem is a moral dilemma studied in psychology. It involves either saving five lives by sacrificing one life through an action (the so-called "utilitarian" response) or refusing to sacrifice one person for the benefit of five (the so-called "deontic" response). Within the Western population, most responses are utilitarian in the so-called *lever* condition, in which you have to pull a lever to deviate the trajectory of the train from the five people onto an alone victim. Whereas in the *bridge* condition, the action is to push an overweight person on the tracks to stop the train. There the majority of the responses are deontic. Instead of the refusal to break a moral code, the deontic choice could be motivated by a strong feeling of responsibility toward the victim we have to kill in order to save the others. A disengagement from this responsibility should lead to more utilitarian responses. We tested this hypothesis by adding two versions in which before acting, the approval of the sacrificed person is requested. Under these conditions, the responses are utilitarian in both scenarios.

Keywords: reasoning · moral decision · deontologism · utilitarianism · responsibility · trolley dilemma

1 Introduction

Moral philosophy usually describes two opposing ways in which we can distinguish what's right from wrong. One focuses on following predetermined moral principles while the other judges the morality of an action by its consequences. The first one has been dubbed *Deontologism* and the other *Consequentialism*. *Utilitarianism* is the most prominent branch of Consequentialism. It considers that a good action is one that tries to maximize the good outcomes for as many

J. Baratgin et al. (Eds.): HAR 2024, LNCS 15504, pp. 403–410, 2025.
https://doi.org/10.1007/978-3-031-84595-6_24

people as it can. Usually, it tries to maximize happiness. We will explore further the distinction between a deontic behavior and a utilitarian one. To illustrate those two moral positions, we can use moral dilemmas. They are thought experiments in which we have to make a decision between a deontic and a utilitarian choice. The most famous of them is *the trolley problem* (Bruers & Braeckman, 2014; Foot, 1967; Thomson, 1976).

Fig. 1. representation of the *lever* scenario of the trolley problem

In the classical version of this dilemma (Fig. 1) you are standing next to some rails and there is a lever next to you. There is a train with broken brakes that is coming your way. Further on its pass, five people are attached to the rails and cannot move out. If the train continues its course it will kill those 5 people. Fortunately, the lever next to you can send the train on a side track before it rides over the five people. Less fortunately, on the side track, there is another person attached to the rails and they will die if you choose to pull the lever. The question is, should you pull the lever and kill one person to save five or do nothing and let five people die? Pushing the lever is of course the utilitarian choice since it maximizes the number of people saved. On the contrary, not doing anything is seen as the deontic choice by following the principle that you should never attempt on someone's life. Hauser et al. (2007) has shown that most people who are put in this hypothetical situation make the utilitarian choice and pull the lever. That said, they also ask some participants about another variant of this dilemma where instead of pulling a lever to deviate the train on a side track, you are standing on a bridge above the track (Fig. 2). In this version, instead of pulling a lever you can push a overweight person standing next to you onto the tracks. The overweight person is heavy enough to stop the train and prevent the other five people from dying at the cost of their own life. Contrary to the *lever* condition, in this *bridge* condition most people choose the deontic option.

First, this dichotomy between those two moral positions has received explanations from the *dual-process theory of moral judgment* (Greene, 2007; Greene

Fig. 2. representation of the *bridge* scenario of the trolley problem

et al., 2001, 2004, 2008). This theory draws from a broader *dual-process theory of reasoning* (Evans, 2003; Kahneman, 2011) which states that there are two ways for us to reason. The type 1 process, also called system 1 or intuitive process, is generally defined as fast and automatic while the type 2 process, sometimes called system 2 or deliberative/reflexive process, is slow, conscious, and requires cognitive resources. The dual-process theory of moral judgment claims that deontic choice always comes from the intuitive process and that deliberative thinking leads to making utilitarian decisions all the time. Indeed the application of the classical dual-process theory to moral judgment seems to be fitting well. It seems logical that the maximization of the good outcomes would require more processing and therefore take more time than the mere application of predefined moral rules. Greene et al. (2008) has shown in that regard that participants gave more deontic responses when they were under cognitive load. Yet other studies point toward other conclusions. Trémolière and Bonnefon (2014) have shown that while giving extreme ratios (e.g. 1 *vs* 500) in problems similar to the trolley one, participants give utilitarian answers even under cognitive load. In the same line, Bago and De Neys (2019) showed that most participants who gave the utilitarian answer after some deliberation were already doing so when they had to respond under cognitive load and time pressure before that. Another element that could lead us to think that there is not a strict association between one moral position and the reasoning process that could produce it is the difference between cultures. Hashimoto et al. (2022) did a study within the Japanese population and found out that in the standard *lever* condition, already half of the participants chose the utilitarian choice when they had to respond under time pressure. Even more surprising, the percentage of utilitarian choices decreased instead of going up after letting the participants think for some time. It decreased even more after participants discussed with each other their opinions on whether to pull the lever or not. Hashimoto et al. (2022) make the hypothesis that the deontic choice might be motivated by a desire to avoid the responsibility of the decision. To explore further this idea, we conducted an experiment within a western population of French participants in which we manipulated their responsibility for the decision in a trolley situation.

2   Experiment

2.1   Participants

80 French native speakers were recruited on social media (Whatsapp, Discord) for this experiment (47 females and 33 males) with an average age of 25 years old ($SD = 8$). They were informed that the experiment was on the basis of voluntary participation and that they could withdraw themselves at any moment.

2.2   Material and Procedure

To manipulate the degree of responsibility that bears on the participant, we modified the standard *lever* and *bridge* scenarios so that the burden of taking the decision to kill falls onto the person who would be sacrificed instead of the participant. To do so participants would have the choice to ask permission for pulling the lever or pushing the person on the tracks instead of directly deciding for them. While in the standard scenarios, participants have a choice between acting and not doing anything, in those new versions, the participant has to choose between asking for permission to act or not doing anything. This way the participant is still deciding whether or not they think they should act but leaves the final word to the concerned person making them responsible for their own death if they decide to sacrifice themselves. Of course, we precised to the participant that whether the person accepts to be sacrificed or not, they will have to respect their decision, but we never reveal the choice they make.

The questionnaires were designed on SoSci Survey. Participants were randomly assigned to one of 4 questionnaires. There were 2 questionnaires for the *lever* scenario and 2 for the *bridge* scenario. The scenarios were either the standard ones or the *ask for permission* versions. All participants in the *lever* condition, whether they received questions in its standard form or its *ask for permission* form were presented with the standard *lever* scenario first. Participants who would receive questions on the *bridge* scenario in either of its forms were, them, presented with the standard *bridge* scenario. Finally, participants were asked to choose what they would do between doing the action (pull the lever or push the person) or not do anything in the following situations for those who were in the standard condition. Those who were in the *ask for permission* condition were asked if instead of directly doing the action in the scenario previously presented to them, they would ask permission to the concerned person to do the action or not do anything. All participants were asked the question 11 times, with each time more victims put on the main track (1, 2, 3, 4, 5, 10, 25, 50, 100, 250 or 500 victims). Each questionnaire finally contained a series of demographic questions. Questions were always presented in this order.

2.3   Results

We used regression models to analyze our data and compared the models to find the one that would best explain the data we observed. We based our evaluation of the best model on three values: AIC (Akaike Information Criterion), which has the tendency to prefer too complex models, BIC (Bayesian Information Criterion), which has a tendency to prefer too simple models, and the Bayes Factor. AIC, BIC, and Bayes factor were computed using multinomial mixed models to take into account the repeated measures per participant. We are looking for the lowest value possible for the AIC and BIC and the highest for the Bayes factor. All models were tested using the lme4 R package (Bates et al., 2015), then compared using the R flexplot (Fife, 2021).

We created six models. Model 0 is the baseline model in which none of the factors has an effect on participants' choice. Model 1 only accounts for the effect of the scenario (*lever* or *bridge*). Model 2 only accounts for the effect of the method (with or without asking permission). Models 3 only consider the effect of the number of victims on the main tracks (1, 2, 3, 4, 5, 10, 25, 50, 100, 250, 500). Model 4 accounts for the interaction effect between the scenario and the method. Model 5 is for the full interaction between the scenario, the method, and the number of victims. Finally, Model 6 takes the main interaction effect between the scenario and the method while also taking into account the effect of the number of victims separately (Table 1).

Model 6 was the best model to explain our data according to the Bayes Factor, AIC and BIC ($BF_{60} = 9.3e + 19, AIC = 587, BIC = 613$). A graphical representation of this model can be found in Fig. 3.

3   Discussion

This experiment confirms that decisions in a moral dilemma, such as the trolley problem, are influenced by the nature of the action the participant has to perform. We see an interaction effect between the scenario (*lever* or *bridge*) that determines which type of action is associated with the utilitarian choice and the method used (Classic or Permission). We found, as expected in the Western population, more utilitarian choices in the *lever* condition than in the *bridge* condition. The transformation of those classical scenarios into their *ask for permission* versions led to an increase in utilitarian choices in both of them, but more strikingly in the *bridge* condition where we can see a clear shift from deontic to utilitarian choices. We also see an independent effect of the number of people on the main tracks, where more potential victims leads to more utilitarian choices.

Table 1. Comparison of the models for participants' responses to the dilemma

Model name	AIC	BIC	BF_{X0}	BF_{6X}	
Model 0 (null)	698	708	1	9.3e+19	
Model 1 (scenario)	683	697	192	4.8e+17	
Model 2 (method)	690	704	7	1.4e+19	
Model 3 (victims)	612	627	4.3e+17	212	
Model 4 (scenario*method)	671	695	485	1.9e+17	
Model 5 (scenario*method*victims)	592	635	5.2e+15	17957	
Model 6 (scenario*method+victims)	587	613	9.3e+19	1	⋆

⋆: Best model
BF_{X0}: Bayes Factor indicating how the given model (X, from 0 to 6) compares to the null model (0).
BF_{6X}: Bayes Factor indicating how much better the best model (6) was compared to the given model (X, from 0 to 6). For example, Model 6 is 17957 times as likely as Model 5 to explain our data.

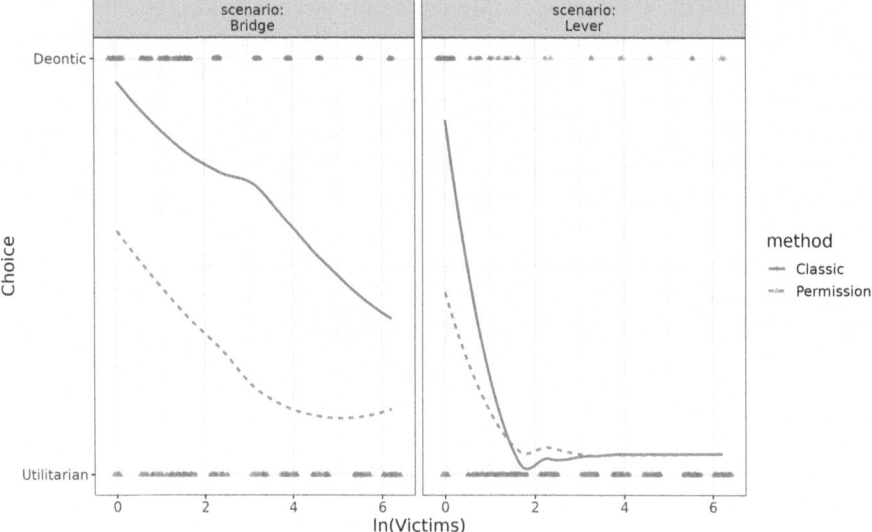

Fig. 3. Graphical representation of model 6 with the choice of the participants on the y-axis and the natural logarithm of the number of victims on the main tracks on the x-axis.

The situation where there is only one victim on the main tracks is quite particular since there is no utilitarian nor deontic choice. In this situation, there is the same number of victims whether the participant choose to act or not. Therefore in the classical scenarios of the *lever* and the *bridge* conditions, almost all participants chose not to act. This was easily predictable since participants would not make the difficult choice of taking an innocent's life only to save another stranger. Similarly, we could have predicted the result in this situation for the *ask for permission condition*. *The percentage of deontic choices decreased* to about 50% which can be explained by the fact that none of the options, letting someone die or asking someone to make the decision to sacrifice themselves, favors whatever moral position the participants might stand for.

Rapidly after this first condition with only one victim on the main tracks, the percentage of deontic choices quickly drops as the number of victims increases for the *lever* scenario. To a point where there is almost only utilitarian choices and this whether it was the classical or the *ask for permission* method. In the classical bridge condition however, the percentage of deontic choices pass under the 50% bar only after a large number of people are put on the main tracks and only reach 66% when there were 500 people on the main tracks. When transformed into the *ask for permission* version we can see that already half of the participants choose to be utilitarians when the number of victim was low and the number of participants endorsing the utilitarian choice keep rising with the number of victims to reach 85%.

In accordance with Hauser et al. (2007), we found more utilitarian choices in the *lever* condition than the *bridge* one. But like in Trémolière and Bonnefon (2014), as the number of victims rises up, participants start to make more and more utilitarian choices, even in the *bridge* condition. However this condition never quite got as many utilitarian choices as in the *lever* condition even in the case where 500 lives were at stake. There are of course many external reasons why we could see this limit in the *bridge* condition. For example, we could be more inclined to refuse to push the overweight person compared to pulling the lever because of the physical proximity with the victim or because we don't want to use them as an instrument to stop the train (Bruers & Braeckman, 2014). The probability of success of the action might also be a factor to take into account (Ryazanov et al., 2023). Here, the participants could have doubt that the sacrifice of the overweight person would be enough to stop the train.

As Hashimoto et al. (2022) hypothesized, our results show that the reluctance to be responsible for the death of a person plays an important role in explaining why participants would endorse the "deontic" choice, even within a Western population. The discharge of responsibility greatly increased the number of utilitarian choices in the *bridge* condition. It also increased in the *lever* condition though it was not as striking as the percentage of utilitarian choices was already so high in this condition even when the number of victims was low. It would be interesting to replicate this experiment in the Japanese population where we could potentially see better effects in the *lever* condition.

Acknowledgements. A special thanks to Baptiste Jacquet for reviewing our statistical analysis.

References

Bago, B., De Neys, W.: The intuitive greater good: testing the corrective dual process model of moral cognition. J. Exp. Psychol. Gener. **148**(5), 1782–1801 (2019). https://doi.org/10.1037/xge0000533

Bates, D., Mächler, M., Bolker, B., Walker, S.: Fitting linear mixedeffects models using LME4. J. Stat. Softw. **67**(1), 1–48 (2015). https://doi.org/10.18637/jss.v067.i01

Bruers, S., Braeckman, J.: A review and systematization of the trolley problem. Philosophia **42**, 251–269 (2014). https://doi.org/10.1007/s11406-013-9507-5

Evans, J.S.B.: In two minds: dual-process accounts of reasoning. Trends Cogn. Sci. **7**(10), 454–459 (2003)

Fife, D.: Flexplot: graphically-based data analysis. Psychol. Methods **27**(4), 477–496 (2021). https://doi.org/10.1037/met0000424

Foot, P.: The problem of abortion and the doctrine of the double effect. Oxford Rev. **5**, 5–15 (1967)

Greene, J.D.: Why are VMPFC patients more utilitarian? A dual-process theory of moral judgment explains. Trends Cogn. Sci. **11**(8), 322–323 (2007)

Greene, J.D., Morelli, S.A., Lowenberg, K., Nystrom, L.E., Cohen, J.D.: Cognitive load selectively interferes with utilitarian moral judgment. Cognition **107**(3), 1144–1154 (2008). https://doi.org/10.1016/j.cognition.2007.11.004

Greene, J.D., Nystrom, L.E., Engell, A.D., Darley, J.M., Cohen, J.D.: The neural bases of cognitive conflict and control in moral judgment. Neuron **44**(2), 389–400 (2004)

Greene, J.D., Sommerville, R.B., Nystrom, L.E., Darley, J.M., Cohen, J.D.: An fMRI investigation of emotional engagement in moral judgment. Science **293**(5537), 2105–2108 (2001)

Hashimoto, H., Maeda, K., Matsumura, K.: Fickle judgments in moral dilemmas: time pressure and utilitarian judgments in an interdependent culture. Front. Psychol. **13** (2022). https://doi.org/10.3389/fpsyg.2022.795732

Hauser, M., Cushman, F., Young, L., Kang-Xing Jin, R., Mikhail, J.: A dissociation between moral judgments and justifications. Mind Lang. **22**(1), 1–21 (2007). https://doi.org/10.1111/j.1468-0017.2006.00297.x

Kahneman, D.: Thinking, Fast and Slow. Farrar, Straus and Giroux (2011)

Ryazanov, A.A., Wang, S.T., Nelkin, D.K., McKenzie, C.R., Rickless, S.C.: Beyond killing one to save five: sensitivity to ratio and probability in moral judgment. J. Exp. Soc. Psychol. **108**, 104499 (2023)

Thomson, J.J.: Killing, letting die, and the trolley problem. Monist **59**(2), 204–217 (1976)

Trémolière, B., Bonnefon, J.-F.: Efficient kill-save ratios ease up the cognitive demands on counterintuitive moral utilitarianism. Pers. Soc. Psychol. Bull. **40**(7), 923–930 (2014). https://doi.org/10.1177/0146167214530436

Author Index

A

Abe, Hirohiko   3
Agbanglanon, Sylvain Luc   308
Almeida, Ana   385
Alves, Bruna   385
Ando, Risako   3

B

Baratgin, Jean   185, 250, 403
Bompy, Raphaël   366
Bourlier, Maxime   185, 403
Brás, Susana   385
Brasdu, Massimo   142

C

Cabral, David Ricardo Galeano   110
Caravona, Laura   323

D

Delsuc, Solène   291
Du, Yaoli   131
Dubois-Sage, Marion   250

E

El Nouty, Charles   18

F

Fernandes, José Maria   385
Filatova, Darya   18

G

Gama, João   385

H

Hashimoto, Hirofumi   55, 403
Hoffrage, Ulrich   196

J

Jacquet, Baptiste   142
Jamet, Frank   250

K

Komis, Vassilis   308

L

Lachaud, Léa   208
Laronze, Florian   291
Lassiter, Daniel   185
Leroux, Cassandra   403
Louis, Jérémy   208

M

Macchi, Laura   323
Maeda, Kaede   55, 403
Martignon, Laura   196
Mineshima, Koji   3
Morishita, Takanobu   3
Mosset-Cancel, Yohann   250
Mura, Alberto   167

N

N'kaoua, Bernard   234
N'Kaoua, Bernard   291

O

Okada, Mitsuhiro   3
Ozeki, Kentaro   3

P

Pais, Daniela   385

Palazzi, Elisa   323
Pardieu, Véronique Salvano   64
Pennequin, Valérie   64
Petturiti, Davide   118
Provasi, Joelle   345

R
Reffad, Sabrina   345
Ribeiro, Rita P.   385
Rosenberger, Julien   35

S
Sabouret, Nicolas   35
Saunier, Julien   35
Sebastião, Raquel   385

Seino, Kai   219
Silva, Catarina   385

T
Tahan, Kerem   234
Tanida, Shigehito   55

U
Uzan, Pierre   91

V
Vantaggi, Barbara   118

Y
Yama, Hiroshi   403